Agricultural Food Production: Water Availability and Management

Agricultural Food Production: Water Availability and Management

Edited by Issac Cook

New York

Published by Syrawood Publishing House,
750 Third Avenue, 9th Floor,
New York, NY 10017, USA
www.syrawoodpublishinghouse.com

Agricultural Food Production: Water Availability and Management
Edited by Issac Cook

International Standard Book Number: 978-1-64740-417-8 (Hardback)

Cataloging-in-publication Data

Agricultural food production : water availability and management / edited by Issac Cook.
p. cm.
Includes bibliographical references and index.
ISBN 978-1-64740-417-8
1. Water-supply, Agricultural. 2. Water in agriculture. 3. Water supply. 4. Water-supply--Management.
5. Food industry and trade. 6. Crop science. 7. Agriculture. I. Cook, Issac.
S494.5.W3 A37 2023
631.7--dc23

TABLE OF CONTENTS

PREFACE

Agricultural food production refers to the activity of producing food. Water and its efficient management is an integral part of agricultural food production, since it enables crop diversification and production intensification. The production of food using processes and systems that are non-polluting and aid in conserving non-renewable energy is known as sustainable food production. It involves ensuring the sustained availability of resources including water and energy. Various challenges that make it difficult to achieve sustainable food production include rising population and demand for food, climate change, decreasing per capita land, and limited water resources. Implementing sustainable water management strategies in agriculture such as adopting best irrigation practices, improving agricultural practices, direct seeding of water intensive crops, landform management, and rainwater harvesting are important for sustainable food production. This book outlines the role of water availability and management in agricultural food production in detail. It includes contributions of experts, which will provide innovative insights into this area of agriculture. The book will serve as a valuable source of reference for graduate and postgraduate students.

This book is a comprehensive compilation of works of different researchers from varied parts of the world. It includes valuable experiences of the researchers with the sole objective of providing the readers (learners) with a proper knowledge of the concerned field. This book will be beneficial in evoking inspiration and enhancing the knowledge of the interested readers.

In the end, I would like to extend my heartiest thanks to the authors who worked with great determination on their chapters. I also appreciate the publisher's support in the course of the book. I would also like to deeply acknowledge my family who stood by me as a source of inspiration during the project.

Editor

1

How Consumers Perceive Water Sustainability (HydroSOStainable) in Food Products and How to Identify it by a Logo

Paola Sánchez-Bravo [1], Edgar Chambers V [2], Luis Noguera-Artiaga [1], Esther Sendra [1], Edgar Chambers IV [2] and Ángel A. Carbonell-Barrachina [1,*]

[1] Research Group "Food Quality and Safety (CSA)", Department of Agro-Food Technology, Escuela Politécnica Superior de Orihuela (EPSO), Universidad Miguel Hernández de Elche (UMH), Ctra. de Beniel, km 3.2, 03312 Orihuela, Alicante, Spain; paola.sanchezb@umh.es (P.S.-B.); lnoguera@umh.es (L.N.-A.); esther.sendra@umh.es (E.S.)

[2] Center for Sensory Analysis and Consumer Behavior, Kansas State University, Manhattan, KS 66502, USA; echambersv@gmail.com (E.C.V); eciv@ksu.edu (E.C.IV)

* Correspondence: angel.carbonell@umh.es

Abstract: Water is the most essential resource for food production and socioeconomic development worldwide. Currently, industry and agriculture are the most water consuming activities, creating high levels of pollution, and intensifying the scarcity of water especially in arid regions. The term "hydroSOStainable products" has been used to define those foodstuffs grown under irrigation strategies that involve optimized water management. A study to understand how consumers perceive options to save water in the food chain and how to identify the water sustainable products by a logo, was conducted in Brazil, China, India, Mexico, Spain and USA, with 600 consumers per country. In all countries, consumers think that the food categories in which it is possible to save the most water are those linked directly to agricultural products: (i) "grains and grain products" and (ii) "vegetables, nuts and beans". Also, consumers do not associate processed products, such as snacks, with high water consumption, even though they come from agricultural products such as grains and require more processing. The logo was positively rated by consumers, especially by young generations. There is a need to properly inform consumers about water sustainability to gain their confidence in the hydroSOS logo.

Keywords: consumer behavior; environmental friendly; food category; hydroSOS brand; logo; water footprint

---

1. Introduction

Different definitions exist for the sustainability concept. Some of the most complete concepts explain sustainability throughout a product life cycle assessment and how it relates to poverty, environment, and the human-, production- and social-capitals left available for future populations [1,2]. In general, discussions of sustainability should include an implementation strategy and this should generate a change that contributes to the development of future generations [3]. Nevertheless, the definition proposed by Moore, Mascarenhas, Bain and Straus [3] could be considered incomplete as they did not consider religion, history, or happiness, as suggested by Shaharir and Alinor [4]. Johnston, et al. [5] highlighted the links between sustainability and ethics.

A focus on environmental issues, due to extreme difficulty in measuring the social and political aspects of global sustainability, makes assessment easier. Consequently, Poore and Nemecek [6] suggested that the term "sustainability" cover measurable features, such as land use, freshwater use weighted against local water scarcity, and greenhouse gas emissions, acidifying, and eutrophying emissions. In addition, climate change will affect the water cycle including rain patterns, availability, and quality. This in turn will affect agricultural production and ecosystems, as it is expected to lead to frequent and severe droughts in the near future [7,8]. Water is already one of, if not the most, essential resource in world food production and socioeconomic development. With the need for increased food production and environmental issues further affecting the availability of water in the future, improving water efficiency is increasingly necessary [9–11].

Currently, industrial companies compete with agriculture for the use of water, which generates a high level of water stress and pollution, intensifying water scarcity in areas where its use is not sustainable [8]. In this sense, a sustainable use of water is that which does not reduce the quantity or quality of freshwater [10]. The term "hydrosustainable or hydroSOStainable" has been branded to fruits and vegetables grown under irrigation strategies that require the use of smaller volumes of irrigation water, for example, regulated deficit irrigation. These products have a solid identify based on the increase of their quality and functionality (increase of secondary metabolites) and their water optimization in their production [10]. There are many scientific studies evaluating in detail the effects of deficit irrigation on the composition, functionality and consumer acceptance of different type of agricultural products, such as pistachios [12–15], pomegranates [16,17], table olives and olive oil [11,18], and almonds [19,20]. All these publications provide information to select hydroSOStainability markers, such as total phenolic content (TPC), oleuropein and oleic acid in extra virgin olive oil and oleic acid in table olives [11], polymeric procyanidins in pistachios [12], anthocyanins, TPC, punicalagin (α and β) and ellagic acid in pomegranates [16,17], etc. In addition, protocols have been prepared to certify whether a product deserves the hydroSOStainable logo, at two consecutive levels (i) the orchard [9] and (ii) the product itself [11].

Another aspect to be considered is the water footprint (WF). This indicator establishes the consumption (direct and indirect) of fresh water used for the production of a product along the supply chain. The WF is divided into three levels: blue, green and gray. The blue one refers to the consumption of surface and underground water; the green one to the consumption of rainwater; and the gray one defines the volume of fresh water that is needed to assimilate the pollution caused [21]. Addressing water efficiency in supply chains, especially commodities, can encourage the consumption of sustainable products and drive sustainable solutions into water management [22]. Prior research suggests that when consumers understand the antecedents of water use, they are more likely to be more aware of the WF and conserve water [23].

Increasingly, consumers demand sustainably produced food and many are trying to bring about change [24]. In 2018, according to the 13th Annual Food and Health Survey, carried out by the Foundation of the International Food Information Council (IFIC), 59% of US consumers positively valued food that was purchased and consumed in a sustainable manner; this percentage represented a significant increase (9%) over the previous year [25,26]. However, consumers still do not have a clear vision or perception of the concepts "sustainability", including "hydroSOStainability", due to the abuse of the term, even in inappropriate scenarios. Various authors [27,28] have suggested that terms such as natural, organic, sustainable, and others are all lumped together by consumers into a single concept, which is not actually true.

The aims of this study were (a) to assess which food categories consumers from six countries (representing the Americas, Asia and Europe) believe it is most possible to save water and (b) determine if the use of a hydroSOStainability logo could provide consumers with a likeable, easy way to identify water sustainability of their products.

2. Materials and Methods

The study was carried out using an online survey, run through the Qualtrics platform (Provo, UT, USA). Six countries were selected (USA, China, Mexico, Brazil, Spain and India) based on availability of databases and to represent large population countries on 4 continents. Qualtrics or its partners maintain proprietary databases of consumers in each country (usually with more than 1 million respondents per country throughout the country and many more in some countries such as the USA). Of course, as with any on-line survey only those consumers who have access to and are available are used. Consumers who are on-line and accessible are an increasingly large part of the population, but still only a portion of the global population. In some parts of the world such testing is impossible, and those sections are missed in on-line testing regardless of the question format used. China, India, the USA, and Brazil had the highest number of internet users in 2015 (Mexico ranked 7th) and Spain had one of the highest percentages of users [29]. However, some individuals are not accessible using this method and, therefore, are excluded from this type of survey.

No specific criteria regarding food habits or behavior towards the environment were used to qualify the respondents. The survey was completed by 3600 consumers (50% self-identified men and women; 600 consumers per country). Four age ranges were selected (25% of participants for each age range), clearly differentiated: 18–23 years (centennials); 24–41 years (millennials); 42–52 years (gen X) and 53–73 years (baby boomers). Respondents did not receive a financial incentive to complete the online survey, but the Qualtrics database has a reward system to compensate respondents for their time and collaboration.

The questionnaire used included queries related to how to save water in the food chain and about the hydroSOStainable logo (Figure 1a), which is registered in Spain (Spanish Patent and Trademark Office) and the European Union (European Union Intellectual Property Office). The meaning of the logo was not explained to the respondents, only its design was evaluated; only the text on questions Q4 and Q5 was presented to consumers. The idea behind the questionnaire was to determine what types of products (e.g., agricultural vs. processed foods) were most likely to contribute to water conservation and how consumers understand hydroSOStainability. This can help in determining whether their perception of water sustainability is similar to the actual concept of the WF, which is an objective marker of the water being used to produce a specific food item. Our working hypothesis was that consumers, in general, understand sustainability as a theoretical and nonconcrete concept but our aim is to transform this subjective image into an objective one, which can be evaluated and measured using objective parameters, as it has already done when controlling hydroSOStainability at orchards [9] and at the final product [11]. This questionnaire was a part of a broader survey on "global" sustainability, which included general questions on sustainability (including questions on: basic statements or definitions, benefits for local communities, sensory quality, price and purchase intention, health effects, and relevance to consumers) and consumer willingness to pay for sustainable products (including up to 29 food categories). Demographic data also was collected to classify consumers according to the factors to be studied (country, gender, age, education, income) (question Q1, Table 1).

The survey was pretested in English to ensure consumers could easily complete the task and then translated into the languages of the participating countries using a modified translation, review, adjudication, retesting, and documentation (TRAPD) approach [30,31] described by Seninde and Chambers IV [32], which includes a pretesting step in each country. The translations were done for Mandarin Chinese, Spanish, Portuguese and Hindi, and the logo also was translated to those languages. The word "hydroSOStainable" was presented together with its proper translation in India and China (Figure 1a). The survey was conducted online and in each country was presented in its most common official language or a choice of languages (English and Hindi in India).

To measure the responses, 2 scales were used: (i) 7-point scale, where 1 meant "dislike very much", 4 was "neither like nor dislike" and 7 was "like very much" (i.e., question Q4, Table 1) and (ii) 7-point scale, where 1 meant "strongly disagree" and 7 was "strongly agree" (i.e., question Q5, Table 1). In addition, 1 ranking question (Q2) in which consumer ranked 11 food categories regarding their belief that the category could be used to help in saving water) and 1 selection question (Q3, in which consumers checked three food categories, from a total of 11, where the most effort should be made to save water) were also included (Table 1). Finally, the noticeability of logo areas was evaluated by having consumers select the area of the logo that was most eye-catching (Q6).

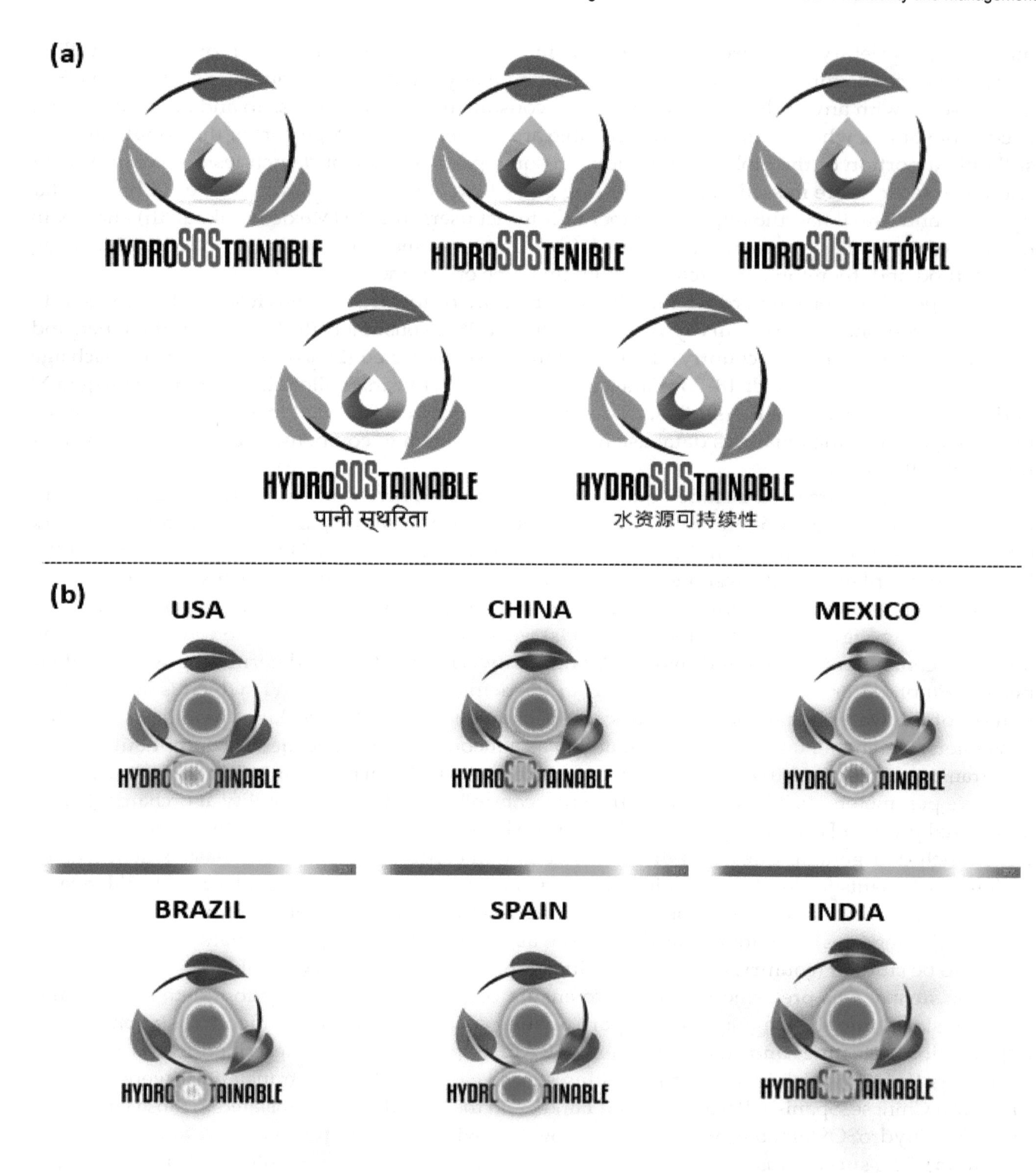

Figure 1. (**a**) HydroSOStainable logo in the different languages (from left to right and top row to bottom row: English, Spanish, Portuguese, Hindi and Mandarin Chinese); (**b**) Areas that were considered as most prominent in the logo in each country.

Table 1. Full questionnaire used in this study.

Number	Question
Q1	Demographics
Q1.1	What is your gender?
Q1.2	What is your age?
Q1.3	What is the highest education level you have completed?
Q1.4	How many adults live in your household including yourself?
Q1.5	How many children under the age of 18 are in your household?
Q1.6	How much is your approximate annual income?
Q2	Assuming we have the same consumption of products we have today, please rank these 11 food categories from 1 to 11 based on what you think is most likely to help in saving water (if we could change growing, production, and/or preparation). Rank 1 for the product you think we could save the most water when growing/making, then 2 for the second, ... to 11 for the one you think we could have the least impact on water use if we changed the growing/making of this. Coffee, tea, & cocoa; Eggs; Fish & seafood; Fruits & juices; Grains & grains products; Meat & meat products; Milk & dairy products; Snack foods; Soft drinks & bottled water; Starchy roots & potatoes; Vegetables, nuts &, beans
Q3	Assuming we have the same consumption of products we have today, please check which 3 categories of products we should work hardest to save water during growing, production, and preparation. Coffee, tea, & cocoa; Eggs; Fish & seafood; Fruits & juices; Grains & grains products; Meat & meat products; Milk & dairy products; Snack foods; Soft drinks & bottled water; Starchy roots & potatoes; Vegetables, nuts &, beans
Q4	We have developed this logo to identify water sustainable products. How much do you like the logo?
Q5	How easy do you think it would be to identify products as more water sustainable using this logo?
Q6	Please, CLICK on the area of the image that stands out most.

Statistical Analysis

A one-way analysis of variance (ANOVA), Tukey's multiple range test and Friedman analysis, with the subsequent LSD (Least Significant Difference) test, were performed for the analysis of the results. For this, software XLSTAT (2016.02.27444 version, Addinsoft) was used. The confidence interval was 95% and the significant difference was defined as $p < 0.05$. Data were not weighted to represent all demographic classes in a country because equal numbers of consumers in each sex and age category were used for comparison purposes within and across countries. The noticeability of logo areas counted and a "heat map" created that shows the areas where consumers had highlighted.

3. Results and Discussion

The rank of food categories perceived to be most likely to help in saving water during their growing, production and/or preparation is shown in Table 2. In this table, the lower the sum of ranges, the more potential has the food category of saving water, according to the consumer opinion. Results showed that the food categories pointed by international consumers as those needed attention to save water, regardless of country, were: (i) grains and grain products and (ii) vegetables, nuts and beans. In contrast, consumers indicated the least possibility of influencing the saving water in the production of snack food, soft drinks and bottled water. In question Q3 (Table 1), consumers were asked to choose the 3 food categories in which more attention and work is needed to save water during growing, production and preparation. As one can see in Table 3, consumers pointed out that "grains and grains products", "fruits and juices" and "vegetables, nuts and beans" were the main 3 products where effort and attention should be paid according to consumers, especially in grains and vegetables. Of course, the use of irrigation in agricultural systems, leads to higher yields and also produces a high impact on water sustainability [33]. However, protein-rich products (e.g., lamb, cheese, pig meat, and peas and nuts), olive oil, and milk are the food items needing the highest volumes of freshwater [6]. In addition, the highest water footprint values in the European Union (EU) are related to the consumption of milk, beef and pork [22].

Table 2. Foods groups that consumer think could save more water as affected by the "country" factor.

	Brazil	China	India	Mexico	Spain	USA
Coffee, tea, & cocoa	3745 cde [†]	4033 ef	3723 cde	3708 d	3688 de	3803 de
Eggs	3716 cde	**3364 ab**	3558 bc	3712 de	3816 e	3696 cd
Fish & seafood	3794 de	3861 de	3945 efg	3935 ef	4082 f	3815 de
Fruits & juices	**3302 a**	3540 bc	**3513 abc**	3402 bc	3548 bcd	3565 c
Grains & grains products	**3305 a**	**3290 a**	**3344 ab**	**3177 ab**	**3326 ab**	**3250 a**
Meat & meat products	3545 bc	3815 de	3973 fg	3561 cd	3665 cde	3692 cd
Milk & dairy products	3583 bcd	3676 cd	**3378 ab**	3637 d	3743 de	3667 cd
Snack foods	4050 f	4223 fg	3809 def	4350 g	4355 g	4219 f
Soft drinks & bottled water	3868 ef	4312 g	4077 g	4154 fg	4100 f	4000 ef
Starchy roots & potatoes	**3460 ab**	**3463 abc**	3623 cd	3523 cd	3457 bc	3710 cd
Vegetables, nuts &, beans	**3364 ab**	**3277 a**	**3317 a**	**2969 a**	**3206 a**	**3305 a**

The values correspond to the ranges sum after the Friedman test, in which 11 food categories were ranked according to the potential help in saving water according to consumer opinion. Note: a sample was ranked 1 for the product in which it is possible to save the most possible water when growing/making, then, 2 for the second, and successively . . . , and finally 11 for the one that will have the lowest impact on water saving. [†] Values followed by different letters, within the same column were significantly different ($p < 0.05$). Food groups with letter "a" (highlighted in red font) had highest potential to safe water according to the consumer opinion.

Table 3. Food groups that consumers think it should work hardest to save water (bold font by ≥30% and red font by ≥45% of consumers of each country Brazil, China, India, Mexico, Spain and USA).

	Brazil (%)	China (%)	India (%)	Mexico (%)	Spain (%)	USA (%)
Coffee, tea & cocoa	19.5	23.3	22.0	16.8	18.8	15.5
Eggs	9.7	18.8	21.5	6.3	9.3	11.2
Fish & seafood	13.0	23.7	14.7	14.2	10.8	20.5
Fruits & juices	**37.8**	**33.7**	**31.5**	**38.0**	**36.2**	**32.3**
Grains & grains products	**46.3**	**42.3**	**53.3**	**37.3**	**50.0**	**46.5**
Meat & meat products	**35.7**	22.0	16.7	**31.8**	**30.2**	**37.5**
Milk & dairy products	22.7	24.5	**31.3**	26.8	20.2	29.2
Snack foods	15.0	22.0	10.7	17.2	15.7	13.5
Soft drinks & bottled water	**34.3**	29.3	26.5	**42.3**	26.0	27.5
Starchy roots & potatoes	29.0	29.8	24.3	25.3	33.8	26.3
Vegetables, nuts & beans	**38.0**	**39.8**	**52.5**	**47.8**	**59.5**	**48.0**

It is important to point out the lack of association between the group "vegetables and fruits" with snacks foods, which were not associated with the term "sustainability" by consumers. This shows that consumers associated water consumption mainly with primary food production, leaving processed food items aside. In fact, in Spain, more than 85% of the organic products consumption is due to fruits and vegetables [34]. There is not a clear connection in the consumers mind linking processed foods, such as snacks, with irrigation water which seems to them only be used for primary products such as fruits and vegetables. That idea is in direct conflict with the WF, which includes much more than just the agricultural input of water [21]; for instance, a large volume of water is needed to produce power energy [35,36]. Therefore, those sectors and products requiring large amounts of electricity and processing are also heavy water users [36]. For example, the water footprint for pasta production may be double (up to 1336–2847 m^3/t) that of wheat (656–1300 m^3/t) from which it is made (Table 4). But it is fully understandable that consumers have difficulties in comparing water saving in producing simple products (such as vegetables) with those products generated after more complex processes including several stages in their production.

Table 4. Examples of water footprint for the evaluated foods groups.

Food Category	Water Footprint	Reference
Coffee, tea, & cocoa	Coffee = 140 dm^3/cup; Tea = 34 dm^3/cup	[37–39]
Eggs	Eggs = 3265 m^3/t	[40]
Fish & seafood	Farmed fishes and crustaceans = 1974 m^3/t	[39,41]
Fruits & juices	Strawberries = 70 m^3/t; Tomatoes = 120–184 m^3/t; Tomato sauce = 195 m^3/t; Dried tomato = 199 m^3/t; Fruits = 962–1000 m^3/t	[39,42–45]
Grains & grains products	Maize = 900–1222 m^3/t; Wheat = 656–1300 m^3/t; Rice = 1325–3000 m^3/t; Pasta = 1336–2847 m^3/t	[39,40,45–48]
Meat & meat products	Chicken = 3900 m^3/t; Pork = 4900 m^3/t; Beef = 15,500 m^3/t	[39,46]
Milk & dairy products	Milk = 1020 m^3/t	[39,49]
Snack foods	-	
Soft drinks & bottled water	-	
Starchy roots & potatoes	Potatoes = 75–324 m^3/t; Sweet potatoes = 823 m^3/t	[39,47,50]
Vegetables, nuts &, beans	Nuts = 9000 m^3/t; Almonds = 10,240–20,820 m^3/t; Soy bean = 1816 m^3/t; Watermelon = 136 m^3/t; Groundnuts = 1330 m^3/t; Olive = 3434 m^3/t (traditional) and 2782 m^3/t (high-density)	[47,51,52]

Besides, to produce tomato sauce or dried tomato, the water footprint increases up to 195 m^3/t and 199 m^3/t, respectively as compared to the 184 m^3/t needed to produce fresh tomatoes [39]. This consumption of water, damages the water natural cycle, affects the natural thermal regime of rivers and, therefore, affects the availability of oxygen and the metabolism of natural biota [36].

Other studies have shown the importance of developing and agree on a tool that identifies those products or items considered as sustainable including those that show a reduced WF [53]. In this sense, a logo for hydroSOStainable products was proposed, and has been registered in Spain (Spanish Patent and Trademark Office) and the European Union (European Union Intellectual Property Office) and international consumers were asked on: (i) how much did they like the logo? (ii) how easy they thought it would be to identify products as more "water sustainable" by using this logo? In addition, they were asked to click on the area of the image that stands out the most. The proposed logo consists of a drop of water surrounded by 3 leaves and at the base the word "hydroSOStainable" (Figure 1a). This logo emerged as an improvement to the first logo proposed for the identification of hydroSOStainable products in Spanish and English-speaking countries [10]. The results showed that the logo was positively rated by most of the international consumers. In India, the logo was rated as being much more attractive to consumers (5.3) than in countries such as Spain and USA (4.1 and 4.3, respectively); although it should be noted that in all countries it received positive ratings (Table 5).

Table 5. Consumers opinion on the hydroSOStainable logo as affected by country, age, gender, income and education factors.

Factor	Q4	Q5
ANOVA Test †		
Country	***	***
Age	***	***
Gender	NS	NS
Income	NS	***
Education	NS	***
Tukey's Test ‡		
Country		
Brazil	4.9 b	5.8 ab
China	4.5 cd	4.8 d
India	5.3 a	5.7 bc
Mexico	4.5 c	6.0 a
Spain	4.1 e	5.8 ab
USA	4.3 de	5.5 c
Age (years)		
18–23	4.6 ab	5.5 b
24–41	4.8 a	5.7 a
42–52	4.6 b	5.7 a
53–73	4.4 c	5.5 b
Gender		
Male	4.6	5.5
Female	4.6	5.6
Income (US dollars)		
25,000 or less	4.6	5.7
25,001–50,000	4.7	5.7
50,001–100,000	4.6	5.5
100,000+	4.6	5.4
Education		
Primary school or less	4.5	4.9 c
High school diploma	4.5	5.5 b
Associate's degree	4.4	5.6 ab
Bachelor's degree	4.6	5.5 b
Graduate degree or higher	4.7	5.8 a

Q4 = How much do you like the logo? Q5 = How easy do you think it would be to identify products as more water sustainable using this logo? † NS, not significant ($p > 0.05$), *** significant differences $p < 0.001$. ‡ Values followed by different letters, within the same column and factor, were significantly different ($p < 0.05$). Age: 18–23 years old (centennials); 24–41 years old (millennials); 42–52 years old (gen X) and 53–73 years old (baby boomers).

In a study on the commercialization of organic products, Prentice, et al. [54] suggested that, for Chinese consumers, the most important factor for the purchase of organic products and that demonstrates their authenticity, is their certification and labeling. Also, although Indian consumers are more likely to accept organic products, they do not trust their authenticity due to fraud in certification systems [55,56]. This logo and the development of a transparent certification procedure could be an important option to make consumers trust hydroSOStainable products and be sure of the quality and origin of the product to be bought. In this sense, indicators of hydroSOStainablibilty have been developed to provide the specific requirements that orchards/farms [9] and products (starting with extra virgin olive oil and processed table olives; [11] must fulfil to be able to be branded under the logo hydroSOStainable. The products that achieve this certification meet the environmental requirements to be sustainable regarding irrigation water. However, farmers, producers and distributors must ensure the other aspects of the whole sustainability concept.

Young generations (centennials, millennials and gen X) rated the logo higher (4.6, 4.8 and 4.6, respectively) than older generations (baby boomers, 4.4). This shows that, in general, the level of satisfaction with the logo decreased as the consumers age increased. Young consumers seem to be more focused on visual information than the old ones, who are more focused on written information; successful case studies supporting this statement are the use of social networks and emoji by youngsters. Also, young generations are more concerned about the environment and sustainability and have greater confidence in food labeling as a source of information [57]. There were no statistical significant differences for the factors gender, annual income, or educational level. Concerning the use of the logo, consumers in Mexico, Brazil and Spain (6.0, 5.8 and 5.8, respectively) were the ones who most agreed that the logo made easier the identification of hydroSOStainable products (question Q5, Table 1), whereas Chinese consumers (4.8) were in the opposite side, although they also agreed with the help of using this logo (Table 5). These results could be due to an important percentage of Chinese consumers not understand English language and making more difficult to understand the term (hydroSOStainable) which was written in English (capital letters) and subtitled in Mandarin Chinese. Probably the expression SOS losses its power when translated into other languages without Latin alphabet. In this way, it should be noted that those countries that saw the greatest utility in the use of the logo were the countries whose language use the Latin alphabet. Nowadays, consumers at the market may find more than 200 logos referring to products with healthy and sustainable aspects. In addition, these labels compete with each other, which creates confusion and distrust among consumers [58]. In this regard, an important work should be done day by day to disseminate the meaning (increased accumulation of bioactive compounds as a response to water stress) and controls (at farm and at the product to be marketed) done on the hydroSOStainable logo to make it widely accepted by international consumers. Although it will be difficult, there is no other way to get consumer confidence and trust in this special type of agronomic products; there is a need to inform consumers based on all scientific findings supporting the use of the hydroSOStainable logo. The most important tool to gain consumers trust would be to ensure complete transparency in the certification process, which must be based on scientific findings and indicators.

Consumers in the middle age groups (millennials and gen X) agreed more with the usefulness of the proposed logo, while consumers on extreme ages (the youngest and the oldest ones) gave lower scores (Table 2). This may be due to the fact that the elderly prefer to read the labeling rather than base their choice on visual aspects, whereas the youngest, despite being the most technologically developed, are still not entirely familiar with the purchasing process. Consumers who believe that their behavior can reduce environmental impact tend to pay more attention to label information and drive their choices accordingly [59]. But in any case, the mean score for this question was 5.6 (with 5 = agree somewhat and 6 = agree) was high and clearly demonstrated that the use of the logo or the implementation of the hydroSOS brand would serve to create a means of differentiation for consumers. Also, the participants level of education had a significant influence on consumer opinion on the logo, with consumers having the lowest education (primary school holders) giving the lowest scores (4.9) contrary to those with associate's degree or higher (mean of 5.6). These results are similar to those obtained by Ditlevsen, et al. [60], who reported that in general highly educated consumers were the biggest buyers of organic food. However, there were no statistically significant differences due to gender and annual income for the logo usefulness. These results agreed with our initial hypothesis regarding the need to provide more information to consumers specifying that they are environment friendly. The development of a hydroSOStainable logo would mean providing the necessary information to the consumer, and with it, providing full transparency, derived from a certification process associated with the logo, that the consumer needs to trust and buy the product [11,55]. This logo would help consumers to easily identify the products in which water has been saved in their production and also in their transport and distribution. This logo provides full information about the product and its farming, and ensures that irrigation water was optimized, water usage was controlled by farmers and final quality of the product was increased by water stress produced on the plant/tree [9,11].

With respect to the area of the logo that stands out the most (Figure 1b), the water drop was getting the attention of most of the consumers in all studied countries. The word "hydro" is well associated with the water drop which is located in the center of the image and therefore plays a key role in the logo. The next most attractive area was the expression "SOS" which was highly relevant in all countries except in China and India, which could be due to the difference in grammar and how easy was to understand the concept of SOS as asking for help (e.g., save our ship or save our souls). Previous studies revealed that consumers prefer simple and easy to understand labels compared to labels more complexed labels although more detailed information is provided [61]. Therefore, understanding consumer perception is essential to understand why some marketing campaigns do not reach their targeted goals.

Water consumption of the products generates an impact on water resources, mostly indirectly, either through packaging, transport, etc. Therefore, it is also necessary to include transport and packaging in the carbon footprint and the water footprint of the product. For this, the longer the distance between the farm where the vegetable is produced and the shop where it is sold, the less hydroSOStainable it will be because its water footprint will be higher [21,22,39,62]. In general, transport is often eliminated from the calculation of the water footprint [21]. However, transport consumes a lot of energy, especially if includes the use of biofuels or hydroelectric energy (energies that have a high water footprint because irrigation water must be used for the first one and dams must be built and water is lost by evaporation); thus, transporting a product to its final destination can generate a considerable contribution to the water footprint of the product [21,63]. As an example, Page, et al. [64] found that the transportation of tomatoes to the market was a key factor in determining the carbon footprint of the product. Therefore, it was important to reduce the energy involved in transportation (reduce the millage between the farm and the shop) to minimize the carbon footprint and water use [21].

4. Conclusions

Given the complexity of the term sustainability and consumer behavior (the fact of expressing an opinion in a survey does not necessarily have to be reflected in their behavior) establishing concrete results is complicated; thus, it is necessary to continue carrying out studies to fully understand both the opinion and commercial actions taken by consumers regarding water-saving products. In all countries, consumers think that the food categories in which more water can be saved during their full production and distribution chain are (i) "grains and grains products" and (ii) "vegetables, nuts and beans". This finding clearly shows that consumers do not associate processed products (e.g., snacks) with significant water consumption. In general, the logo proposed for hydroSOStainable products was positively rated, especially by young generations, and it was considered useful for the identification of these sustainable foods. However, spreading its meaning and to provide confidence in the hydroSOS brand is essential. The development of a certification will guarantee the quality and origin of the product to consumers, helping them to easily identify them in their markets.

Author Contributions: Formal analysis, P.S.-B. and E.C.V; Investigation, P.S.-B.; Methodology, L.N.-A.; Project administration, E.C.IV; Supervision, E.S., E.C.IV and Á.A.C.-B.; Writing—original draft, P.S.-B., E.C.V and L.N.-A.; Writing—review & editing, E.S., E.C.IV and Á.A.C.-B. All authors have read and agreed to the published version of the manuscript.

References

1. Jørgensen, A.; Herrmann, I.T.; Bjørn, A. Analysis of the link between a definition of sustainability and the life cycle methodologies. *Int. J. Life Cycle Assess.* **2013**, *18*, 1440–1449.
2. WCED. *Our Common Future*; Oxford University Press: Oxford, NY, USA, 1987.
3. Moore, J.E.; Mascarenhas, A.; Bain, J.; Straus, S.E. Developing a comprehensive definition of sustainability. *Implement. Sci.* **2017**, *12*, 110. [CrossRef] [PubMed]
4. Shaharir, B.M.Z.; Alinor, M.B.A.K. The need for a new definition of sustainability. *J. Indones. Econ. Bus.* **2013**, *28*, 251–268.

5. Johnston, P.; Everard, M.; Santillo, D.; Robèrt, K.H. Reclaiming the definition of sustainability. *Environ. Sci. Pollut. Res. Int.* **2007**, *14*, 60–66. [PubMed]
6. Poore, J.; Nemecek, T. Reducing food's environmental impacts through producers and consumers. *Science* **2018**, *360*, 987–992. [CrossRef]
7. Collins, R.; Kristensen, P.; Thyssen, N. Water Resources across Europe-Confronting Water Scarcity and Drought. Ph.D. Thesis, Fakulteta za Kmetijstvo in Biosistemske Vede, Univerza v Mariboru, Maribor, Slovenia, 2009.
8. FAO. *The Future of Food and Agriculture—Trends and Challenges*; Food and Agriculture Organisation: Rome, Italy, 2017.
9. Corell, M.; Martín-Palomo, M.J.; Sánchez-Bravo, P.; Carrillo, T.; Collado, J.; Hernández-García, F.; Girón, I.; Andreu, L.; Galindo, A.; López-Moreno, Y.E.; et al. Evaluation of growers' efforts to improve the sustainability of olive orchards: Development of the hydrosostainable index. *Sci. Hortic.* **2019**, *257*, 108661. [CrossRef]
10. Noguera-Artiaga, L.; Lipan, L.; Vázquez-Araújo, L.; Barber, X.; Pérez-López, D.; Carbonell-Barrachina, Á.A. Opinion of spanish consumers on hydrosustainable pistachios. *J. Food Sci.* **2016**, *81*, S2559–CS2565. [CrossRef]
11. Sánchez-Bravo, P.; Collado-González, J.; Corell, M.; Noguera-Artiaga, L.; Galindo, A.; Sendra, E.; Hernández, F.; Martín-Palomo, M.J.; Carbonell-Barrachina, Á.A. Criteria for hydrosos quality index. Application to extra virgin olive oil and processed table olives. *Water* **2020**, *12*, 555. [CrossRef]
12. Noguera-Artiaga, L.; Pérez-López, D.; Burgos-Hernández, A.; Wojdyło, A.; Carbonell-Barrachina, Á.A. Phenolic and triterpenoid composition and inhibition of α-amylase of pistachio kernels (pistacia vera l.) as affected by rootstock and irrigation treatment. *Food Chem.* **2018**, *261*, 240–245. [CrossRef]
13. Noguera-Artiaga, L.; Salvador, M.D.; Fregapane, G.; Collado-González, J.; Wojdyło, A.; López-Lluch, D.; Carbonell-Barrachina, Á.A. Functional and sensory properties of pistachio nuts as affected by cultivar. *J. Sci. Food Agric.* **2019**, *99*, 6696–6705. [CrossRef]
14. Noguera-Artiaga, L.; Sánchez-Bravo, P.; Hernández, F.; Burgos-Hernández, A.; Pérez-López, D.; Carbonell-Barrachina, Á.A. Influence of regulated deficit irrigation and rootstock on the functional, nutritional and sensory quality of pistachio nuts. *Sci. Hortic.* **2020**, *261*, 108994. [CrossRef]
15. Noguera-Artiaga, L.; Sánchez-Bravo, P.; Pérez-López, D.; Szumny, A.; Calin-Sánchez, Á.; Burgos-Hernández, A.; Carbonell-Barrachina, Á.A. Volatile, sensory and functional properties of hydrosos pistachios. *Foods* **2020**, *9*, 158. [CrossRef] [PubMed]
16. Galindo, A.; Calín-Sánchez, Á.; Griñán, I.; Rodríguez, P.; Cruz, Z.N.; Girón, I.F.; Corell, M.; Martínez-Font, R.; Moriana, A.; Carbonell-Barrachina, A.A.; et al. Water stress at the end of the pomegranate fruit ripening stage produces earlier harvest and improves fruit quality. *Sci. Hortic.* **2017**, *226*, 68–74. [CrossRef]
17. Galindo, A.; Collado-González, J.; Griñán, I.; Corell, M.; Centeno, A.; Martín-Palomo, M.J.; Girón, I.F.; Rodríguez, P.; Cruz, Z.N.; Memmi, H.; et al. Deficit irrigation and emerging fruit crops as a strategy to save water in mediterranean semiarid agrosystems. *Agric. Water Manag.* **2018**, *202*, 311–324. [CrossRef]
18. Sánchez-Rodríguez, L.; Lipan, L.; Andreu, L.; Martín-Palomo, M.J.; Carbonell-Barrachina, Á.A.; Hernández, F.; Sendra, E. Effect of regulated deficit irrigation on the quality of raw and table olives. *Agric. Water Manag.* **2019**, *221*, 415–421. [CrossRef]
19. Lipan, L.; García-Tejero, I.F.; Gutiérrez-Gordillo, S.; Demirbaş, N.; Sendra, E.; Hernández, F.; Durán-Zuazo, V.H.; Carbonell-Barrachina, A.A. Enhancing nut quality parameters and sensory profiles in three almond cultivars by different irrigation regimes. *J. Agric. Food Chem.* **2020**, *68*, 2316–2328. [CrossRef] [PubMed]
20. Lipan, L.; Martín-Palomo, M.J.; Sánchez-Rodríguez, L.; Cano-Lamadrid, M.; Sendra, E.; Hernández, F.; Burló, F.; Vázquez-Araújo, L.; Andreu, L.; Carbonell-Barrachina, Á.A. Almond fruit quality can be improved by means of deficit irrigation strategies. *Agric. Water Manag.* **2019**, *217*, 236–242. [CrossRef]
21. Hoekstra, A.Y.; Chapagain, A.K.; Aldaya, M.M.; Mekonnen, M.M. *The Water Footprint Assessment Manual—Setting the Global Standard*; Routledge: London, UK, 2011.
22. Vanham, D.; Bidoglio, G. A review on the indicator water footprint for the eu28. *Ecol. Indic.* **2013**, *26*, 61–75. [CrossRef]
23. Gómez-Llanos, E.; Durán-Barroso, P.; Robina-Ramírez, R. Analysis of consumer awareness of sustainable water consumption by the water footprint concept. *Sci. Total Environ.* **2020**, *721*, 137743. [CrossRef]

24. Malochleb, M. Sustainability: How food companies are turning over a new leaf. *Food Technol.* **2018**, *72*, 32–43.
25. IFIC, International Food Information Council Foundation. One-Third of Americans Are Dieting, Including One in 10 Who Fast ... While Consumers also Hunger for Organic, "Natural" and Sustainable. Available online: https://foodinsight.org/ (accessed on 21 February 2020).
26. IFIC, International Food Information Council Foundation. Food & Health Survey. Available online: https://foodinsight.org/2018-food-and-health-survey/ (accessed on 21 February 2020).
27. Chambers, V.E.; Tran, T.; Chambers IV, E. Natural: A $75 billion word with no definition—why not? *J. Sens. Stud.* **2019**, *34*, e12501. [CrossRef]
28. Meas, T.; Hu, W.; Batte, M.T.; Woods, T.A.; Ernst, S. Substitutes or complements? Consumer preference for local and organic food attributes. *Am. J. Agric. Econ.* **2015**, *97*, 1044–1071. [CrossRef]
29. Roser, M.; Ritchie, H.; Ortiz-Ospina, E.; Hasell, J. Coronavirus Pandemic (COVID-19). Our World in Data. Available online: https://ourworldindata.org/coronavirus (accessed on 21 July 2020).
30. Curtarelli, M.; van Houten, G. Questionnaire translation in the european company survey: Conditions conducive to the effective implementation of a trapd-based approach. *Transl. Interpret.* **2018**, *10*, 34–54. [CrossRef]
31. Harkness, J.A.; van de Vijver, F.J.; Mohler, P.P.; Wiley, J. *Cross-Cultural Survey Methods*; Wiley-Interscience: Hoboken, NJ, USA; New York, NY, USA, 2003; Volume 325.
32. Seninde, D.R.; Chambers IV, E. Comparing four question formats in five languages for on-line consumer surveys. *Methods Protoc.* **2020**, *3*, 49. [CrossRef] [PubMed]
33. Darré, E.; Cadenazzi, M.; Mazzilli, S.R.; Rosas, J.F.; Picasso, V.D. Environmental impacts on water resources from summer crops in rainfed and irrigated systems. *J. Environ. Manag.* **2019**, *232*, 514–522. [CrossRef] [PubMed]
34. Cerdeño, V.J.M. Alimentos ecológicos: Oferta y demanda en españa. *Distrib. Consumo* **2010**, *20*, 49–60.
35. EEA, European Environment Agency. Water Use in Europe—Quantity and Quality Face Big Challenges. Available online: https://www.eea.europa.eu/signals/signals-2018-content-list/articles/water-use-in-europe-2014 (accessed on 15 March 2020).
36. EEA, European Environment Agency. Use of Freshwater Resources in Europe. Available online: https://www.eea.europa.eu/data-and-maps/indicators/use-of-freshwater-resources-3/assessment-4 (accessed on 15 March 2020).
37. Chapagain, A.K.; Hoekstra, A.Y. *The Water Needed to Have the Dutch Drink Coffee*; Unesco-IHE Institute for Water Education: Delft, The Netherlands, 2003.
38. Chapagain, A.K.; Hoekstra, A.Y. The water footprint of coffee and tea consumption in the netherlands. *Ecol. Econ.* **2007**, *64*, 109–118. [CrossRef]
39. Lovarelli, D.; Bacenetti, J.; Fiala, M. Water footprint of crop productions: A review. *Sci. Total Environ.* **2016**, *548*, 236–251. [CrossRef]
40. Mekonnen, M.M.; Hoekstra, A.Y. *The Green, Blue and Grey Water Footprint of Farm Animals and Animal Products*; UNESCO-IHE: AX Delft, The Netherlands, 2010; Volume 1.
41. Pahlow, M.; van Oel, P.R.; Mekonnen, M.M.; Hoekstra, A.Y. Increasing pressure on freshwater resources due to terrestrial feed ingredients for aquaculture production. *Sci. Total Environ.* **2015**, *536*, 847–857. [CrossRef]
42. García Morillo, J.; Rodríguez Díaz, J.A.; Camacho, E.; Montesinos, P. Linking water footprint accounting with irrigation management in high value crops. *J. Clean. Prod.* **2015**, *87*, 594–602. [CrossRef]
43. Manzardo, A.; Mazzi, A.; Loss, A.; Butler, M.; Williamson, A.; Scipioni, A. Lessons learned from the application of different water footprint approaches to compare different food packaging alternatives. *J. Clean. Prod.* **2016**, *112*, 4657–4666. [CrossRef]
44. Ramírez, T.; Meas, Y.; Dannehl, D.; Schuch, I.; Miranda, L.; Rocksch, T.; Schmidt, U. Water and carbon footprint improvement for dried tomato value chain. *J. Clean. Prod.* **2015**, *104*, 98–108. [CrossRef]
45. Chapagain, A.K.; Hoekstra, A.Y. The blue, green and grey water footprint of rice from production and consumption perspectives. *Ecol. Econ.* **2011**, *70*, 749–758. [CrossRef]
46. Chapagain, A.K.; Hoekstra, A.Y. *Water footprints of nations In Value of Water Research Report Series*; Unesco-IHE Institute for Water Education: Delft, The Netherlands, 2004; Volume 16, p. 80.
47. Huang, J.; Zhang, H.-L.; Tong, W.-J.; Chen, F. The impact of local crops consumption on the water resources in beijing. *J. Clean. Prod.* **2012**, *21*, 45–50. [CrossRef]

48. Ruini, L.; Marino, M.; Pignatelli, S.; Laio, F.; Ridolfi, L. Water footprint of a large-sized food company: The case of barilla pasta production. *Water Resour. Ind.* **2013**, *1–2*, 7–24. [CrossRef]
49. Palhares, J.C.P.; Pezzopane, J.R.M. Water footprint accounting and scarcity indicators of conventional and organic dairy production systems. *J. Clean. Prod.* **2015**, *93*, 299–307. [CrossRef]
50. Hess, T.M.; Lennard, A.T.; Daccache, A. Comparing local and global water scarcity information in determining the water scarcity footprint of potato cultivation in great britain. *J. Clean. Prod.* **2015**, *87*, 666–674. [CrossRef]
51. Fulton, J.; Norton, M.; Shilling, F. Water-indexed benefits and impacts of california almonds. *Ecol. Indic.* **2019**, *96*, 711–717. [CrossRef]
52. Pellegrini, G.; Ingrao, C.; Camposeo, S.; Tricase, C.; Contò, F.; Huisingh, D. Application of water footprint to olive growing systems in the apulia region: A comparative assessment. *J. Clean. Prod.* **2016**, *112*, 2407–2418. [CrossRef]
53. De Laurentiis, V.; Hunt, D.V.L.; Lee, S.E.; Rogers, C.D.F. Eats: A life cycle-based decision support tool for local authorities and school caterers. *Int. J. Life Cycle Assess.* **2019**, *24*, 1222–1238. [CrossRef]
54. Prentice, C.; Chen, J.; Wang, X. The influence of product and personal attributes on organic food marketing. *J. Retail. Consum. Serv.* **2019**, *46*, 70–78. [CrossRef]
55. Boobalan, K.; Nachimuthu, G.S. Organic consumerism: A comparison between india and the USA. *J. Retail. Consum. Serv.* **2020**, *53*, 101988. [CrossRef]
56. Misra, R. An analysis of factors affecting growth of organic food: Perception of consumers in delhi-ncr (india). *Br. Food J.* **2016**, *118*, 2308–2325. [CrossRef]
57. Bollani, L.; Bonadonna, A.; Peira, G. The millennials' concept of sustainability in the food sector. *Sustainability* **2019**, *11*, 2984. [CrossRef]
58. Institute of organic agriculture FiBL; IFOAM. *The World of Organic Agriculture: Statistics and Emerging Trends 2019*; Research Institute of Organic Agriculture FiBL and IFOAM Organics International: Rheinbreitbach, Germany, 2019.
59. Dascher, E.D.; Kang, J.; Hustvedt, G. Water sustainability: Environmental attitude, drought attitude and motivation. *Int. J. Consum. Stud.* **2014**, *38*, 467–474. [CrossRef]
60. Ditlevsen, K.; Denver, S.; Christensen, T.; Lassen, J. A taste for locally produced food—values, opinions and sociodemographic differences among 'organic' and 'conventional' consumers. *Appetite* **2020**, *147*, 104544. [CrossRef]
61. Leach, A.M.; Emery, K.A.; Gephart, J.; Davis, K.F.; Erisman, J.W.; Leip, A.; Pace, M.L.; D'Odorico, P.; Carr, J.; Noll, L.C.; et al. Environmental impact food labels combining carbon, nitrogen, and water footprints. *Food Policy* **2016**, *61*, 213–223. [CrossRef]
62. Vanham, D.; Hoekstra, A.Y.; Bidoglio, G. Potential water saving through changes in european diets. *Environ. Int.* **2013**, *61*, 45–56. [CrossRef]
63. Mekonnen, M.M.; Hoekstra, A.Y. The blue water footprint of electricity from hydropower. *Hydrol. Earth Syst. Sci.* **2012**, *16*, 179–187. [CrossRef]
64. Page, G.; Ridoutt, B.; Bellotti, B. Carbon and water footprint tradeoffs in fresh tomato production. *J. Clean. Prod.* **2012**, *32*, 219–226. [CrossRef]

2

Modeling the Effects of Irrigation Water Salinity on Growth, Yield and Water Productivity of Barley in Three Contrasted Environments

Zied Hammami [1,*], Asad S. Qureshi [1], Ali Sahli [2], Arnaud Gauffreteau [3], Zoubeir Chamekh [2,4], Fatma Ezzahra Ben Azaiez [2], Sawsen Ayadi [2] and Youssef Trifa [2]

[1] International Center for Biosaline Agriculture (ICBA), Dubai P.O. Box 14660, UAE; a.qureshi@biosaline.org.ae
[2] Laboratory of Genetics and Cereal Breeding, National Agronomic Institute of Tunisia, Carthage University, 43 Avenue Charles Nicole, 1082 Tunis, Tunisia; sahli_inat_tn@yahoo.fr (A.S.); zoubeirchamek@gmail.com (Z.C.); zahra.azaiez@gmail.com (F.E.B.A.); sawsen.ayadi@gmail.com (S.A.); youssef.trifa@gmail.com (Y.T.)
[3] INRA–INA-PG–AgroParisTech, UMR 0211, Avenue Lucien Brétignières, F-78850 Thiverval Grignon, France; arnaud.gauffreteau@inrae.fr
[4] Carthage University, National Agronomic Research Institute of Tunisia, LR16INRAT02, Hédi Karray, 1082 Tunis, Tunisia
* Correspondence: z.hammami@biosaline.org.ae

Abstract: Freshwater scarcity and other abiotic factors, such as climate and soil salinity in the Near East and North Africa (NENA) region, are affecting crop production. Therefore, farmers are looking for salt-tolerant crops that can successfully be grown in these harsh environments using poor-quality groundwater. Barley is the main staple food crop for most of the countries of this region, including Tunisia. In this study, the AquaCrop model with a salinity module was used to evaluate the performance of two barley varieties contrasted for their resistance to salinity in three contrasted agro-climatic areas in Tunisia. These zones represent sub-humid, semi-arid, and arid climates. The model was calibrated and evaluated using field data collected from two cropping seasons (2012–14), then the calibrated model was used to develop different scenarios under irrigation with saline water from 5, 10 to 15 dS m^{-1}. The scenario results indicate that biomass and yield were reduced by 40% and 27% in the semi-arid region (KAI) by increasing the irrigation water salinity from 5 to 15 dS m^{-1}, respectively. For the salt-sensitive variety, the reductions in biomass and grain yield were about 70%, respectively, although overall biomass and yield in the arid region (MED) were lower than in the KAI area, mainly with increasing salinity levels. Under the same environmental conditions, biomass and yield reductions for the salt-tolerant barley variety were only 16% and 8%. For the salt-sensitive variety, the biomass and grain yield reductions in the MED area were about 12% and 43%, respectively, with a similar increase in the salinity levels. Similar trends were visible in water productivities. Interestingly, biomass, grain yield, and water productivity values for both barley varieties were comparable in the sub-humid region (BEJ) that does not suffer from salt stress. However, the results confirm the interest of cultivating a variety tolerant to salinity in environments subjected to salt stress. Therefore, farmers can grow both varieties in the rainfed of BEJ; however, in KAI and MED areas where irrigation is necessary for crop growth, the salt-tolerant barley variety should be preferred. Indeed, the water cost will be reduced by 49% through growing a tolerant variety irrigated with saline water of 15 dS m^{-1}.

Keywords: salinity; environments; AquaCrop model; water productivity; scenarios; tolerant

1. Introduction

The world food supply is affected by environmental abiotic stresses, which damages up to 70% of food crop yields [1–3]. In the Near East and North Africa (NENA) region, physical water scarcity is already affecting food production [4]. The NENA region is characterized by an arid climate with a total annual rainfall much lower than the evapotranspiration of the field crops. In the Arab World, more than 85% of the available water resources are used for agriculture [5]. Despite this high-water allocation for the agriculture sector, about 50% of food requirements are imported [4]. Crop irrigation uses poor quality groundwater, which is saline in nature. The uninterrupted application of groundwater for irrigation is replete, which leads to a severe increase in soil salinity and reduction in crop yields. Climate changes, namely the increase in global temperatures and the decline in rainfall, exacerbate soil salinization, resulting in loss of production in arable lands [6]. According to recent estimates, one-fifth of the irrigated lands in the world are affected by salinity. Every day, on average, 2000 ha of irrigated land in arid and semi-arid areas is adversely affected by salinity problems [7]. The annual economic loss due to these increases in soil salinity is about USD 27.3 billion [8].

Cereals are the main crops in the Mediterranean and NENA regions, contributing to food security and social stability. Barley is one of these staple crops in the area. However, its production is constrained by abiotic factors, such as the arid climate, low and erratic rainfall, and soil and water salinity. The anticipated climate changes will further increase the negative impacts of these factors in the future [9]. Barley (*Hordeum vulgare* L.) is a drought- and salt-tolerant crop with considerable economic importance in Mediterranean and NENA regions since it is a source of stable farm income [10]. Indeed, barley is a staple food for over 106 countries in the world [11]. Barley is characterized by its high adaptability from humid to arid and even Saharan environments. Barley is grown in many areas of the world and is used for feed, food, and malt production [2,12].

To improve barley production in these regions, plant scientists have adopted a strategy to identify tolerant genotypes for maintaining reasonable yield on salt-affected soils [13]. Crops physiologists and breeders are working to assess how efficient a genotype is in converting water into biomass or yield. To do so, they use production parameters, with which measurement in field experiments is difficult and time-consuming. However, these complex parameters can be determined with the help of crop growth simulation models [6,13]. Dynamic simulation models describe the growth and development of crops based on the interaction with soil, water, and climate parameters. Models can be used to simulate soil and water salinity and crop management practices on the growth and yield of crops under different agro-climatic conditions [6].

Models were used to test the impact of salinity on crops under different environmental conditions and different fertilization practices [14,15].

AquaCrop is a water-driven dynamic model (Vanuytrecht et al., 2014). AquaCrop is a simulation model to study crops' water productivity. As crop-water-productivity is affected by climatic conditions, it is crucial to understand water productivity's response to changing rainfall and temperature patterns [9].

Among the available models, AquaCrop is preferred due to its robustness, precision, and the limited number of variables to be introduced [16]. It uses a small number of explicit and intuitive parameters that require simple calculation [16]. AquaCrop is a software system developed by the Land and Water Division of FAO to estimate water use efficiency and improve agricultural systems' irrigation management practices [17,18].

Water productivity (WP) can be described as the ratio of crops' net benefits, including both rain and irrigation.

According to [19], irrigation management organizations are interested in the yield per unit of irrigation water applied, as they have to improve the yield through human-induced irrigation processes. However, the downside is that not all irrigation water is used to generate crop production. Therefore, FAO defines water productivity as a ratio between a unit of output and a unit of input. Here, water productivity is used exclusively to indicate the amount or value of the product over the volume or value of water that is depleted or diverted [20].

This model was developed by the Food and Agriculture Organization (FAO) [16,21]. AquaCrop simulates the response of crop yield to water and is particularly suited to regions where water is the main limiting factor for agricultural production. The model is based on the concepts of crops' yield response to water developed by Doorenbos and Kassam [22]. The AquaCrop model (v4.0) published in 2012 can estimate yield under salt stress conditions.

The AquaCrop model has been used to predict crop yields under salt stress conditions in different parts of the world [23,24]. Kumar et al. [23] successfully used the AquaCrop model to predict the water productivity of winter wheat under different salinity irrigation water regimes. Mondal et al. [24] used AquaCrop to evaluate the potential impacts of water, soil salinity, and climatic parameters on rice yield in the coastal region of Bangladesh. The AquaCrop model has also been widely used to simulate yields of various crops under diverse environments. For example, barley (*Hordeum vulgare* L.) [5,25,26], teff (*Eragrostis teff* L.) [5], cotton (*Gossypium hirsutum* L.) [27], maize (*Zea mays* L.) [28] wheat (*Triticum aestivum* L.) [3].

In this study, the AquaCrop model (v4.0) is used to assess the performance of two barley genotypes under three contrasted agro-ecosystems (soil, salinity, and climate). In these areas, groundwater is primarily used for irrigation. The salinity of irrigation water ranges from 3 to 15 dS m^{-1}. Farmers do not know which barley variety is most tolerant to producing a reasonable yield under these saline environments. Furthermore, model simulations were also performed to evaluate the impact of three irrigation water salinity levels (5, 10, and 15 dS m^{-1}) on the barley yield. A cost–benefit analysis was performed to determine the economic returns of each level of salinity water irrigation and genotype tolerance based on model simulation results. Those results should help recommend the farmers of saline areas to enhance barley yield and economic return.

2. Materials and Methods

Description of Field Trial Sites

Field experiments were conducted during the 2012–2014 period in three contrasting locations (Beja, Kairouan, Medenine) of Tunisia. The Beja site (36°44′01.13″ N; 9°08′14.30″ E) is sub-humid, Kairouan (35°34′34.97″ N; 10°02′50.88″ E) is located in the semi-arid area of central Tunisia, and Medenine (33°26′54″ N, 10°56′31″ E) is part of the South East arid region of Tunisia (Figure 1). Two barley varieties (Konouz from Tunisia and Batini 100/1 B from Oman) were used for field experiments. The Konouz variety is salt-sensitive [29,30], whereas Batini 100/1 B is salt-tolerant [29,31].

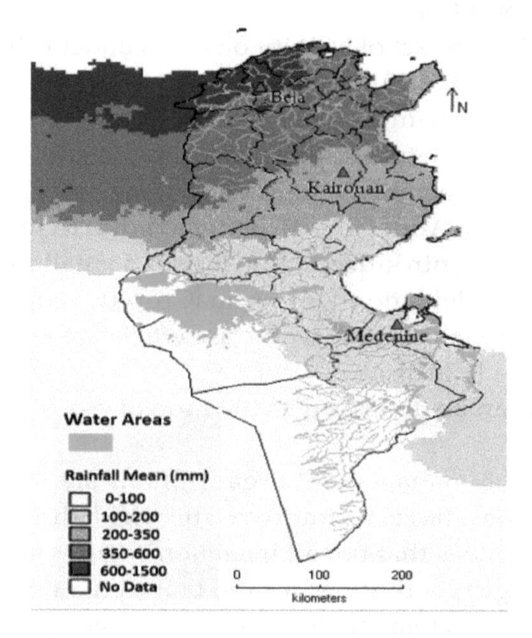

Figure 1. Location of field trial sites in different agro-climatic zones of Tunisia.

In Kairouan (KAI) and Medenine (MED) field trial sites were divided into two sub-plots. Each subplot was irrigated by one water salinity treatment (EC = 2 and 13 dS m^{-1}). Three blocks were defined perpendicularly to the sub-plots so that both treatments were observed in each block. As Beja is located in the rainfed cereal growing area of Tunisia, no irrigation was applied.

The weather data characterize the trials sites related to temperature, and rain was described by [29]. The irrigation water applied and reference evapotranspiration (ET_o) registered in the trials during the two growing seasons are presented in Table 1. The collected data from each site were used to estimate the reference evapotranspiration (ET) according to the Penman-Monteith Evapotranspiration FAO-56 Method, and then the total water supplied was determined for each site to obtain the water barley requirement. Irrigation was applied using a drip system. To ensure water supply homogeneity, line source emitters were installed at each planting row and 33-cm spacing between emitters on the same row.

Table 1. Rainfall, irrigation water applied and evapotranspiration (ET_o) in three trial sites.

Growing Season	Rainfall (mm)			Irrigation Water Applied (mm)			ET_o (mm)		
	Sites			Sites			Sites		
	Beja	KAI	MED	Beja	KAI	MED	Beja	KAI	MED
2012–2013	472.2	151.9	81.1	0	360	455	393.8	364.7	327.6
2013–2014	413.5	180.0	156.1	0	360	405	390.1	363.7	328.4

Soil samples were taken from the trial sites, and physico-chemical analyses were performed. The site's soil characteristics are diverse, from soil rich in clay and organic matter in BEJ to sandy soil with impoverished organic matter continent in MED (Table 2).

Table 2. Soil properties in three field trial sites.

Site	Sand (%)	Clay (%)	Silt (%)	OM (%)	Na^+ Content (ppm)	K^+ Content (ppm)	Ca^{2+} Content (ppm)	PWP (% vol)	FC (% vol)
Beja	15.0	57.5	27.5	4.7	10–20	250–300	100–110	32.0	50.0
KAI	14.8	45.1	40.1	4.0	230–270	390–550	90–140	23.0	39.0
MED	55.5	20.5	24.0	0.9	120–200	30–70	30–55	6.0	13.0

(OM: organic matter, PWP: permanent wilting point; FC: field capacity).

Crops were sown during the last week of November. Seeds were hand sown at the rate of 200 viable grains per m^2. Nitrogen, potassium and phosphorus were applied separately at 85, 50, and 50 kg/ha rates, respectively.

At the five different stages, plants for each genotype, from three small areas (25 × 25 cm) were taken from each experimental unit and used to determine the biomass. At a final harvest stage, plot (1 × 2 m) was used for biomass and grain yield assessment. Water productivity (WP) was calculated as the ration between the collected yield expressed in kg ha^{-1} and the daily transpiration simulated by the model.

3. Description of the AquaCrop Model

The model describes soil, water, crop, and atmosphere interactions through four sub-model components: (i) the soil with its water balance; (ii) the crop (development, growth, and yield); (iii) the atmosphere (temperature, evapotranspiration, and rainfall), and carbon dioxide (CO_2) concentration; and (iv) the management, such as irrigation and crop fertilization soil fertility.

The AquaCrop model is based on the relationship between the relative yield and the relative evapotranspiration [22] as follows

$$\frac{Y_x - Y_a}{Y_x} = K_y\left(\frac{ET_x - ET_a}{ET_x}\right) \tag{1}$$

where Y_x is the maximum yield, Y_a is the actual yield, ET_x is the maximum evapotranspiration, ET_a is the actual evapotranspiration, and K_y is the yield response factor between the decrease in the relative yield and the relative reduction in evapotranspiration.

The AquaCrop model does not take into account the non-productive use of water for separating evapotranspiration (ET) into crop transpiration (T) and soil evaporation (E)

$$ET = E + Tr \tag{2}$$

where ET = actual evapotranspiration, E = soil evaporation and Tr = the sweating of crop.

At a daily time step, the model successively simulates the following processes: (i) groundwater balance; (ii) development of green canopy (CC); (iii) crop transpiration; (iv) biomass (B); and (v) conversion of biomass (B) to crop yield (Y). Therefore, through the daily potential evapotranspiration (ET_o) and productivity of water (WP*), the daily transpiration (Tr) is converted into vegetal biomass as follows

$$B_i = WP^*\left(\frac{T_{ri}}{ETo_i}\right) \tag{3}$$

where WP* is the normalized water productivity [32,33] relative to Tr. After the normalization of water productivity for different climatic conditions, its value can be converted into a fixed parameter [34]. The estimation and prediction of performance are based on the final biomass (B) and harvest index (HI). This allows a clear distinction between impact of stress on B and HI, in response to the environmental conditions

$$Y = HI * (B) \tag{4}$$

where: Y = final yield; B = biomass; HI = harvest index.

During the calibration and testing of the model, we calculated water productivity (WP) as presented by Araya et al. [5]

$$WP = \left[\frac{Y}{\sum Tr}\right] \tag{5}$$

where Y is the yield expressed in kg ha^{-1} and Tr is the daily transpiration simulated by the model.

3.1. Crop Response to Soil Salinity Stress

The electrical conductivity of saturation soil-past extracts from the root zone (ECe) is commonly used as an indicator of the soil salinity stress to determine the total reduction in biomass production, determines the value for soil salinity stress coefficient ($K_{s,\ salt}$).

The coefficient of soil salinity stress (Ks_{salt}) varied between 0 (full effect of stress of soil salinity) and 1 (no effect). The following equation determined the reduction in biomass

$$B_{rel} = 100\ (1 - Ks_{salt}) \tag{6}$$

B_{rel} represents the expected biomass production under given salinity stress relative to the biomass produced in the absence of salt stress. The coefficient is adjusted daily to the average ECe in the root zone [35].

Then, the thresholds values are given for the sensitive and tolerant barley genotype and expressed in dS m^{-1}. This allows the estimation of the lower limit (EC_{en}) to which the soil salinity stress begins to affect the production of biomass and the upper threshold (EC_{ex}), in which soil salinity stress has reached its maximum effect.

3.2. Soil Salinity Calculation

AquaCrop adopts the calculation procedure presented in BUDGET [36] to simulate the movement and retention of salt in the soil profile. The salts enter the soil profile as solutes after irrigation with saline water or through capillary rise from a shallow groundwater table (vertical downward and upward salt movement). The average ECe in the compartments of the effective rooting depth determines the effects of soil salinity on biomass production.

To explain the movement and retention of soil water and salts in the soil profile, AquaCrop divides the soil profile into 2 to 11 soil compartments called "cells", depending on the type of soil in each horizon (clay, sandy horizon) and its saturated hydraulic conductivity (Ksat in mm/day). The salt diffusion between two adjacent cells (cell j and cell j+1) is determined by the differences in salt concentration and expressed by the electrical conductivity (EC) of soil water.

AquaCrop determines the vertical salt movement in response to soil evaporation, considering the amount of water extracted from the soil profile by evaporation and the wetness of the upper soil layer. The relative soil water content of the topsoil layer determines the fraction of the dissolved salts that moves with the evaporating water.

AquaCrop determines the vertical salt movement because of the capillary rise. Finally, the salt content of a cell is determined by

$$\mathrm{Salt_{cell}} = 0.64\ \mathrm{W_{cell} EC_{cell}} \tag{7}$$

$\mathrm{Salt_{cell}}$ is the salt content expressed in grams salts per m^2 soil surface, Wcell its volume expressed in liter per m^2 (1 mm = 1 L/m^2), and 0.64 a global conversion factor used in AquaCrop to convert dS/m to g/L. The electrical conductivity of the soil water (ECsw) and of the electrical conductivity of saturation soil-past extract (ECe) at a particular soil depth (soil compartment) is calculated as

$$\mathrm{EC_{sw}} = \frac{\sum_{j=1}^{n} \mathrm{Salt_{cell.j}}}{0.64\ (1000\ \theta\Delta_z)\left\{1 - \frac{\mathrm{Vol\%_{gravel}}}{100}\right\}} \tag{8}$$

$$\mathrm{EC_{e}} = \frac{\sum_{j=1}^{n} \mathrm{Salt_{cell.j}}}{0.64\ (1000\ \theta_{sat}\Delta_z)\left\{1 - \frac{\mathrm{Vol\%_{gravel}}}{100}\right\}} \tag{9}$$

where n is the number of cells in each soil compartment; θ is the soil water content (m^3/m^3); θ_{sat} is the soil water content (m^3/m^3) at saturation; Δz (m) is the thickness of the soil compartment and Vol% gravel is the volume percentage of the gravel in the soil horizon of each compartment.

4. Model Calibration

4.1. Input and Output Variables of the Model

The model was calibrated using data from the growing season of 2012–2013 and evaluated using data from 2013–2014. Determining parameters for crop development and production, as well as water and salinity stress, was fundamental for calibrating the AquaCrop model. The parameters of climate, soil, and crop management used for the model calibration are presented in Table 3.

Table 3. Climate, soil and crop parameters used for the simulation model AquaCrop.

Climate		- Daily rainfall, daily ET_o, daily temperatures - CO_2 concentration
Crop	Limited set	Crop development and production parameters which include phenology and life cycle
	Crop parameters	- Harvest index - Root zone threshold at the end of the canopy expansion - Threshold root zone depletion for early senescence - Time for the maximum canopy cover - Maximum vegetation - Flowering time - Initial vegetative cover - Depletion threshold root zone for stomata closure - Extraction of water

Table 3. *Cont.*

	Field	- Soil fertility, mulch - Field practices (surface runoff presence, ground bond)
Soil	Soil profile	Characteristics of soil horizon (no of soil horizon, thickness, Permanent Wilting Point (PWP), Field Capacity (FC), Soil saturation (SAT), Ksat); soil surface (runoff, evaporation); Restrictive soil layer capillary rise).
	Soil water and groundwater	Constant depth; variable depth; water quality.

4.2. Statistical Parameters Used for the Calibration and Evaluation of Model

Several statistical indices were used to evaluate the performance of the model on the field measured data. These include Percentage Error (PE), Root Mean Square Error (RMSE), Model Efficiency (ME) and Coefficient of Determination (R^2).

Percentage Error (PE) was determined using the following equation

$$EP = \frac{(S_i - O_i)}{O_i} \times 100 \tag{10}$$

where S_i and O_i are simulated and observed values, respectively.

The root means square error (RMSE) [37] is presented by the following equation

$$RMSE = \sqrt{\frac{1}{n}\sum_{i=1}^{n}(S_i - O_i)^2} \tag{11}$$

with the values of RMSE close to zero indicate the best model fit.

The model efficiency (ME) [38] was applied to assess the effectiveness of the model. The ME indicator compares the variability of prediction errors by the model to those of collected data from the field. If the prediction errors are greater than the data error, then the indicator becomes negative. The upper ME bound is at 1.

$$ME = \frac{\sum_{i=1}^{n}(O_i - MO)^2 - \sum_{i=1}^{n}(S_i - O_i)^2}{\sum_{i=1}^{n}(O_i - MO)^2} \tag{12}$$

The coefficient of determination (R^2), as a result of regression analysis, is the proportion of the variance in the dependent variable (predict value) that is predictable from the independent variable (observed value) and is computed according to [35]

$$R^2 = \left\{ \frac{\sum_{i=1}^{n}(O_i - \overline{O})(S_i - \overline{S})}{\left[\sum_{i=1}^{n}(O_i - \overline{O})^2\right]^{0.5}\left[\sum_{i=1}^{n}(S_i - \overline{S})^2\right]^{0.5}} \right\}^2 \tag{13}$$

R^2 is between 0 and 1.

4.3. Parameters Used for Model Calibration

In total, 26 input parameters were used for the model calibration (Table 4). Out of these, 14 parameters were considered as "conservative" because they do not change with salinity and are independent of limiting or non-limiting conditions. These parameters include normalized crop water productivity and crop transpiration coefficient. The remaining 12 are site-specific (climate, water, and soil salinity) and crop-specific (tolerant or sensitive). These input parameters were adjusted during the calibration process to obtain better adequacy between the measured and simulation values.

Table 4. Final values of different model input parameters obtained after calibration for two genotypes under different salinity levels (S1 = 2 dS m^{-1}; S2 = 13 dS m^{-1}).

		Batini-100/1 B (Salt-Tolerant)			Konouz (Salt-Sensitive)			Remarks
		BEJ	KAI	MED	BEJ	KAI	MED	
Base temperature (°C)	S1	0	0	0	0	0	0	Conservative
	S2	-	0	0	-	0	0	
Upper temperature (°C)	S1	30	30	30	30	30	30	Conservative
	S2	-	30	30	-	30	30	
Initial canopy cover, CC0 (%)	S1	1.5	1.5	1.5	1.5	1.5	1.5	Conservative
	S2	-	1.5	1.5	-	1.5	1.5	
Canopy cover per seeding (cm^2/plant)	S1	0.75	0.75	0.75	0.75	0.75	0.75	Conservative
	S2	-	0.75	0.75	-	0.75	0.75	
Maximum coefficient for transpiration, KcTr, x	S1	0.90	0.90	0.90	0.90	0.90	0.90	Conservative
	S2	-	0.90	0.90	-	0.90	0.90	
Maximum coefficient for soil evaporation, Kex	S1	0.4	0.4	0.4	0.4	0.4	0.4	Conservative
	S2	-	0.4	0.4	-	0.4	0.4	
Upper threshold for canopy expansion, Pexp, upper	S1	0.30	0.30	0.30	0.20	0.20	0.20	Varietal effect
	S2	-	0.30	0.30	-	0.20	0.20	
Lower threshold for canopy expansion, Pexp, lower	S1	0.65	0.65	0.65	0.55	0.55	0.55	Varietal effect
	S2	-	0.65	0.65	-	0.55	0.55	
Leaf expansion stress coefficient curve shape	S1	4.5	4.5	4.5	4.5	4.5	4.5	Conservative
	S2	4.5	4.5	4.5	4.5	4.5	4.5	
Upper threshold for stomatal closure, Psto, upper	S1	0.6	0.6	0.6	0.55	0.55	0.55	Varietal effect
	S2	-	0.6	0.6	-	0.55	0.55	
Leaf expansion stress coefficient curve shape	S1	4.5	4.5	4.5	4.5	4.5	4.5	Conservative
	S2	4.5	4.5	4.5	4.5	4.5	4.5	
Canopy senescence stress coefficient, Psen, upper	S1	0.65	0.65	0.65	0.55	0.45	0.45	Varietal effect and site effect for the sensitive
	S2	-	0.65	0.65	-	0.45	0.45	
Senescence stress coefficient curve shape	S1	4.5	4.5	4.5	4.5	4.5	4.5	Conservative
	S2	4.5	4.5	4.5	-	4.5	4.5	

Table 4. *Cont.*

		Batini-100/1 B (Salt-Tolerant)			Konouz (Salt-Sensitive)			Remarks
		BEJ	KAI	MED	BEJ	KAI	MED	
Reference harvest index, HI0 (%)	S1	40	40	41	41	42	45	Varietal and salt stress effect
	S2		40	45		41	41	
Normalized crop water productivity, WP* (g/m^2)	S1	14	14	14	14	14	14	Conservative
	S2	14	14	14	14	14	14	
Time from sowing to emergence (day)	S1	7	7	7	7	7	7	Conservative
	S2	7	7	7	7	7	7	
Time from sowing to maximum CC (jours)	S1	60	60	60	62	60	57	Varietal and salt stress effect
	S2	-	60	58	-	59	55	
Time from sowing to maximum CC (day)	S1	145	145	145	145	145	145	Conservative
	S2	-	145	145	145	145	145	
Time from sowing to maturity (day)	S1	178	157	157	178	157	157	Varietal and salt stress effect
	S2	-	157	157	-	157	157	
Maximum canopy cover, CCx (%)	S1	87	87	87	87	75	63	Varietal and salt stress effect
	S2	-	87	70	-	60	40	
Canopy growth coefficient, CGC (%/day)	S1	12.5	12.5	12.5	12	12	12	Varietal effect
	S2		12.5	12.5		12	12	
Canopy decline coefficient, CDC (%/day)	S1	6	6	6	6	6	6	Conservative
	S2	-	6	6	-	6	6	
Maximum effective rooting depth, Zx (m)	S1	0,9	0.75	0.75	0,9	0.75	0.75	Site effect
	S2	-	0.75	0.75	-	0.75	0.75	
Salinity stress, lower threshold, ECen ($dS\ m^{-1}$)	S1	3	3	3	1	1	1	Varietal effect
	S2	3	3	3	1	1	1	
Salinity stress, upper threshold, ECex ($dS\ m^{-1}$)	S1	22	22	22	18	18	18	Varietal effect
	S2	22	22	22	18	18	18	
Shape factor for salinity stress coefficient curve	S1	1	1	1	1	1	1	Conservative
	S2	1	1	1	1	1	1	

For Bej only rainfall; for Kai, two levels of water salinity (S1 = 1.2 $dS\ m^{-1}$; S2 = 13 $dS\ m^{-1}$ (S2)); for Med, two levels of water salinity (S1 = 2 $dS\ m^{-1}$; S2 = 13 $dS\ m^{-1}$ (S2)).

4.4. Development of Different Scenarios

After calibration and evaluation, the model was used to assess the performance of two barley varieties under three water salinity conditions scenarios i.e., 5, 10, and 15 dS m^{-1}, using the weather data for the growing season 2013–2014.

4.5. The Economic Gain from the Use of a Unit of Water Consumed in the Tow Barley Varieties under Different Climatic Conditions

The economic productivity of two barley varieties was estimated using the average unit cost of one water cubic meter in Tunisia and the water use predicted by AquaCrop. The crop water economic productivity of the tolerant and the sensitive barley varieties as the measure of the biophysical and then economic gain from the use of a unit of water consumed were estimated by AquaCrop model in grain yield production [20]. This is expressed in productive crop units of kg/m^3 and money unit/m^3.

5. Results

5.1. Biophysical Environments Variability of Experimental Sites

The experiments are conducted in adaptability trials set up in three contrasting biophysical environments (from the sub-humid to the arid interior). These sites, namely Beja, Kairouan and Medenine, were selected on a North–South transect (Figure 1). The soils of the trial sites are very diverse, from soil rich in clay and poor in organic matter in BEJ to sandy soil with poor organic matter continent in MED (Table 2). Beja's sub-humid site received annual rainfall of 472 and 413 mm respectively during the two cropping seasons. However, in the semi-arid and arid sites, low rainfall was registered. The arid site of MED received an annual rainfall of 81 mm during the first cropping season and 156 mm during the second. At Kairouan, the rainfall for the 2012/2013 and 2013/2014 seasons was 152 and 180 mm, respectively (Table 1). As Beja is located in the rainfed cereal-growing area of Tunisia, no irrigation was applied. KAI and MED field trial sites, two different salinities (EC = 2 and 13 dS m^{-1}) of water were used for irrigation.

Soil calcium and potassium content was higher in KAI as compared to MED. Soil sodium content changes during the different experimentation period following irrigation with saline water in KAI and MED, where sodium is the dominant element present in the saline irrigation water. The variation between sites might be explained by the variation in the cationic exchange capacity of the sandy soil and torrential character of the rainfall in this area (Table 2).

5.2. Biomass, Grain Yield, and Water Use Efficiency

The correlation between grain yield, biomass, and water productivity values for two barley genotypes showed that the observed and simulated values are closely co-related, as evidenced by the high R^2 values, i.e., 0.91, 0.93, and 0.89 for grain yield, biomass, and water productivity, respectively (Figure 2).

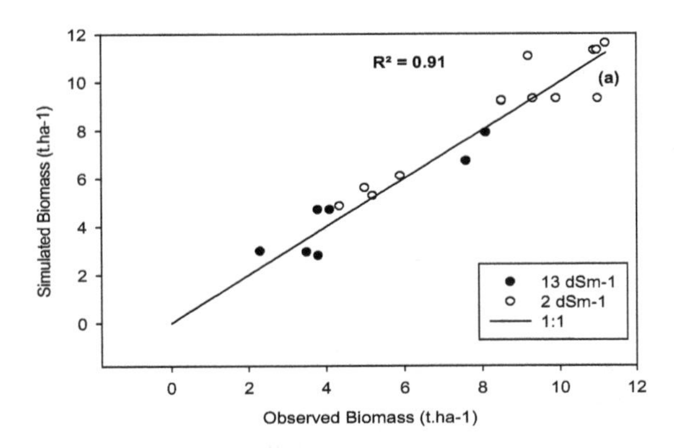

Figure 2. *Cont.*

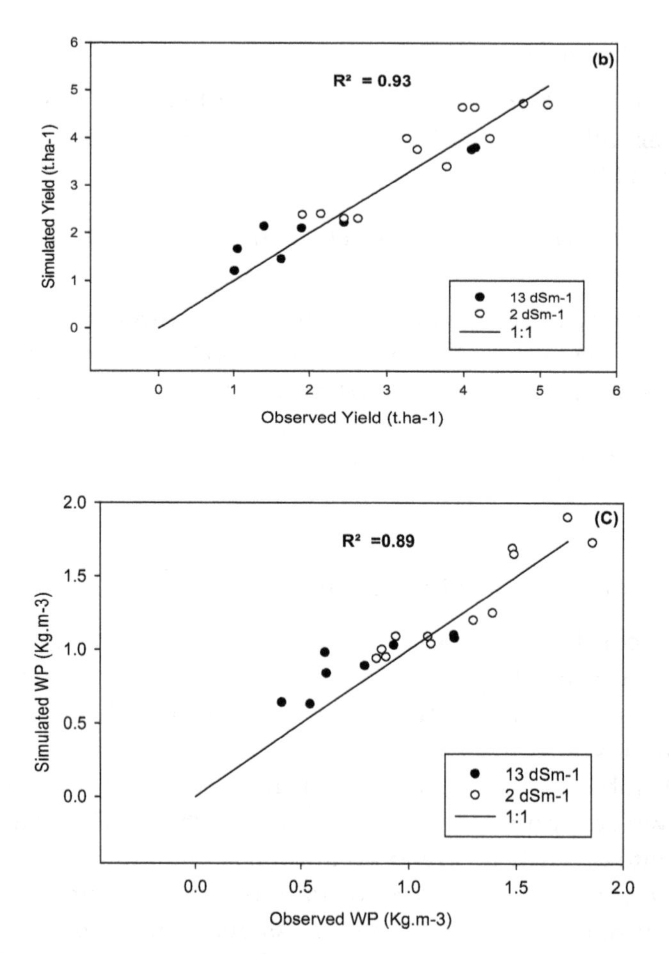

Figure 2. Correlation between observed and simulated (**a**) biomass yield; (**b**) grain yield; and (**c**) water productivity compared with 1:1 line.

The correlation between observed and simulated values of biomass yield for two barley genotypes at three locations showed proximity (Figure 3), which indicates the excellent ability of the AquaCrop model to predict biomass yield under different agro-climatic conditions. The results also show that the sensitive barley variety at MED produces the lowest biomass for both irrigation water qualities. Similar trends were observed for grain yield, where the tolerant barley variety performed better than the sensitive variety regardless of the location and the quality of irrigation water.

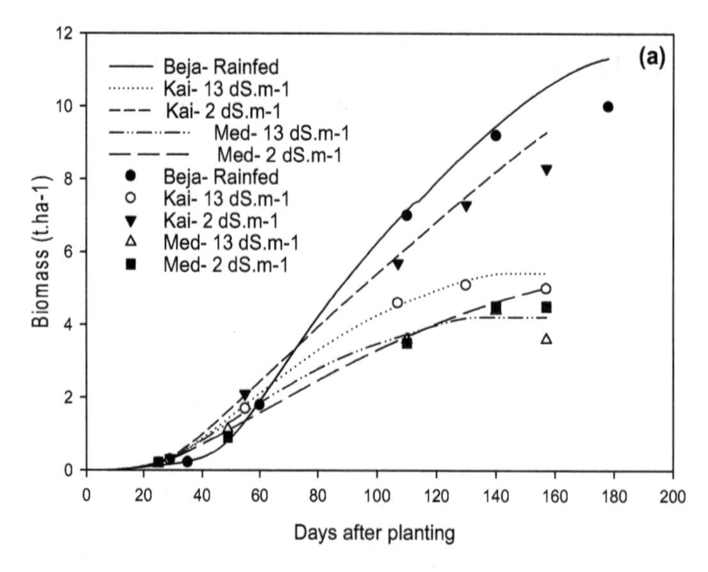

Figure 3. *Cont.*

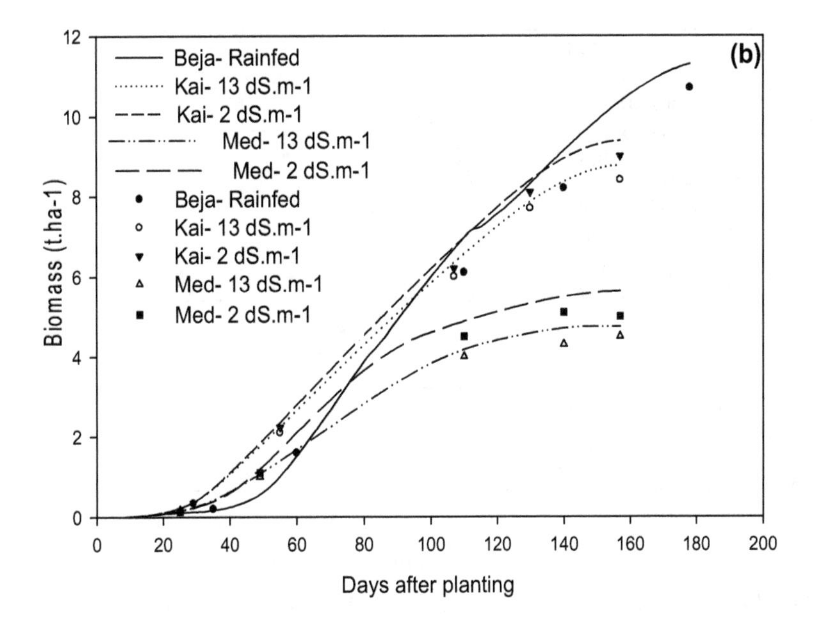

Figure 3. Simulated and observed biomass of (**a**) tolerant and (**b**) sensitive barley genotypes (dots represent observations; simulations are represented by lines).

5.3. Canopy Cover (CC)

The maximum and minimum CC were 85% and 30% in the sub-humid and arid areas, respectively. The salinity induces a 10% reduction in the CC in the sub-humid environment and 5–30% in the dry climate of MED. CC reduction under saline irrigation water is less noticeable in the tolerant variety than the sensitive variety for both salinity levels. However, in the rainfed area of Beja, the growth of both varieties was comparable.

Figure 4 shows a strong correlation between measured and simulated CC values for both varieties of barley ($R^2 = 0.91$ and $R^2 = 0.93$). In general, a good match between the observed and the simulated CC was observed in all three locations. However, the model somewhat over-estimated CC in the rainfed environment of Beja and slightly under-estimated it in the other two situations.

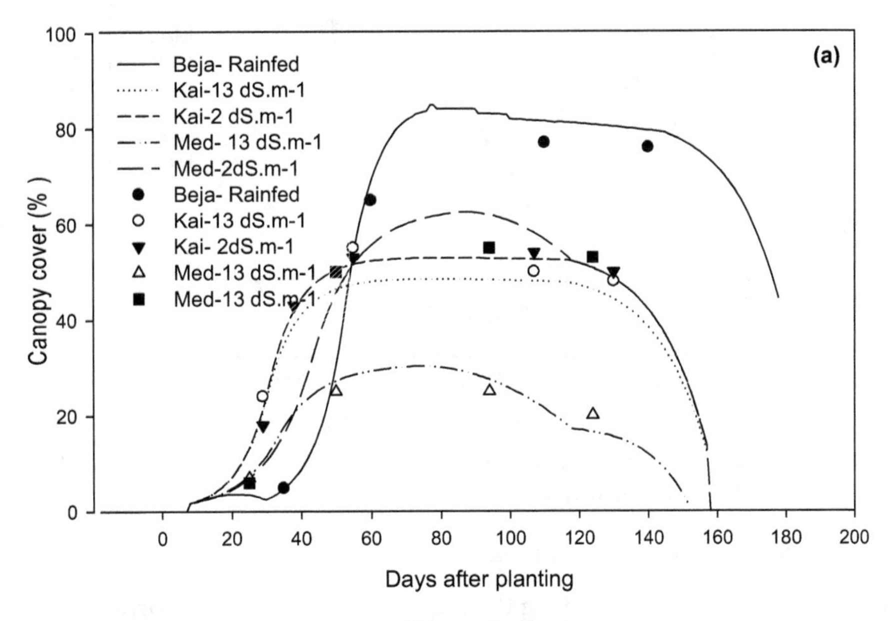

Figure 4. *Cont.*

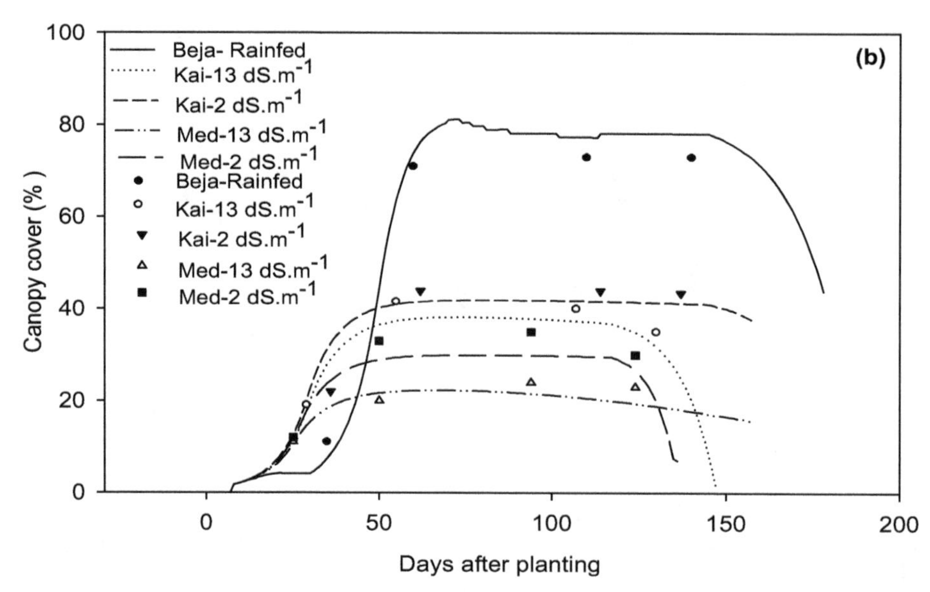

Figure 4. Simulated and observed canopy cover for (**a**) tolerant and (**b**) sensitive barley varieties.

5.4. *Effects of Soil Salinity*

The maximum soil salinity was in the arid and semi-arid areas irrigated with saline water, respectively. The soil salinisation dynamic depends on the salinity of irrigation water. However, in the rainfed area of Beja, we noted the absence of any salty issue.

Figure 5 shows that the simulated soil salinity trend in the root zone (up to a depth of 0.7 m) corresponds very well with the measured values under different saline water regimes across different environments throughout the growing season. The observed and modeled soil salinity correlated well, with an R^2 of 0.96. Figure 5 shows that the model reliably simulated average root zone salinity when the crop is irrigated with low-salinity water (2 dS m^{-1}). However, it slightly underestimated soil salinity under higher saline water conditions (13 dS m^{-1}), particularly for the late growing season.

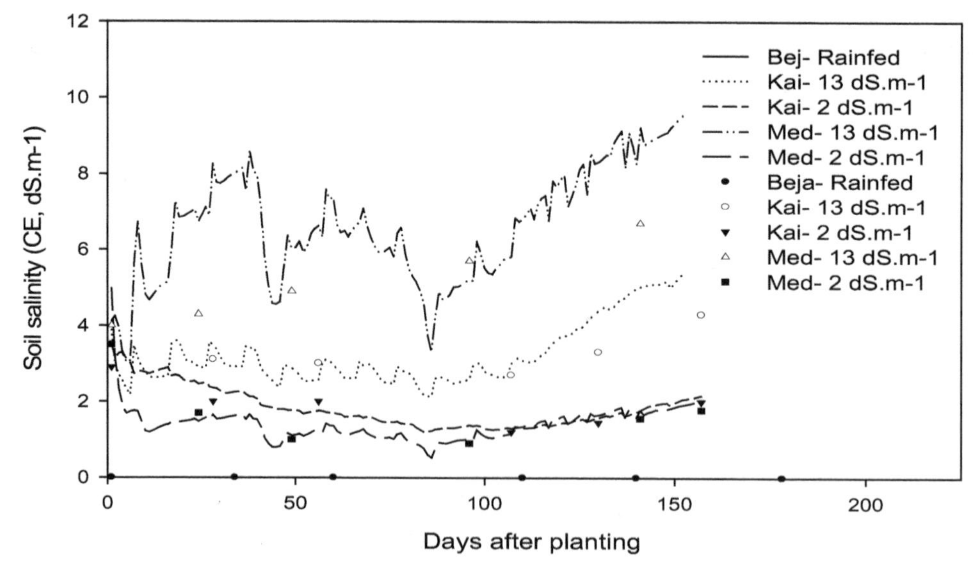

Figure 5. Simulated and observed soil salinity in the testing-cropping season under different saline water regimes and across different environments.

5.5. *Statistical Indices for AquaCrop Model Evaluation*

The statistical indices derived for evaluating the AquaCrop model's performance in predicting soil water content, yield, canopy cover percent, biomass, and water productivity (*WP*) of barley genotypes under different saline water regimes across different environments are given in Table 5. All statistical parameters depict a strong correlation between simulated and observed values for model calibration and evaluation periods. The correlation between all statistical parameters remained almost the same for the calibration and evaluation period, which indicates the robustness of the model prediction. Based on the model calibration and evaluation results, the model was found robust enough to calculate different scenarios.

Table 5. Statistical indices values for different parameters obtained for model calibration.

	Variable	RMSE	ME	R^2
Calibration	Grain yield (t ha^{-1})	0.40	0.89	0.91
	Biomass (t ha^{-1})	0.87	0.96	0.93
	water productivity (kg ha^{-1} mm^{-1})	0.15	0.84	0.89
	Soil salinity	0.34	0.91	0.95
	Canopy cover percent	1.5	0.89	0.91
Evaluation	Grain yield (t ha^{-1})	0.45	0.87	0.89
	Biomass (t ha^{-1})	0.89	0.86	0.87
	water productivity (kg ha^{-1} mm^{-1})	0.13	0.91	0.84
	Soil salinity	1.25	0.91	0.96
	Canopy cover percent	2.25	0.89	0.91

6. Development of Different Scenarios

Due to a shortage of surface water, farmers of KAI and MED regions have no option than to use groundwater for irrigation. The quality of groundwater ranges from 4 to15 dS m^{-1} in these two regions. Farmers are interested to know which barley varieties would be most suitable to grow under these groundwater quality conditions. The calibrated and evaluated model was used to assess the performance of two barley varieties under three water salinity conditions i.e., 5, 10, and 15 dS m^{-1}, and the results are presented in Table 6.

Table 6. Predicted values of biomass, yield, and water productivity of two barley varieties for different scenarios.

	BEJ	KAI			MED		
	Rainfed	5 dS m^{-1}	10 dS m^{-1}	15 dS m^{-1}	5 dS m^{-1}	10 dS m^{-1}	15 dS m^{-1}
			Tolerant genotype				
Biomass (t ha^{-1})	11.30	9.07	8.36	5.48	5.60	4.74	4.70
Yield (t ha^{-1})	4.70	3.65	3.44	2.20	2.29	2.13	2.10
WP (kg m^{-3})	1.73	1.29	1.19	0.85	1.25	1.18	1.00
			Sensitive genotype				
Biomass (t ha^{-1})	11.33	6.62	4.60	1.90	3.18	3.03	2.80
Yield (t ha^{-1})	4.64	2.70	1.90	0.80	1.40	1.30	0.80
WP (kg m^{-3})	1.65	1.12	0.85	0.45	0.74	0.72	0.51

The performance of both barley varieties in the KAI area is predicted to be much higher than MED area under all salinity levels due to prevailing climatic conditions. In the KAI area, biomass and grain yield reductions are much higher with the increasing water salinity for both varieties. For example, the biomass and yield reductions in the KAI area were about 40%with an increase in salinity from 5 to 10 and 15 dS m^{-1}. For the sensitive genotype, the biomass and yield reductions in the KAI area would be above 72% with a similar increase in the salinity levels. Although overall biomass and grain yields in the MED area were lower than in the KAI area, biomass and yield reductions for the salt-tolerant barley variety were only 16% and 8%, with an increase in salinity from 5 to 15 dSm^{-1}, respectively.

However, for the sensitive genotype, reductions in biomass and yield were 12% and 43%, respectively, with a similar increase in salinity levels. Similar trends are obtained for water productivities.

Without salt stress, both varieties have the same performance. However, the tolerant variety performs better than the sensitive variety under salt stress. This is because it has better potential. Therefore, farmers can grow both varieties in the rainfed areas of BEJ, while, in KAI and MED areas where irrigation is necessary for crop growth, the salt-tolerant barley variety should be preferred. The cultivation of the salt-sensitive barley variety in the MED area will be risky, as the yields will be low, and the development of soil salinity over time will remain a challenge. This situation will be very critical for long-term sustainable crop production in the area.

7. Economic Productivity of Barley Varieties under Different Climatic Conditions

The economic productivity of two barley varieties was estimated using the average unit cost of one water cubic meter in Tunisia and the water use predicted by AquaCrop. The results show that the production cost of 1 kg of barley is lowest in the BEJ area compared to those areas where it is irrigated with saline water.

In the KAI region, the cost will be reduced by 13.28% 28.72% and 47.19% by growing the tolerant variety irrigated with saline water of 5, 10, and 15 dS m^{-1}, respectively. In the arid region of MED, the benefit will be reduced by 40%, 38%, and 49% by growing the tolerant barley variety by irrigating with saline water of 5, 10 and 15 dS m^{-1}, respectively (Figure 6). However, in the sub-humid region of BEJ, there is no significant difference between susceptible and tolerant genotypes. The results show the economic interest for arid region farmers to grow the tolerant barley variety. This stresses the need for appropriate breeding programs for the saline environments for optimizing crop production instead of targeting potential yields.

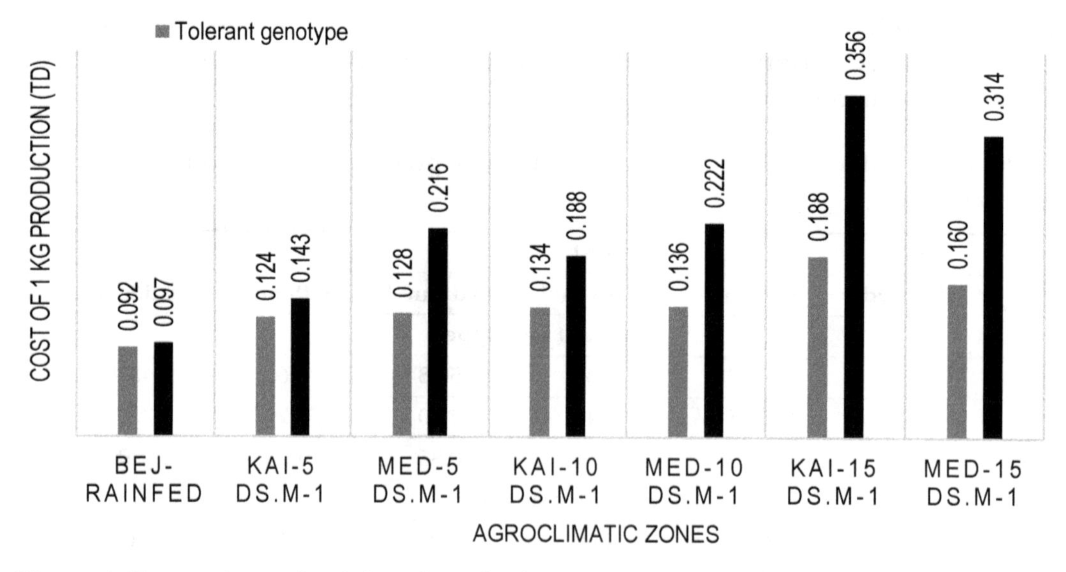

Figure 6. Economic productivity of two barley varieties under different climatic conditions.

8. Discussion

We evaluated the AquaCrop model for two barley varieties under contrasting environments and different water salinity levels. The simulated model values were close to the field measurements concerning biomass, yield and soil salinity. ME and R^2 parameters were close to 0.9, showing the model's ability to simulate the behavior of sensitive and resistant cultivars in contrasting environments and irrigation practices. Araya et al. [5] reported R^2 values of 0.80 when simulating barley biomass and grain yield using AquaCrop. El Mokh et al. [25] reported R^2 values of 0.88 when simulating barley yield under different irrigation regimes in a dry environment using AquaCrop. Mondal et al. [24] reported a 0.12 t ha^{-1} root mean square error after simulating the yield response of rice to salinity stress with the AquaCrop model. Our results also show a correct prediction with an RMSE of 0.45 t ha^{-1} (Table 5). This shows that the AquaCrop model simulates biomass production for all environments with an acceptable accuracy level.

AquaCrop model produces consistent simulation results for CC with an R^2 of 0.89 and RMSE of 2.25 (Table 5). The model also simulated soil salinity satisfactorily for all environments (R^2 = 0.96) for all situations. The R^2 values exceeding 0.8 are considered excellent for model performance [39]. The ability of AquaCrop to predict yield depends on the appropriate calibration of the canopy cover curve [1,40]. Indeed, after simulation of soil water balance at a daily time step, the model simulates CC and then simulates the transpiration of a crop, biomass above the soil, and converts biomass into yield. Therefore, it is essential to make accurate predictions of the canopy cover by the proper calibration of crop traits.

Therefore, through proper calibration, models can be used for additional solutions for the quantification of salinity build-up in the root zone [41].

We also noted the overestimation of the soil salinity at the end of the growing season when saline water is used for irrigation (Figure 5). This could be due to the excessive leaching of salts from the soil profile through irrigation, as reported by Mohammadi et al. [42]. Over- or underestimation at the end of the season could be the simplification of soil salt transport calculations in the model based on some empirical functions, including the parameters of Ks and the drainage coefficient for vertical downward salt movement. Furthermore, the occasional leaching of salts from the root zone using relatively better-quality water is also recommended. Changing cropping patterns is also a useful strategy for the rehabilitation and management of saline soils, especially when only saline water is available for irrigation.

The AquaCrop model was also capable of predicting water productivity under sub-humid, semi-arid, and arid environments and the effect of salinity. Plants subjected to salinity stress show a varying response in *WP*. The sensitive genotype was more exposed to varying responses in *WP*. Besides, heat stress induced by increased temperatures and the water deficit also decreases productivity, as demonstrated by Hatfield [43]. The observed and predicted water productivities were directly affected by climate aridity and the salinity of the irrigation water. However, the tolerant barley variety was less affected by these factors. These results are in agreement with the earlier studies [16,44].

Water scarcity is already hampering agricultural production in the MENA region. Therefore, the adoption of integrated management strategies will be useful for growing tolerant genotypes under saline water conditions and increasing the water use efficiency. For the sustainable management of crop growth in saline environments, soil-crop-water management interventions consistent with site-specific conditions need to be adopted [41]. These may include cyclic or conjunctive saline water use and freshwater through proper irrigation scheduling to avoid salinity development.

There are several traits available for screening genetic material for enhanced production and *WP* under different climate scenarios. This study shows that, under different water salinity conditions, sensitive barley genotype is more affected by the increasing water salinity than the tolerant barley genotype. The crop yields for both genotypes under all water salinity levels were higher in KAI area compared to the MED area. Therefore, this study recommends that farmers with higher salinity water for irrigation should grow tolerant barley genotypes, allowing them to reduce the cost, on average, by 30% (Figure 6). However, from a sustainability point of view, irrigation amounts should be kept to a minimum to optimize crop yields instead of targeting potential yields [45]. This exercise will help there be less accumulation of salts in the root zone. Besides, the occasional leaching of salts from the root zone using relatively better-quality water is also recommended. Changing cropping patterns is also regarded as a useful strategy for the rehabilitation and management of saline soils, especially when only saline water is available for irrigation [46,47].

9. Conclusions

The AquaCrop model with a salinity module was used to evaluate the agronomic performance of two barley varieties for the three different agro-climatic zones in Tunisia. These zones represent sub-humid, semi-arid, and arid climates. The model was calibrated and evaluated using field data from two years (2012 and 2014). The excellent correlation between the simulated and measured data of biomass, yield, and soil salinity confirms the ability of AquaCrop model to simulate crop growth under different climatic conditions. The scenario results using the calibrated model indicate that farmers with higher salinity water for irrigation should grow tolerant barley genotypes. However, from a sustainability point of view, irrigation amounts should be kept to a minimum to optimize crop yields instead of targeting potential yields.

Author Contributions: Conceptualization, Z.H.; Data curation, F.E.B.A. and S.A.; Formal analysis, Z.H.; Methodology, Z.H., A.S. and Z.C.; Supervision, Y.T.; Writing—Original draft, Z.H.; Writing—Review and editing, A.S.Q. and A.G. All authors have read and agreed to the published version of the manuscript.

References

1. Abi Saab, M.T.; Albrizio, R.; Nangia, V.; Karam, F.; Rouphael, Y. Developing scenarios to assess sunflower and soybean yield under different sowing dates and water regimes in the Bekaa valley (Lebanon): Simulations with Aquacrop. *Int. J. Plant Prod.* **2014**, *8*, 457–482.
2. Ahmed, M.; Goyal, M.; Asif, M. Silicon the non-essential beneficial plant nutrient to enhanced drought tolereance in wheat. In *Crop Plant*; Goyal, A., Ed.; Intech Publication House: London, UK, 2012; pp. 31–48.
3. Andarzian, B.; Bannayan, M.; Steduto, P.; Mazraeh, H.; Barati, M.E.; Barati, M.A.; Rahnama, A. Validation and testing of the AquaCrop model under fulland deficit irrigated wheat production in Iran. *Agric. Water Manag.* **2011**, *100*, 1–8. [CrossRef]
4. Araya, A.; Habtub, S.; Hadguc, K.; Kebedea, A.; Dejened, T. Test of AquaCrop model in simulating biomass and yield of water deficient and irrigated barley (Hordeum vulgare). *Agric. Water Manag.* **2010**, *97*, 1838–1846. [CrossRef]
5. Araya, A.; Keesstra, S.D.; Stroosnijder, L. Simulating yield response to water of Tef (Eragrostistef) with FAO's AquaCrop model. *Field Crop. Res.* **2010**, *116*, 1996–2204. [CrossRef]
6. Chauhdarya, J.N.; Bakhsh, A.; Ragab, R.; Khaliq, A.; Bernard, A.; Engeld, M.R.; Shahid, M.N.; Nawaz, Q. Modeling corn growth and root zone salinity dynamics to improve irrigation and fertigation management under semi-arid conditions. *Agric. Water Manag.* **2020**, *230*, 105952. [CrossRef]
7. FAO (Food and Agriculture Organization of the United Nations). *Advances in the Assessment and Mmonitoring of Salinization and Status of Biosaline Agriculture*; FAO: Rome, Italy, 2010.
8. Qadir, M.; Quillérou, E.; Nangia, V.; Murtaza, G.; Singh, M.; Thomas, R.J.; Drechsel, P.; Noble, A.D. Economics of salt-induced land degradation and restoration. *Nat. Resour. Forum* **2014**, *38*, 282–295. [CrossRef]
9. Hatfield, J.L.; Dold, C. Water-Use efficiency: Advances and challenges in a changing climate. *Front. Plant Sci.* **2019**. [CrossRef]
10. Zhou, G.; Johnson, P.; Ryan, P.R. Quantitative trait loci for salinity tolerance in barley (*Hordeum vulgare* L.). *Mol. Breed.* **2012**, *29*, 427–436. [CrossRef]
11. Newton, A.C.; Flavell, A.J.; George, T.S.; Leat, P.; Mullholland, B.; Ramsay, L.; RevoredoGiha, C.; Russell, J.; Steffenson, B.J.; Swanston, J.S.; et al. Crops that feed the world 4. Barley: A resilient crop? Strengths and weaknesses in the context of food security. *Food Secur.* **2011**, *3*, 141–178. [CrossRef]
12. Zhang, H.; Han, B.; Wang, T.; Chen, S.; Li, H.; Zhang, Y.; Dai, S. Mechanisms of Plant Salt Response: Insights from Proteomics. *J. Proteome Res.* **2012**, *11*, 49–67. [CrossRef]
13. Negrao, S.; Schmockel, S.M.; Tester, M. Evaluating physiological responses of plants to salinity stress. *Ann. Bot.* **2017**, *119*, 1–11. [CrossRef] [PubMed]
14. Heng, L.K.; Hsiao, T.; Evett, S.; Howell, T.; Steduto, P. Validating the FAO AquaCrop model for irrigated and water deficient field maize. *Agron. J.* **2009**, *101*, 488–498. [CrossRef]

15. Verma, A.K.; Gupta, S.K.; Isaac, R.K. Use of saline water for irrigation in monsoon climate and deep water table regions: Simulation modelling with SWAP. *Agric. Water Manag.* **2012**, *115*, 186–193. [CrossRef]
16. Soothar, R.K.; Wenying, Z.; Yanqing, Z.; Moussa, T.; Uris, M.; Wang, Y. Evaluating the performance of SALTMED model under alternate irrigation using saline and fresh water strategies to winter wheat in the North China Plain. *Environ. Sci. Pollut. Res.* **2019**, *26*, 34499–34509. [CrossRef] [PubMed]
17. Steduto, P.; Hsiao, T.C.; Raes, D.; Fereres, E. AquaCrop—The FAO crop model to simulate yield response to water. I. Concepts and underlying principles. *Agron. J.* **2009**, *101*, 426–437. [CrossRef]
18. Van Gaelen, H. *AquaCrop Training Handbooks–Book II Running AquaCrop*; Food and Agriculture Organization of the United Nations: Rome, Italy, 2016.
19. Doorenbos, J.; Kassam, A.H. *Yield Response to Water*; FAO Irrigation and Drainage Paper No. 33; FAO: Rome, Italy, 1979.
20. Kumar, A.; Sarangi, D.K.; Singh, R.; Parihar, S.S. Evaluation of aquacrop model in predicting wheat yield and water productivity under irrigated saline regimes. *Irrig. Drain.* **2014**, *63*, 474–487. [CrossRef]
21. Mondal, M.S.; Fazal, M.A.; Saleh, M.D.; Akanda, A.R.; Biswas, S.K.; Moslehuddin, Z.; Sinora, Z.; Attila, N. Simulating yield response of rice to salinity stress with the AquaCrop model. *Environ. Sci. Process. Impacts* **2015**, *17*, 1118–1126. [CrossRef]
22. El Mokh, F.; Nagaz, K.; Masmoudi, M.M.; Mechlia, N.B.; Fereres, E. *Calibration of AquaCrop Salinity Stress Parameters for Barley under Different Irrigation Regimes in a Dry Environment*; Springer: Cham, Germany, 2017. [CrossRef]
23. Hellal, F.; Mansour, H.; Mohamed, A.H.; El-Sayed, S.; Abdelly, C. Assessment water productivity of barley varieties under water stress by AquaCrop model. *AIMS Agric. Food* **2019**, *4*, 501–517. [CrossRef]
24. Tan, S.; Wang, Q.; Zhang, J.; Chen, Y.; Shan, Y.; Xu, D. Performance of AquaCrop model for cotton growth simulation under film-mulched drip irrigation in southern Xinjiang, China. *Agric. Water Manag.* **2018**, *196*, 99–113. [CrossRef]
25. Hammami, Z.; Gauffreteau, A.; BelhajFraj, M.; Sahlia, A.; Jeuffroy, M.H.; Rezgui, S.; Bergaoui, K.; McDonnell, R.; Trifa, Y. Predicting yield reduction in improved barley (*Hordeum vulgare* L.) varieties and landraces under salinity using selected tolerance traits. *Field Crop. Res.* **2017**, *211*, 10–18. [CrossRef]
26. Sbei, H.; Sato, K.; Shehzad, T.; Harrabi, M.; Okuno, K. Detection of QTLs for salt tolerance in Asian barley (*Hordeum vulgare* L.) by association analysis with SNP markers. *Breed. Sci.* **2014**, *64*, 378–388. [CrossRef] [PubMed]
27. Jaradat, A.A.; Shahid, M.; Al-Maskri, A.Y. Genetic diversity in the Batini barley landrace from Oman: Spike and seed quantitative and qualitative traits. *Crop. Sci.* **2014**, *44*, 304–315. [CrossRef]
28. Raes, D.; Steduto, P.; Hsiao, T.C.; Fereres, E. *Crop Water Productivity. Calculation Procedures and Calibration Guidance. AquaCrop Version 3.0. FAO*; Land and Water Development Division: Rome, Italy, 2009.
29. Trombetta, A.; Iacobellis, V.; Tarantino, E.; Gentile, F. Calibration of the AquaCrop model for winter wheat using MODIS LAI images. *Agric. Water Manag.* **2015**, *164*. [CrossRef]
30. Hanks, R.J. Yield and water-use relationships. In *Limitations to Efficient Water Use in Crop Production*; Taylor, H.M., Jordan, W.R., Sinclair, T.R., Eds.; ASA, CSSA, and SSSA: Madison, WI, USA, 1983; pp. 393–411.
31. Tanner, C.B.; Sinclair, T.R. Efficient water use in crop production: Research or re-search? In *Limitations to Efficient Water Use in Crop Production*; Taylor, H.M., Jordan, W.R., Sinclair, T.R., Eds.; ASA, CSSA, and SSSA: Madison, WI, USA, 1983; pp. 1–27.
32. Steduto, P.; Hsiao, T.C.; Fereres, E. On the conservative behavior of biomass water productivity. *Irrig. Sci.* **2007**, *25*, 189–207. [CrossRef]
33. Raes, D.; Steduto, P.; Hsiao, T.C.; Fereres, E. *AquaCrop Version 5.0 Reference Manual*; Food and Agriculture Organization of the United Nations: Rome, Italy, 2016.
34. Loague, K.; Green, R.E. Statistical and graphical methods for evaluating solute transport models: Overview and application. *J. Contam. Hydrol.* **1991**, *7*, 51–73. [CrossRef]
35. Minhas, P.S.; Tiago, B.; AlonBen-Gal, R.; Pereira, L.S. Coping with salinity in irrigated agriculture: Crop evapotranspiration and water management issues. *Agric. Water Manag.* **2020**, *227*, 105832. [CrossRef]
36. Pereira, L.S.; Paredes, P.; Rodrigues, G.C.; Neves, M. Modeling barley water use and evapotranspiration partitioning in two contrasting rainfall years. Assessing SIMDualKc and AquaCrop models. *Agric. Water Manag.* **2015**, *159*, 239–254. [CrossRef]

37. Iqbal, M.A.; Shen, Y.; Stricevic, R.; Pei, H.; Sun, H.; Amiri, E.; Rio, S. Evaluation of FAO Aquacrop model for winter wheat on the North China plain under deficit from field experiment to regional yield simulation. *Agric. Water Manag.* **2014**, *135*, 61–72. [CrossRef]
38. Mohammadi, M.; Ghahraman, B.; Davary, K.; Ansari, H.; Shahidi, A.; Bannayan, M. Nested validation of AquaCrop model for simulation of winter wheat grain yield soil moisture and salinity profiles under simultaneous salinity and water stress. *Irrig. Drain.* **2016**, *65*, 112–128. [CrossRef]
39. Hatfield, J.L. Increased temperatures have dramatic effects on growth and grain yield of three maize hybrids. *Agric. Environ. Lett.* **2016**, *1*, 1–5. [CrossRef]
40. Hsiao, T.C.; Heng, L.; Steduto, P.; Roja-Lara, B.; Raes, D.; Fereres, E. AquaCrop—The FAO model to simulate yield response to water: Parametrization and testing for maize. *Agron. J.* **2009**, *101*, 448–459. [CrossRef]
41. Wiegmanna, M.; William, T.B.; Thomasb, H.J.; Bullb, I.; Andrew, J.; Flavellc, J.; Annette, Z.; Edgar, P.; Klaus, P.; Andreas, M. Wild barley serves as a source for biofortification of barley grains. *Plant Sci.* **2019**, *283*, 83–94. [CrossRef] [PubMed]
42. Roberts, D.P.; Mattoo, A.K. Sustainable agriculture—Enhancing environmental benefits, food nutritional quality and building crop resilience to abiotic and biotic stresses. *Agriculture* **2018**, *8*, 8. [CrossRef]
43. Tavakoli, A.R.; Moghadam, M.M.; Sepaskhah, A.R. Evaluation of the AquaCrop model for barley production under deficit irrigation and rainfed condition in Iran. *Agric. Water Manag.* **2015**, *161*, 136–146. [CrossRef]
44. Eisenhauer, J.G. Regression through the origin. *Teach. Stat.* **2003**, *25*, 76–80. [CrossRef]
45. Teixeira, A.D.C.; Bassoi, L.H. HBassoi Crop Water Productivity in Semi-arid Regions: From Field to Large Scales. *Ann. Arid Zone* **2009**, *48*, 1–13.
46. FAO. The Irrigation Challenge. In *Increasing Irrigation Contribution To Food Security through Higher Water Productivity from Canal Irrigation Systems*; Issue paper; FAO: Rome, Italy, 2003.
47. Barnston, A. Correspondence among the Correlation [root mean square error] and Heidke Verification Measures; Refinement of the Heidke Score. *Notes Corresp. Clim. Anal. Cent.* **1992**, *7*, 699–709.

3

Water use and Rice Productivity for Irrigation Management Alternatives in Tanzania

Stanslaus Terengia Materu [1], Sanjay Shukla [2,*], Rajendra P. Sishodia [2], Andrew Tarimo [1] and Siza D. Tumbo [1]

[1] Department of Engineering Sciences and Technology, Sokoine University of Agriculture, P. O. Box 3003, Chuo Kikuu, Morogoro, Tanzania; stanslaus_materu@yahoo.com (S.T.M.); andrewtarimo2@yahoo.co.uk (A.T.); siza.tumbo@gmail.com (S.D.T.)

[2] Agricultural and Biological Engineering Department, University of Florida, 2685 State Road 29 N, Immokalee, FL 34142, USA; rpsishodia@ufl.edu

* Correspondence: sshukla@ufl.edu

Abstract: Rice production is important for global food security but given its large water footprint, efficient irrigation management strategies need to be developed. Expansion of rice growing area is larger than any other crop in Africa due to increasing demand for rice. Three rice irrigation management alternatives with the system of rice intensification (SRI) were field-evaluated against the conventional continuously flooded system (CF) in Tanzania. Production systems included: (1) CF (50 mm ponding depth for the entire season); (2) SRI (40 mm ponding for 3 days and no irrigation for next 5 days); (3) 80% SRI (80% of the SRI ponding); and (4) 50% SRI (50% of the SRI ponding). Experimental evaluation of the four systems was conducted for both wet and dry seasons. For the dry season, the SRI and 80% SRI produced higher yields of 9.68 tons/ha and 11.45 tons/ha and saved 26% and 35% of water, respectively compared to the CF (8.69 tons/ha). The yield advantage of the 80% SRI and SRI over the CF was less during the wet season with 6.01 tons/ha and 5.99 tons/ha of production, and water savings of 30% and 14%, respectively compared to the CF (5.64 tons/ha). The 50% SRI had lowest yield of all for both seasons, 7.48 tons/ha and 4.99 tons/ha for the dry and wet seasons, respectively. Statistically, the 80% SRI treatment outperformed all other treatments over the two seasons with an additional yield of 1.57 tons/ha and 33% (345 mm) water savings compared to the CF. Economic productivity of water (US$/ha-cm) over two seasons was highest for the 80% SRI ($20.27/ha-cm), while it was lowest for the CF ($12.89/ha-cm). Water saved by converting from the CF to the 80% SRI (1.98 million ha-cm) can support a 50% expansion in the current rice irrigated area in Tanzania. Even without irrigation expansion, the 80% SRI can increase rice production by 1.5 million tons annually while enhancing water availability for industrial and environmental uses (e.g., ecological preserves) and help achieve food security in Tanzania and the greater sub-Saharan Africa.

Keywords: Africa; deficit irrigation; food security; system of rice intensification; water conservation; water productivity

---

1. Introduction

Water is a valuable resource that is becoming increasingly scarce due to growing population and intensifying agriculture [1]. Water scarcity is challenging the ability of countries to meet the increasing food demand [2]. Globally, agriculture is the largest consumer (≈70%) of freshwater accounting for 90% of consumptive water use [3,4]. Of the three main food crops (maize, wheat, and rice), rice is the most important crop especially in developing countries [5]. Given its large water footprint, practices that can reduce water inputs for rice production such as deficit irrigation need to be explored. Deficit irrigation is a technique used to minimize water losses and increase water efficiency, especially in areas where there is insufficient water supply for irrigation. Deficit irrigation management involves inducing marginal stress, except in critical growth stages where crop yield might be negatively affected [6].

Expansion of rice growing area is larger than any other crop in Africa due to its increasing demand. [7]. Tanzania is the largest (947,303 km^2) country in East Africa and accounts for 9% (2.6 million ton) of African rice production (30.8 million ton) [8]. However, due to a rising gap between production and consumption, many African countries, including Tanzania, are becoming increasingly dependent on rice imports [9,10]. At the same time, increasing irrigation withdrawals and spatial and temporal variability in rainfall and surface flows are causing water scarcity in many parts of Tanzania such as the Pangani and Rufiji River basin [11,12]. The Pangani and Rufiji rivers support majority of irrigated agriculture in Tanzania and support almost entire hydroelectric generation in Tanzania (Mtera, Kidatu and Kihansi plants) [12]. Growing population, increasing food demands, and increased rainfall variability due to changed climate is likely to exacerbate water availability in the future. There is a need to develop alternative farming systems that can increase or sustain rice yields with reduced water footprint to ensure the food security in Tanzania.

Field water use for rice typically ranges from 1000–2000 mm [13], which is 2–3 times of other cereal crops. In rice production systems, a large quantity of water is lost through evapotranspiration, surface runoff, seepage, and deep percolation [14]. Several water-saving irrigation techniques have been developed for rice [15]. For instance, in Asia, the most widely adopted water-saving practice is aerobic rice production system. Although the aerobic rice system reduces water use it also results in lower yields compared to lowland flooded rice [16]. This practice also has some limitations related to soil type, rice variety, and socio-economic constraints [17]. Other strategies being pursued to reduce rice water requirements include alternate wetting and drying (AWD) and saturated soil culture [18]. Studies show that AWD can reduce crop water requirements while maintaining or even increasing the yield as compared to the conventional flooded system [19,20]. The AWD is an irrigation practice where water is applied to attain certain depth of ponding after which the field is left unirrigated for some time (e.g., 5 to 7 days) to dry out or drain. In the traditional continuously flooded (CF) system, water is applied at a frequency that will maintain a certain depth (e.g., 5 cm) of ponding throughout the season. Under the CF, more than 50% of irrigation water is lost through seepage, deep percolation, and excessive unproductive evaporation [21].

The system of rice intensification (SRI) is a relatively new production practice that has been adopted by many farmers in Asia and Sub-Saharan Africa [22–25]. The SRI is a combination of agronomic practices comprising of land preparation, seed selection, nursery establishment, transplanting of young age seedlings (8 to 12 days), wider plant spacing, AWD, and frequent weeding [22,26]. The SRI practice has been reported to substantially increase the yields as compared to the CF system [27]. The AWD irrigation technique used in the SRI production system can reduce water use by minimizing evaporation and deep percolation losses. The SRI combined with AWD has a potential to reduce water application and yet increase or sustain current yields in Tanzania. A range of AWD regimes are possible with SRI. However, limited research has been conducted on the evaluation of SRI in combination with different AWD regimes (e.g., ponding depth) in Tanzania. The current SRI practice lacks specific information regarding the irrigation management needed to achieve optimum yield. The goal of this study was to find water sustainable rice production systems in Tanzania by comparing water and yield metrics of conventional continuously flooded rice with SRI production system under different AWD regimes during wet (February-June) and dry (September-January) seasons. Specifically, the study attempts to answer the following questions: (a) Can SRI-AWD combination with reduced ponding depth significantly increase crop yield and reduce irrigation requirement compared to CF? (b) Which irrigation management strategy (scheduling and ponding depth) provides highest water productivity?

2. Methods and Materials

2.1. Study Area

The experiment was conducted in Morogoro, located 200 km south west of the city of Dar es Salaam. The experimental fields were located at the research farm of Sokoine University of Agriculture (SUA) (latitude = 37°39′26″ E and longitude = 6°51′5″ S) at an altitude of 510 m above mean sea level. The average annual temperature at the site is 23 °C with a minimum of 15 °C in July and a maximum of 32 °C in November and December. The mean relative humidity (1971–2000) for the area is 73%. Rice is grown during two seasons, dry (September–January) and wet (February–June) seasons. These two growing seasons have different irrigation needs due to differences in rainfall (1971–2000 average wet season rainfall = 53 cm and dry season = 38 cm) and evaporative demands. The seasonal mean (1971–2000) relative humidity is 66% and 78% for the dry and wet seasons, respectively. Humidity, wind speed, solar radiation, and temperature data, measured at the SUA meteorological station, were used to calculate reference evapotranspiration (ETo) using FAO Penman-Monteith method [28] (Figure 1). The average ETo for the wet and dry seasons are 52 and 64 cm, respectively. Given this weather variability, the optimum irrigation strategy for the dry season is likely to be different from the wet season. Although water availability limits large-scale production, better yields promote rice production during the dry season.

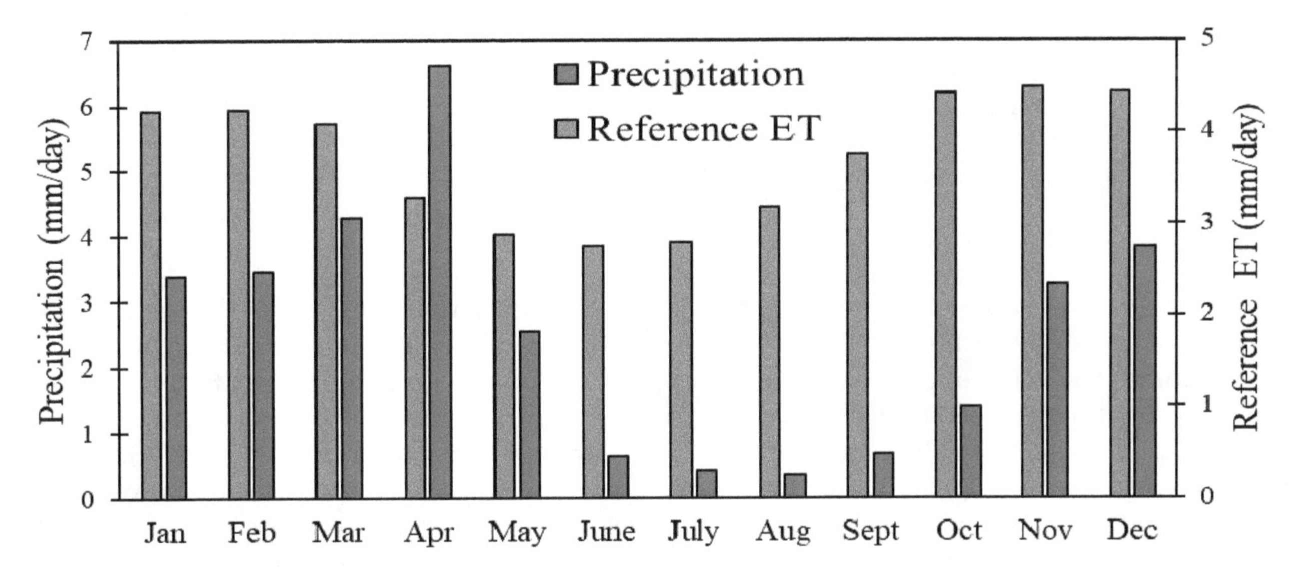

Figure 1. Average daily rainfall and reference ET (1971–2000) for Morogoro, Tanzania.

The experiment was conducted during the dry and wet seasons of 2012–2013. For 2013 wet season, the experimental site was moved to another plot within the same research farm due to land availability issues. Annual rainfall was measured at the site using a standard rain gauge. Another set of rainfall measurements were taken at the meteorological station at SUA. The rainfall system is bimodal, characterized by two rainfall peaks with short rains from October to December and long rains from March to May (Figure 1). Measured rainfall at the site was 489 mm and 1379 mm during the dry and wet season, respectively. High rainfall in the wet season disrupted the AWD cycle for all the SRI treatments. The top soil (0–30 cm) at the research site is a dark brown, clay loam (47% clay, 7% silt and 46% sand). The soil is acidic with a pH of 5.6. The volumetric water content at field capacity and permanent wilting point is 40.1% and 28.7%, respectively for the top soil (0–30 cm). The saturated hydraulic conductivity of the top 30 cm soil is 0.35 mm/h. Soil texture (sand, silt and clay percent) and bulk density data obtained from collected soil samples were used to estimate soil water retention and saturated hydraulic conductivity parameters [29].

2.2. Experimental Design

The layout of the experiment was a complete randomized block design (CRBD) with factorial arrangement of four treatments with three replications. A total of 12 experimental fields (area each field = 40 m^2) were identified within a rice field. Four production systems (treatments) with differential irrigation volumes evaluated in this study included: (1) continuous flooding (CF, Control) (maintain 50 mm ponding depth for the entire season); (2) SRI (maintain 40 mm ponding depth for three consecutive days followed by no irrigation for five days); (3) 80% SRI (maintain 80% of the SRI ponding depth i.e., 32 mm for three consecutive days followed by no irrigation for five days); and (4) 50% SRI (maintain 50% of the SRI ponding depth i.e., 20 mm for three consecutive days followed by no irrigation for five days). All the SRI systems involved alternate wetting and drying (AWD) during the initial stages. After the panicle initiation stage, continuous flooding was practiced in all the SRI treatments to maintain 20 mm ponding depth until plant senescence to ensure better grain filling. Important dates for the two seasons are shown in Table 1.

Table 1. Crop management and development stages for dry and wet seasons.

Event	Period (DAT *)	Dry Season	Wet Season
Field Preparations	-	15 September 2012	8 February 2013
Nursery	-	24 September 2012	18 February 2013
Transplanting	-	6 October 2012	1 March 2013
Tillering	0–46	21 November 2012	16 April 2013
Panicle initiation	47–59	3 December 2012	29 April 2013
Flowering	60–72	15 December 2012	13 May 2013
Grain filling	73–90	1 January 2013	27 May 2013
Harvesting	111 (Wet season) 113 (Dry season)	26 January 2013	19 June 2013

*: DAT-days after transplanting.

The SRI treatments involved transplanting 12-day seedlings, with two leaves. The seedlings were transplanted carefully and quickly to minimize seedling damage. The number of seedling per hill was one for all the SRI treatments; this allowed optimum growth without competition for nutrients. For the CF treatment, based on the conventional practice in Tanzania, three seedlings per hill were planted. For both the CF and SRI treatments, fertilizer was applied at a rate of 50 kg/ha (N), 50 kg/ha (K_2O) and 50 kg/ha (P_2O_5) before the last puddling event. Weeding was done manually at 12-day intervals using a spike-toothed harrow. To encourage greater root and canopy growth, plant-to-plant and row-to-row spacing was maintained at 25 cm for all the SRI and CF treatments.

2.3. Sri Management

Each experimental plot was leveled to allow uniform water ponding. The soil was kept saturated for five days, and then rotavated. The field was harrowed twice at an interval of three days to ensure proper soil-water mixture. Twelve-days-old seedlings were transplanted before the emergence of a third leaf. Care was taken to separate the seedling from the seedbed to avoid damage to the young root. One seedling was planted per hill at a depth of two cm on the 25 cm square grid. Between the transplanting and appearance of panicles, three to five days irrigation cycle was followed, i.e., the field was irrigated for three consecutive days and then left to dry for five days. The goal was to keep the soil moist but not saturated to allow air to get into the soil for improved soil health and root growth. After panicle initiation, irrigation was applied to maintain 20 mm ponding for all three SRI treatments.

2.4. Irrigation and Soil Moisture Measurements

Irrigation water volume for each treatment was measured using a propeller type flow meter. Soil moisture was measured every 15-min using a capacitance probe (EnviroScan, Sentek Technologies,

Stepney, Australia; sensors at 10, 20, 30, 60, 80, and 90 cm below the soil surface) in one of the plots for each of the four treatments. Manufacturer provided calibration equation for clay loam soil was used to measure the soil moisture content at multiple depths. Capacitance-type soil moisture probes provide reasonable soil moisture measurements even without site-specific calibration (3–4% accuracy) and therefore, could be used for irrigation scheduling [30,31]. Measured soil moisture data from capacitance probes have been successfully used to determine irrigation water requirements and scheduling for agricultural crops [32,33]. Furthermore, measured average maximum soil moisture (saturation moisture) within and below the root zone at the study site was 48–60% which is close to the estimated saturation moisture content of 45–50%. Therefore, measured soil moisture was assumed to adequately represent the actual soil moisture content and its variation during the growing seasons. The capacitance probes were connected to a CR206 datalogger (Campbell Scientific Inc., Logan, UT, USA) to store the data. For the dry season, soil moisture was measured from 3 November 2012 to 30 January 2013. During the wet season, soil moisture data could only be measured from 4 March–6 May 2013 due to theft of the datalogger.

2.5. Plant and Yield Observations

Five plants were randomly selected from each plot to measure plant height and number of tillers during each development stage. At grain maturity, the field was drained to allow the soil to dry before harvesting. Three one m quadrants were selected in each plot for yield measurements. Dry biomass (oven dried for 24 h or more until no change in weight) of different plant organs (stem, leaves and panicles) was weighed. Length and width of leaves were measured manually to estimate leaf area index (LAI) using the method by Yin et al. (2000) [34].

2.6. Data Analysis

All statistical analyses were conducted using SAS [35]. The data were analyzed using Tukey-Kramer test for comparing pair wise differences of means [36]. Variables compared included water applied (mm), crop yield (kg/ha), soil moisture (%vol.), LAI, above ground biomass (kg/ha) and economic productivity of water (US$/ha-cm). The economic productivity of water was calculated using the income (I, $) from crop yield and volume of water applied (ha-cm, irrigation plus rainfall) [37] as:

$$Economic\ productivity\ of\ water = \frac{I(\$)}{Water\ applied\ (\mathrm{Ha-cm})}$$

Average farm gate paddy price of 550 Tanzanian Shillings TZS/kg [38] and prevailing exchange rate (1 US$ = 1600 TZS) during 2013 was used to calculate the income (I).

3. Results and Discussion

3.1. Plant Growth

Plant height for the CF and SRI treatments were significantly ($p < 0.05$) higher than both the 80% SRI and 50% SRI for both seasons (Table 2). Plant height for the CF was 23% and 63% higher than the 80% SRI and 50% SRI, respectively during the dry season. Plant height in the wet season for the CF was 11% and 30% greater than the 80% SRI and 50% SRI, respectively. More ponding depth or higher water availability is the likely reason for higher plant height in the CF as it can increase plant nutrient uptake and plant height [39,40]. Similar plant heights for the CF and SRI are likely due to similar soil moisture or water availability in the root zone.

The number of tillers for the 80% SRI was significantly higher ($p < 0.05$) than rest of the three treatments in both seasons (Table 2). Based on the statistical analyses results, the number of tillers can be arranged as 80% SRI > SRI = CF > 50% SRI (Table 2). For the dry season, the 80% SRI had 40% more tillers than the CF and 93% more tillers than the 50% SRI. Shortening of the vegetative stage duration has been shown to result in increased tillers [41]. High number of tillers per hill for the 80% SRI indicates higher potential yield than the rest of the treatments. Because panicles are attached to tillers, the number of tillers are usually an indicator of yield; the higher the number of tillers, the higher the potential for increased yield. The advantage of the SRI method in enhancing tiller numbers was observed by many researchers [42,43]. Transplanting of younger seedlings, higher plant spacing and soil aeration due to wetting and drying cycle promotes root growth and tillers under the SRI system [44]. Results for the wet season were similar to the dry season however the numerical differences between the treatments were much lower. Part of this difference was due to rainfall that masked the effect of differential irrigation input. Overall, the 80% SRI outperformed all other treatments in the number of tillers indicating higher yield potential than other treatments.

Biomass (Figure 2) and LAI (Figure 3) followed similar trends. Based on the results from statistical analyses, the order for the dry and wet seasons biomass were 80% SRI > CF > SRI > 50% SRI and 80% SRI > SRI > CF > 50% SRI (Figure 2), respectively. Higher biomass leads to higher accumulation of non-structural carbohydrate in the culms and leaf cover which can rapidly be trans-located to the panicle during the initial stage of grain filling and can increase the potential for higher crop yield [45]. Overall, key plant growth parameters such as biomass, LAI and number of tillers indicate the best plant performance for the 80% SRI treatment followed by the SRI for both seasons.

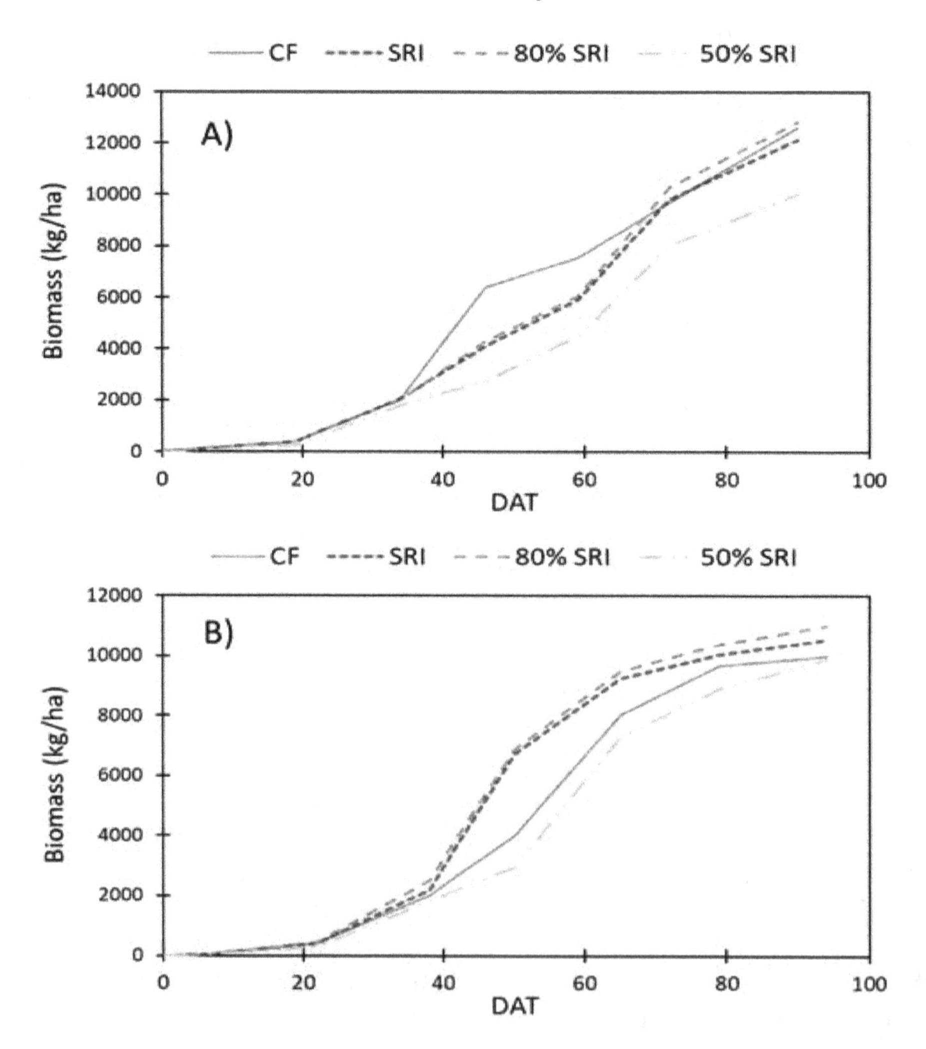

Figure 2. (**A**) Dry season (October 2012 to January 2013) and (**B**) wet season (February 2013 to June 2013) total biomass for continuous flooding (CF), system of rice intensification (SRI), 80% SRI, and 50% SRI treatments. The 80% SRI and 50% SRI refers to 80% and 50% of the SRI ponding depth, respectively. 0 to 46, 47 to 60, 61 to 75, and 77 to 94 days after transplanting (DAT) corresponds to vegetative, flowering, panicle initiation, and senescence stages, respectively.

Table 2. Plant height and number of tillers for the dry and wet seasons.

Treatments *	Plant Height (m)		Number of Tillers	
	Dry Season	Wet Season	Dry Season	Wet Season
CF	0.49 [a]	0.52 [a]	40 [a]	28 [a]
SRI	0.44 [a]	0.48 [a]	42 [a]	31 [a]
80% SRI	0.40 [b]	0.47 [b]	56[b]	38 [b]
50% SRI	0.30 [c]	0.40 [c]	29 [c]	25 [c]

* CF—Continuously flooded, SRI—System of Rice Intensification, 80% SRI—80% of SRI ponding depth, and 50% SRI—50% of SRI ponding depth. Note: Treatments with different letters (superscripts) were significantly different at 0.05 significance level.

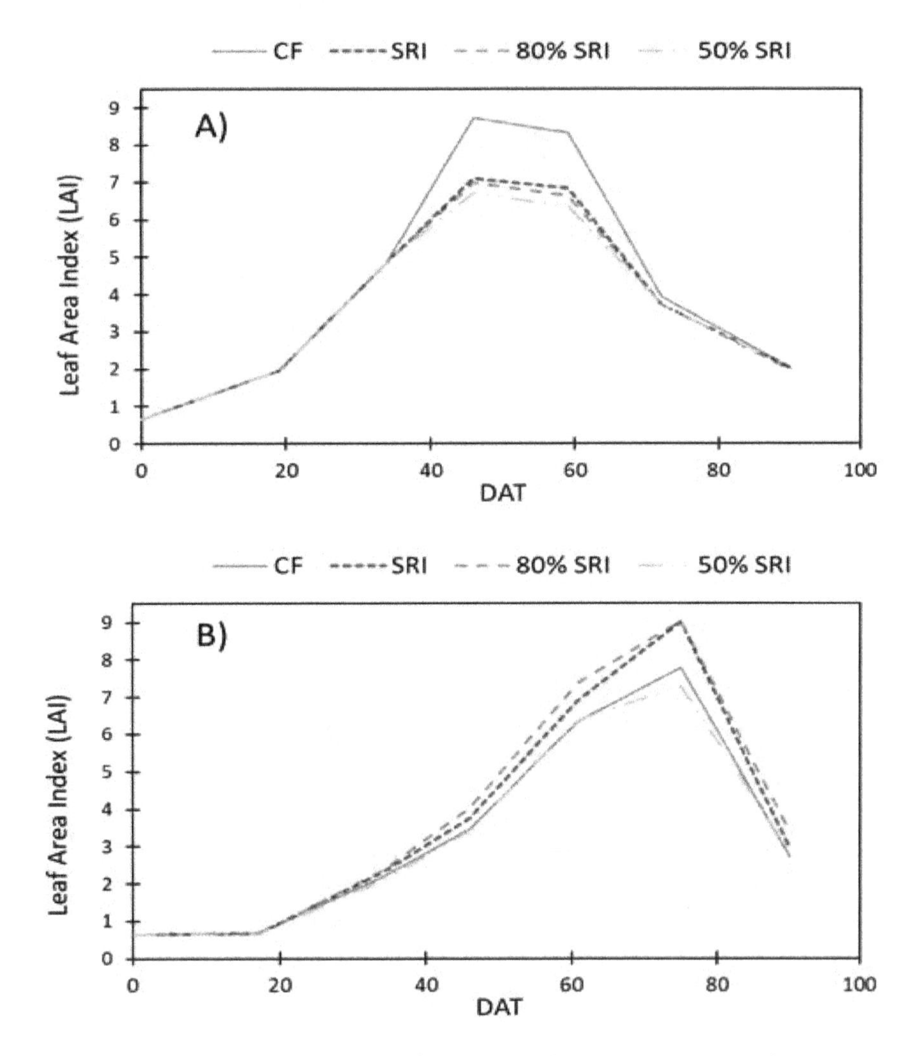

Figure 3. (**A**) Dry season (October 2012 to January 2013) and (**B**) wet season (February 2013 to June 2013) leaf area index (LAI) for continuous flooding (CF), system of rice intensification (SRI), the 80% SRI, and 50% SRI treatments. The 80% SRI and 50% SRI refers to 80% and 50% of the SRI ponding depth, respectively. Days after transplanting (DAT) 0 to 46, 47 to 60, 61 to 75, and 77 to 90 corresponds to vegetative, panicle initiation, flowering, and senescence stages, respectively.

3.2. Yield

For both seasons, the 80% SRI had statistically higher yield than other treatments while the 50% SRI had the lowest yield ($p < 0.05$; Figure 4). Following comparisons were statistically significant ($p < 0.05$) for both seasons: 80% SRI > CF, CF > 50% SRI, SRI > 50% SRI. For the dry season, yield for the 80% SRI was significantly higher than the SRI ($p = 0.01$). However, frequent rainfall events between transplanting and panicle initiation during the wet season (March–May 2013) resulted in similar soil moisture (Figures 5 and 6) in the root zone for all the SRI treatments which is the likely reason for small yield differences between the SRI, 50% SRI, and 80% SRI.

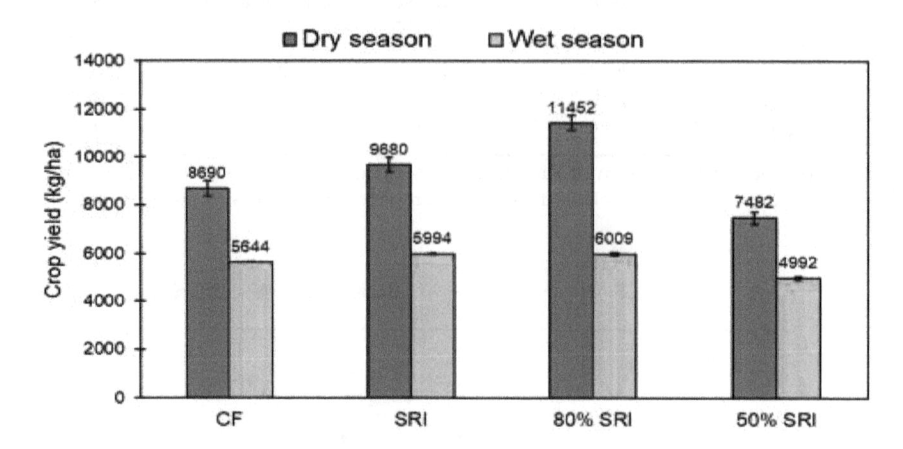

Figure 4. Dry and wet seasons rice crop yields for continuous flooding (CF), system of rice intensification (SRI), 80% SRI, and 50% SRI. 80% SRI and 50% SRI refers to 80% and 50% of the SRI ponding depth, respectively.

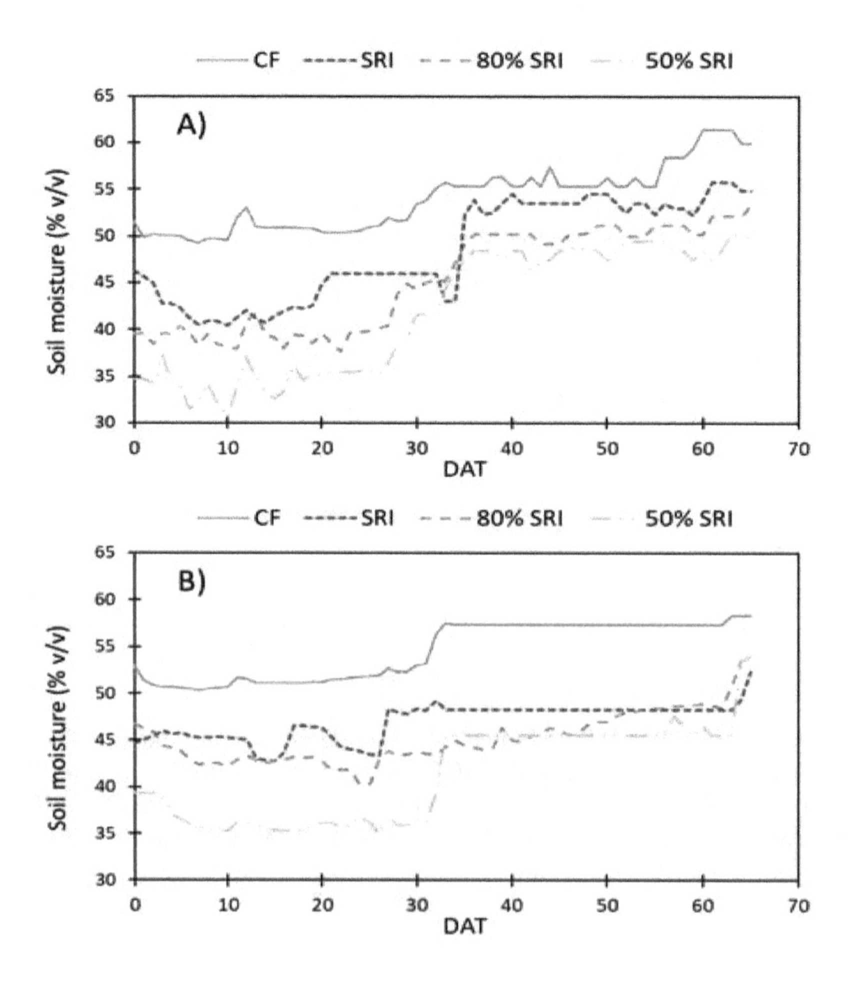

Figure 5. Daily soil moisture at 30 cm depth during the (**A**) dry season (October 2012 to January 2013) and (**B**) wet season (February 2013 to June 2013). CF is continuous flooding, SRI is system of rice intensification, 80% SRI and 50% SRI refers to 80% and 50% of the SRI ponding depth, respectively DAT is days after transplanting.

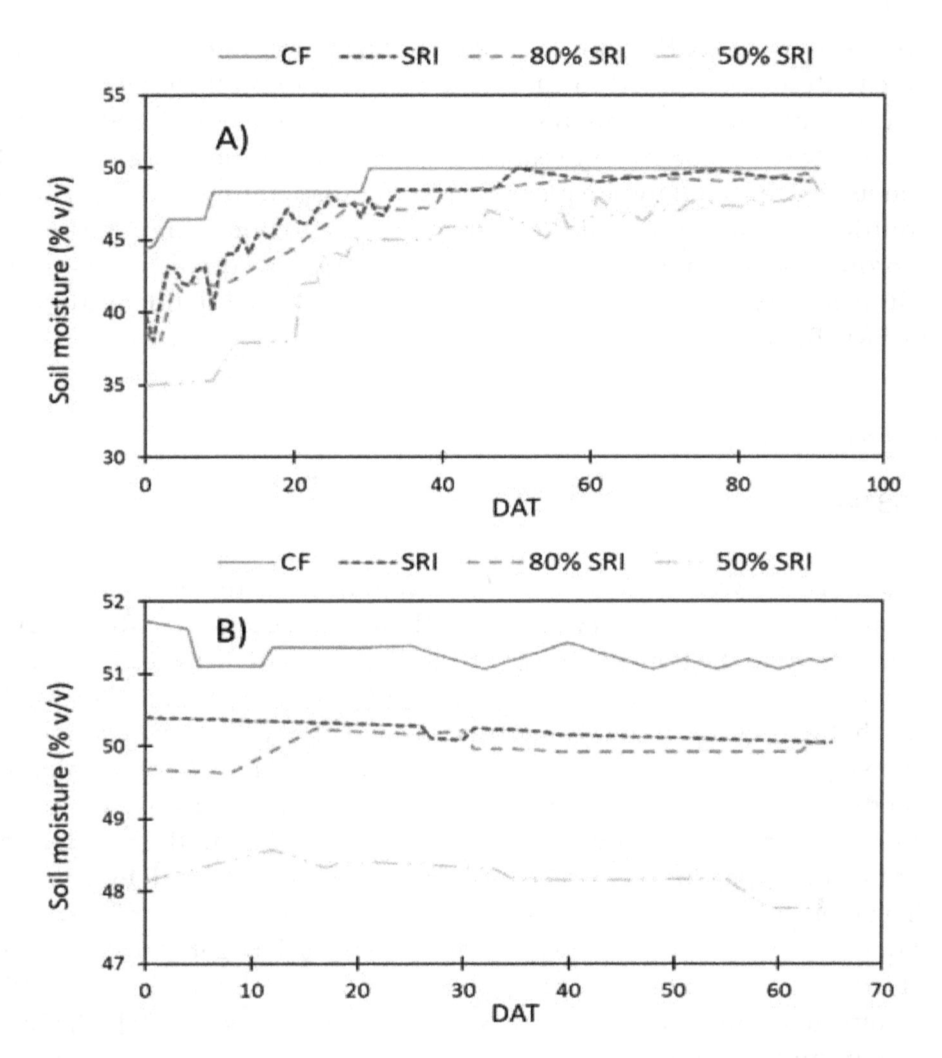

Figure 6. (**A**) Dry season (October 2012 to January 2013) and (**B**) wet season (February 2013 to June 2013) daily average soil moisture below the root zone (60 cm) against days after transplanting (DAT). The CF is continuous flooding, SRI is system of rice intensification, and 80% SRI and 50% SRI refers to 80% and 50% of the SRI ponding depth, respectively.

The wet season yields were lower than dry season mainly due to frequent rainfall between March and May 2013 resulting in saturated to near saturated soil moisture conditions during wet season for all treatments (Figures 5 and 6). The lowest yield observed for the 50% SRI in the dry season was higher than the highest yield for the 80% SRI during the wet season. Very little rainfall during November–December 2012 helped maintain the target soil moisture for all three SRI treatments (Figures 5 and 6) resulting in higher yields during the dry season as compared to the wet season.

The 80% SRI treatment produced 32% and 6% more yield than the CF for the dry and wet seasons, respectively. Although the 80% SRI treatment had almost the same yield as the SRI for the wet season, it had 18% more yield than the SRI treatment for the dry season. Higher yield for SRI is in agreement with observations from other studies [42,46,47] that noted higher grain yield when younger seedlings (8 to 12 days old) are transplanted at spacing ranging from 25 cm × 25 cm to 30 cm × 30 cm under non-flooded conditions. In this study, younger seedling and AWD irrigation were the main synergistic factors that increased the tillering ability (per hill and per area), panicle length, number of filled grains, and finally high yield for the 80% SRI treatment followed by the SRI compared to the CF.

Rice yield for the wet season was low with small differences among the treatments due to heavy and frequent rainfall at the beginning of the wet season (March and May) which resulted in sustained saturation/flooding during the wet period and prevented implementation of the SRI treatments.

Similar results were observed by Stoop et al. (2002) [22], who noted that it was not possible to attain higher yields with SRI compared to CF due to frequent rainfall events. Prolonged root zone saturation due to frequent rainfall events restrict root zone aeration under SRI thereby negatively affecting root and tiller growth. For the dry season, higher yield for the 80% SRI and SRI were due to: (1) adequate soil moisture required by the plant between transplanting and panicle initiation stages which enhances nutrients uptake; (2) reduced plant stress due to non-waterlogged conditions in the root zone which promotes healthier root growth; and (3) improved soil aeration which increases microbial metabolism activity. These factors resulted in better number of tillers and yield observed in this study for 80% SRI (Table 2).

3.3. *Water Use and Productivity*

3.3.1. Irrigation and Soil Moisture

Measured soil moisture within (30 cm) and below root zone (60 cm) indicates the success of treatment implementation. During the dry season, soil moisture of all the SRI treatments fluctuated around soil field capacity from the transplanting to the panicle initiation stage after which soil moisture for all treatments was similar due to sustained flooding (Figures 5 and 6). During the dry season, the only treatment that was allowed to fall below field capacity and, at times, close to wilting point during the tillering stage was 50% SRI (Figure 5). After panicle initiation stage, the soil moisture at 60 cm depth for all the SRI treatments for the dry season was similar to the CF because of continuous ponding maintained for all the treatments. Unlike dry season, soil in all the SRI treatments was near saturation during the initial part of the wet season due to frequent rainfall events (Figures 5 and 6). The desired SRI irrigation cycle (wetting for 3 consecutive days and drying for 5 days) was interrupted by heavy tropical rainfall at 30 days after transplanting (DAT) which negated the effects of the irrigation management and led to similar soil moisture in the root zone for all treatments resulting in reduced yield for the SRI treatments. Similar observations were made by Kombe (2012) [48]. Given that the 80% SRI resulted in maximum yield, it can be inferred that the desired soil moisture from transplanting to panicle initiation stage within the root zone (0–30 cm) is 44–48% (vol.) which falls between field capacity to saturation.

There was a large difference in the amount of water applied between the two seasons. In the dry season, total rainfall was 489 mm and irrigation volumes applied to the 80% SRI and CF treatments were 830 mm and 1286 mm, respectively (Figure 7A). On the other hand, during the wet season, there was a total of 1379 mm rainfall and the irrigation volume applied to the 80% SRI and CF were 554 mm and 787 mm, respectively (Figure 7). Frequent rainfall and variability in deep percolation losses resulted in similar water application for the 80% SRI and 50% SRI treatments (Figure 7). The irrigation volume varied depending on the rainfall, ET and deep percolation losses.

Results show that the 80% SRI can save 35% water compared to the CF during the dry season (Figure 7B). The water savings during the wet season will vary depending on rainfall. Tabbal et al. (2002) also reported that maintaining the soil moisture by alternate wetting and drying reduced irrigation volume by about 40–70%, compared with the traditional CF, without any significant loss in yield. In Ruaha basin of Tanzania, water abstracted from rivers for irrigation accounts for 56% of the wet season river flows and 93% of the dry season river flows [49]. Results from this study indicate that practicing the 80% SRI system for both rice growing seasons can achieve significant water savings which can be used for other purposes and/or help maintain environmental flows.

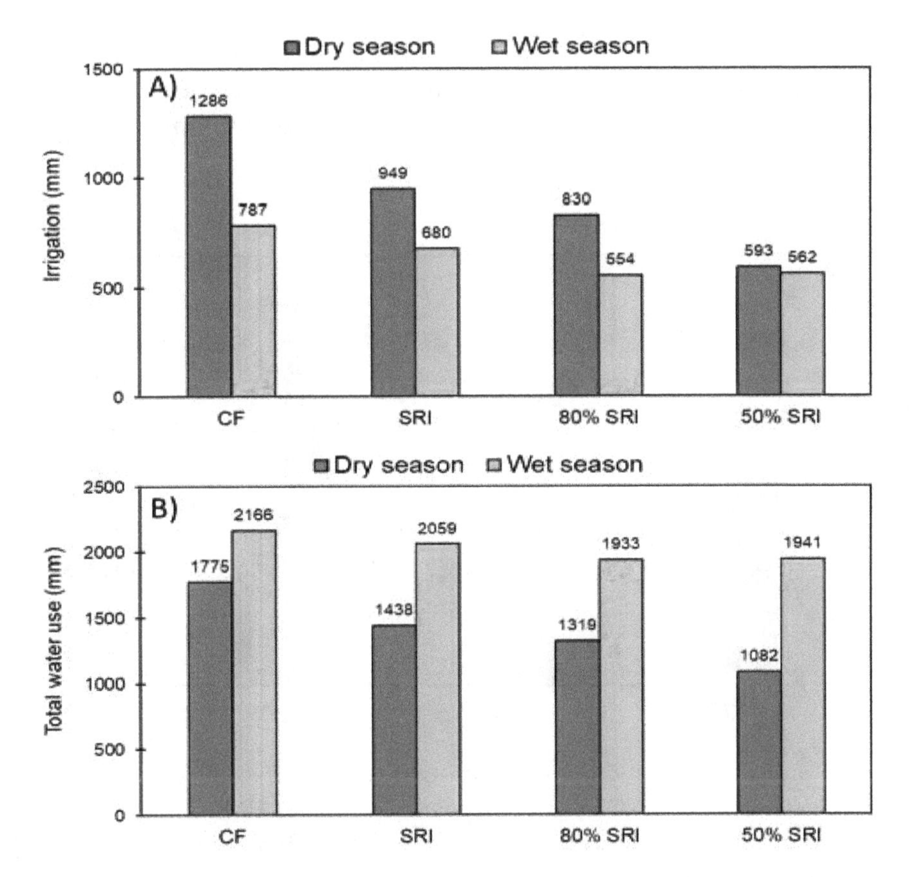

Figure 7. (**A**) Irrigation applied and (**B**) total water (irrigation + rainfall) input for continuous flood (CF), system of rice intensification (SRI), 80% SRI and 50% SRI treatments during the dry and wet seasons. 80% SRI and 50% SRI refers to 80% and 50% of the SRI ponding depth, respectively.

3.3.2. Economic Productivity of Water

All three SRI treatments had higher economic productivity than the CF production system indicating higher "income per drop of water" under SRI. The 80% SRI had the highest economic productivity during both dry and wet seasons (Figure 8). Dry season economic productivity for the 80% SRI was almost twice that of the CF (Figure 8). The 80% SRI treatment was 29% more productive than the SRI during the dry season (Figure 8). As compared to the dry season, wet season economic productivity showed little differences between the treatments. Despite 14–30% (107–233 mm) of difference in irrigation applied between the CF and SRI treatments, high wet season rainfall (1379 mm) masked the treatment effects on economic productivity (Figure 8). However, the 80% SRI was still 19% more beneficial than the CF during the wet season (Figure 8). Considering the yield advantage from the 80% SRI for both seasons, it is a better irrigation management strategy compared to the SRI and the CF (Figure 8). Although the economic productivity for the 50% SRI was higher than the CF for the dry season, the yield loss from this treatment is not likely to result in its acceptance over the 80% SRI. However, for areas with limited water supply the 50% SRI is still a viable option because yield reductions from the 50% SRI were only 13% compared to the CF over the two seasons.

In 2014, 957,218 ha of paddy area was harvested in Tanzania [8] almost all utilizing the CF production system. The yield advantage and water savings from the 80% SRI are likely to vary depending on rainfall amount and distribution, soil properties, water availability, and management strategies. For example, light textured soils (sandy loam) typically require higher irrigation volume than heavy texture soils (e.g., clay) mainly due to higher soil hydraulic conductivity that results in higher deep percolation losses. Assuming that the results from this study are applicable to the entire rice production area in Tanzania, implementation of the 80% SRI will result in annual water savings of 3.29 billion cubic meters and achieve additional production of 1.5 million tons of rice. If used, this water saved from the CF to the 80% SRI conversion can support 50% increase in the current rice irrigated area in Tanzania. Achieving these water savings and yield benefits is likely to increase the sustainability of rice production system in Tanzania and create additional water supplies for industry, environment, and other users.

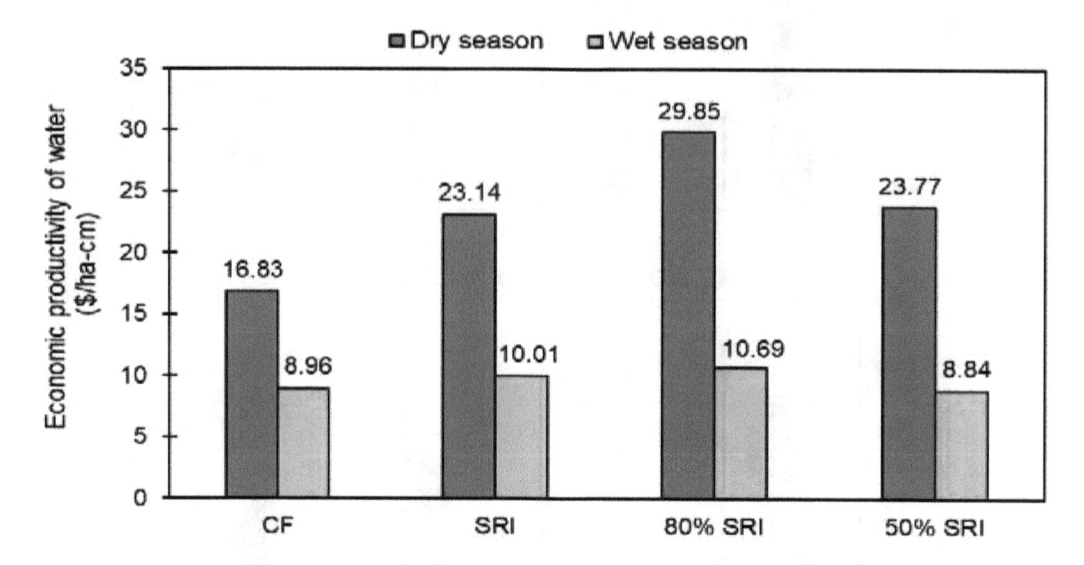

Figure 8. Economic productivity of water (US$/ha-cm) under different water management options during the dry and wet seasons. The CF is continuous flooding and SRI is system of rice intensification. 80% SRI and 50% SRI refers to 80% and 50% of the SRI ponding depth, respectively.

4. Conclusions

The traditional continuously flooded (CF) rice production system has no yield advantage over the SRI-AWD production system for both growing seasons in Tanzania. Consistently, highest yields were obtained from the 80% SRI system for both seasons which indicates that it is possible to increase yields while reducing the total irrigation volume. The 80% SRI system outperformed the CF system by 2762 kg/ha and 365 kg/ha with water savings of 456 mm and 233 mm during the dry and wet seasons, respectively. The water savings from the 80% SRI are 30–35% compared to the CF system. Given that 50% SRI system produced almost 90% yield compared to the CF during the wet season, the farmers in limited water supply regions are still likely to achieve a viable yield. For the farms that grow rice in both seasons, the annual water savings from the 80% SRI will be 689 mm with an additional production of 3127 kg/ha over the CF system. Considering that water savings from the 80% SRI accounts for 74% of annual rainfall (935 mm/year) in southern Tanzania, this irrigation management system has important implications for maintaining water supply and environmental flows in rivers.

To achieve large-scale yield and water saving benefits, there is a need to develop easy-to-understand water management recommendations for farmers in Tanzania. Maintaining 30 mm ponding depth (80% SRI) for three days followed by no irrigation for five days during transplanting to panicle initiation is an easy-to-follow recommendation. After panicle initiation, 20 mm ponding depth can be maintained to achieve increased yield with reduced irrigation. Farmers in the major river basins such as Ruaha are already experiencing water cuts which are likely to become more frequent or permanent in the future considering low reservoir levels for hydropower generation (e.g., Mtera and Kidatu power generation plants) [50]. Basin-scale implementation of the 80% SRI will not only help farmers sustain or improve the yields under current and future water cuts but also help maintain current power generation. One of the avenues for promoting large-scale implementation of the SRI system is to conduct on-farm demonstration studies. Furthermore, socio-economic factors including market prices, soil type, water availability, and existing irrigation infrastructure will have to be considered for wide-scale acceptance of the 80% SRI in Tanzania. Given the large-scale production of rice in Tanzania and projected water stress by 2025 [12], the 80% SRI has the potential to improve the well-being of farmers and contribute to food security in Tanzania.

Author Contributions: S.T.M., S.S., A.T., and S.D.T. conceived and designed the study, S.T.M. collected the data under the supervision of S.S., A.T. and S.D.T., S.T.M., S.S., and R.P.S. conducted the data analysis and wrote the manuscript.

Acknowledgments: The authors gratefully acknowledge D. N. Kimaro, Minde, and D. Kryabill for their helpful suggestions to improve this manuscript. We also thank Macha for data collection, Epignosis and Naza for their encouragement.

References

1. Rijsberman, F.R. Water scarcity: Fact or fiction? *Agric. Water Manag.* **2006**, *80*, 5–22. [CrossRef]
2. Hanjra, M.A.; Qureshi, M.E. Global water crisis and future food security in an era of climate change. *Food Policy* **2010**, *35*, 365–377. [CrossRef]
3. AQUASTAT. *Water Resources Development and Management Service*; Food and Agriculture Organization of the United Nations: Rome, Italy, 2010; Available online: http://www.fao.org/nr/water/aquastat/main/index.stm (accessed on 30 June 2016).
4. Siebert, S.; Burke, J.; Faures, J.M.; Frenken, K.; Hoogeveen, J.; Döll, P.; Portmann, F.T. Groundwater use for irrigation—A global inventory. *Hydrol. Earth Syst. Sci.* **2010**, *14*, 1863–1880. [CrossRef]
5. Nguyen, N.V. (Ed.) *Global Climate Changes and Rice Food Security*; FAO: Rome, Italy, 2002; pp. 24–30.
6. Geerts, S.; Raes, D. Deficit irrigation as an on-farm strategy to maximize crop water productivity in dry areas. *Agric. Water Manag.* **2009**, *96*, 1275–1284. [CrossRef]
7. Balasubramanian, V.; Sie, M.; Hijmans, R.J.; Otsuka, K. Increasing rice production in sub-Saharan Africa: Challenges and opportunities. *Adv. Agron.* **2007**, *94*, 55–133.
8. FAOSTAT. *Statistical Databases*; Food and Agriculture Organization of the United Nations: Rome, Italy, 2014.
9. Africa Rice Center. *Africa Rice Trends: Overview of Recent Developments in the Sub-Saharan Africa Rice Sector*; Africa Rice Center: Cotonou, Benin, 2007.
10. Nasrin, S.; Lodin, J.B.; Jirström, M.; Holmquist, B.; Djurfeldt, A.A.; Djurfeldt, G. Drivers of rice production: Evidence from five Sub-Saharan African countries. *Agric. Food Secur.* **2015**, *4*, 12. [CrossRef]
11. United Republic of Tanzania (URT). *State of the Environment Report—2006*; Division of Environment, Vice President's Office: Dar es Salaam, Tanzania, 2006; ISBN 9987-8990.
12. World Bank. *United Republic of Tanzania, Water Resources Assistance Stretagy, Improving Water Security for Sutaining Livelihoods and Growth*; Report No. 35327-TZ; Water and Urban Unit 1, Africa Region; World Bank: Washington, DC, USA, 2006; Available online: http://documents.worldbank.org/curated/en/378981468117562281/Tanzania-Water-resources-assistance-strategy-improving-water-security-for-sustaining-livelihoods-and-growth (accessesd on 6 July 2017).
13. Bouman, B.A.M.; Tuong, T.P. Field water management to save water and increase its productivity in irrigated rice. *Agric. Water Manag.* **2001**, *49*, 11–30. [CrossRef]
14. Guerra, L.C. *Producing More Rice with Less Water from Irrigated Systems*; International Water Management Institute (IWMI): Colombo, Sri Lanka, 1998; Volume 5.
15. Tuong, T.P.; Bouman, B.A.M. Rice production in water-scarce environments. *Water Prod. Agric. Limits Oppor. Improv.* **2003**, *1*, 13–42.
16. Bouman, B.A.M.; Peng, S.; Castaneda, A.R.; Visperas, R.M. Yield and water use of irrigated tropical aerobic rice systems. *Agric. Water Manag.* **2005**, *74*, 87–105. [CrossRef]
17. Nie, L.; Peng, S.; Chen, M.; Shah, F.; Huang, J.; Cui, K.; Xiang, J. Aerobic rice for water-saving agriculture. A review. *Agron. Sustain. Dev.* **2012**, *32*, 411–418. [CrossRef]
18. Tabbal, D.F.; Bouman, B.A.M.; Bhuiyan, S.I.; Sibayan, E.B.; Sattar, M.A. On-farm strategies for reducing water input in irrigated rice; case studies in the Philippines. *Agric. Water Manag.* **2002**, *56*, 93–112. [CrossRef]
19. Zhang, H.; Xue, Y.; Wang, Z.; Yang, J.; Zhang, J. An Alternate Wetting and Moderate Soil Drying Regime Improves Root and Shoot Growth in Rice. *Crop Sci.* **2009**, *49*, 2246–2260. [CrossRef]

20. Lampayan, R.M.; Rejesus, R.M.; Singleton, G.R.; Bouman, B.A. Adoption and economics of alternate wetting and drying water management for irrigated lowland rice. *Field Crops Res.* **2015**, *170*, 95–108. [CrossRef]
21. Bouman, B.A.M.; Feng, L.; Tuong, T.P.; Lu, G.; Wang, H.; Feng, Y. Exploring options to grow rice using less water in northern China using a modelling approach: II. Quantifying yield, water balance components, and water productivity. *Agric. Water Manag.* **2007**, *88*, 23–33. [CrossRef]
22. Stoop, W.A.; Uphoff, N.; Kassam, A. A review of agricultural research issues raised by the system of rice intensification (SRI) from Madagascar: Opportunities for improving farming systems for resource-poor farmers. *Agric. Syst.* **2002**, *71*, 249–274. [CrossRef]
23. Kassam, A.; Stoop, W.; Uphoff, N. Review of SRI modifications in rice crop and water management and research issues for making further improvements in agricultural and water productivity. *Paddy Water Environ.* **2011**, *9*, 163–180. [CrossRef]
24. Mati, B.M.; Wanjogu, R.; Odongo, B.; Home, P.G. Introduction of the system of rice intensification in Kenya: Experiences from mwea irrigation scheme. *Paddy Water Environ.* **2011**, *9*, 145–154. [CrossRef]
25. Thakur, A.K.; Uphoff, N.T.; Stoop, W.A. Scientific underpinnings of the system of rice intensification (SRI): What is known so far? In *Advances in Agronomy*; Academic Press: Cambridge, MA, USA, 2016; Volume 135, pp. 147–179.
26. Uphoff, N. Agroecological implications of the system of rice intensification (SRI) in Madagascar. *Environ. Dev. Sustain.* **1999**, *1*, 297–313. [CrossRef]
27. Zhao, L.; Wu, L.; Li, Y.; Lu, X.; Zhu, D.; Uphoff, N. Influence of the system of rice intensification on rice yield and nitrogen and water use efficiency with different N application rates. *Exp. Agric.* **2009**, *45*, 275–286. [CrossRef]
28. Allen, R.G.; Pereira, L.S.; Raes, D.; Smith, M. *Crop Evapotranspiration-Guidelines for Computing Crop Water Requirements—FAO Irrigation and Drainage Paper 56*; FAO: Rome, Italy, 1998; Volume 300, D05109.
29. Schaap, M.G.; Leij, F.J.; Van Genuchten, M.T. Rosetta: A computer program for estimating soil hydraulic parameters with hierarchical pedotransfer functions. *J. Hydrol.* **2001**, *251*, 163–176. [CrossRef]
30. Cobos, D.R.; Chambers, C. *Calibrating ECH2O Soil Moisture Sensors*; Application Note; Decagon Devices: Pullman, WA, USA, 2010.
31. Leib, B.G.; Jabro, J.D.; Matthews, G.R. Field evaluation and performance comparison of soil moisture sensors. *Soil Sci.* **2003**, *168*, 396–408. [CrossRef]
32. Dukes, M.D.; Simonne, E.H.; Davis, W.E.; Studstill, D.W.; Hochmuth, R. Effect of sensor-based high frequency irrigation on bell pepper yield and water use. In Proceedings of the 2nd International Conference on Irrigation and Drainage, Phoenix, AZ, USA, 12–15 May 2003; pp. 12–15.
33. Zotarelli, L.; Dukes, M.D.; Scholberg, J.M.S.; Femminella, K.; Munoz-Carpena, R. Irrigation scheduling for green bell peppers using capacitance soil moisture sensors. *J. Irrig. Drain. Eng.* **2010**, *137*, 73–81. [CrossRef]
34. Yin, X.; Schapendonk, A.H.; Kropff, M.J.; van Oijen, M.; Bindraban, P.S. A generic equation for nitrogen-limited leaf area index and its application in crop growth models for predicting leaf senescence. *Ann. Bot.* **2000**, *85*, 579–585. [CrossRef]
35. SAS Institute. *SAS/IML 9.3 User's Guide*; SAS Institute: Cary, NC, USA, 2011.
36. Somerville, P.N. On the conservatism of the Tukey-Kramer multiple comparison procedure. *Stat. Probab. Lett.* **1993**, *16*, 343–345. [CrossRef]
37. Gleick, P.H.; Christian-Smith, J.; Cooley, H. Water-use efficiency and productivity: Rethinking the basin approach. *Water Int.* **2011**, *36*, 784–798. [CrossRef]
38. Wilson, R.T.; Lewis, I. *The Rice Value Chain in Tanzania, a Report from the Southern Highlands Food Systems Programme*; Food and Agriculture Organization of the United Nations: Rome, Italy, 2015; pp. 1–15.
39. Chaudhary, D.K. Effect of water regimes and NPK levels on mid duration rice. (*Oryza sativa* L.). Master's Thesis, Rajendra Agricultural University, Pusa, Bihar, 2003.
40. Parihar, S.S. Effect of crop-establishment method, tillage, irrigation and nitrogen on production potential of rice (*Oryza sativa*)-wheat (*Triticum aestivum*) cropping system. *Indian J. Agron.* **2004**, *49*, 1–5.
41. Panda, S.C.; Rath, B.S.; Tripathy, R.K.; Dash, B. Effect of water management practices on yield and nutrient uptake in the dry season rice. *Oryza* **1997**, *34*, 51–53.
42. Gani, A.; Rahman, A.; Rustam, D.; Hengsdijk, H. Water management experiments in Indonesia. In Proceedings of the International Symposium on Water Wise Rice Production, IARI, New Delhi, India, 2–3 November 2003; pp. 29–37.

43. Thakur, A.K.; Rath, S.; Patil, D.U.; Kumar, A. Effects on rice plant morphology and physiology of water and associated management practices of the system of rice intensification and their implications for crop performance. *Paddy Water Environ.* **2011**, *9*, 13–24. [CrossRef]
44. Uphoff, N. Higher yields with fewer external inputs? The system of rice intensification and potential contributions to agricultural sustainability. *Int. J. Agric. Sustain.* **2003**, *1*, 38–50. [CrossRef]
45. Takai, T.; Matsuura, S.; Nishio, T.; Ohsumi, A.; Shiraiwa, T.; Horie, T. Rice yield potential is closely related to crop growth rate during late reproductive period. *Field Crops Res.* **2006**, *96*, 328–335. [CrossRef]
46. Vijayakumar, M.; Ramesh, S.; Chandrasekaran, B.; Thiyagarajan, T.M. Effect of system of rice intensification (SRI) practices on yield attributes yield and water productivity of rice (*Oryza sativa* L.). *Res. J. Agric. Biol. Sci.* **2006**, *2*, 236–242.
47. Krishna, A.; Biradarpatil, N.K.; Channappagoudar, B.B. Influence of system of rice intensification (SRI) cultivation on seed yield and quality. *Agric. Sci.* **2008**, *21*, 369–372.
48. Kombe, E. The System of Rice Intensification (SRI) as a Strategy for Adapting to the Effects of Climate Change and Variability: A Case Study of Mkindo Irrigation Scheme in Morogoro, Tanzania. Unpublished Master's Thesis, Department of Agricultural Engineering and Land Planning, Sokoine University of Agriculture, Morogoro, Tanzania, 2012.
49. Mwakalila, S. Water resource use in the Great Ruaha Basin of Tanzania. *Phys. Chem. Earth Parts A/B/C* **2005**, *30*, 903–912. [CrossRef]
50. Makoye, K. *Farmers to Lose Water Access as Tanzania's Hydropower Runs Dry*; Reuters: London, UK, 2015; Available online: http://www.reuters.com/article/tanzania-water-hydropower/farmers-to-lose-water-access-as-tanzanias-hydropower-runs-dry-idUSL8N0ZC0X320150626 (accessed on 19 June 2018).

4

Spatial Distribution of Salinity and Sodicity in Arid Climate following Long Term Brackish Water Drip Irrigated Olive Orchard

John Rohit Katuri [1,2], Pavel Trifonov [1] and Gilboa Arye [1,*]

[1] French Associates Institute for Agriculture and Biotechnology of Drylands, Jacob Blaustein Institutes for Desert Research, Ben Gurion University of the Negev, Sede Boqer Campus, Midreshet Ben-Gurion 84990, Israel; john.katuri@gmail.com (J.R.K.); paveltri@post.bgu.ac.il (P.T.)

[2] Department of Agronomy, Directorate of Crop Management, Tamil Nadu Agricultural University, Coimbatore, Tamil Nadu 641003, India

* Correspondence: aryeg@bgu.ac.il

Abstract: The availability of brackish groundwater in the Negev Desert, Israel has motivated the cultivation of various salinity tolerant crops, such as olives trees. The long term suitability of surface drip irrigation (DI) or subsurface drip irrigation (SDI) in arid regions is questionable, due to salinity concerns, in particular, when brackish irrigation water is employed. Nevertheless, DI and SDI have been adopted as the main irrigation methods in olive orchards, located in the Negev Desert. Reports on continued reduction in olive yields and, essentially, olive orchard uprooting are the motivation for this study. Specifically, the main objective is to quantify the spatial distribution of salinity and sodicity in the active root-zone of olive orchards, irrigated with brackish water (electrical conductivity; EC = 4.4 dS m^{-1}) for two decades using DI and subsequently SDI. Sum 246 soil samples, representing 2 m^2 area and depths of 60 cm, in line and perpendicular to the drip line, were analyzed for salinity and sodicity quantities. A relatively small leaching-zone was observed below the emitters depth (20 cm), with EC values similar to the irrigation water. However, high to extreme EC values were observed between nearby emitters, above and below the dripline. Specifically, in line with the dripline, EC values ranged from 10 to 40 dS m^{-1} and perpendicular to it, from 40 to 120 dS m^{-1}. The spatial distribution of sodicity quantities, namely, the sodium adsorption ratio (SAR, (meq $L^{-1})^{0.5}$) and exchangeable sodium percentage (ESP) resembled the one obtained for the EC. In line with the dripline, from 15 to 30 (meq $L^{-1})^{0.5}$ and up to 27%, in perpendicular to the drip line from 30 to 60 (meq $L^{-1})^{0.5}$ and up to 33%. This study demonstrates the importance of long terms sustainable irrigation regime in arid regions in particular under DI or SDI. Reclamation of these soils with gypsum, for example, is essential. Any alternative practices, such as replacing olive trees and the further introduction of even high salinity tolerant plants (e.g., jojoba) in this region will intensify the salt buildup without leaving any option for soil reclamation in the future.

Keywords: arid region; brackish water; sub surface drip irrigation (SDI); salinity; sodicity; olives trees

---

1. Introduction

Salinity and drought are the major abiotic stress factors limiting yield in arid regions [1]. To counteract these limitations, advanced irrigation management practices, such as drip irrigation (DI), were introduced and soon hailed as a breakthrough in agricultural efficiency [2]. Additionally, advanced breeding methods and genetic engineering tools have been developed to confer abiotic stress tolerance in different crops, with emphasis on enhanced tolerance to drought and high soil salinity [3]. With the advent of these technologies, saline water agriculture has gained importance and facilitated cultivation in arid environments. Due to drought conditions (low precipitation) in arid regions soil, salinity often increases, impeding plant water uptake. The initial plant responses to salinity and drought stress are fundamentally identical across species and are often complex [4,5]. Plant root adaptations play a key role in coping with these stresses [6]. For the successful management of arid agriculture choice of crop, cultivar and irrigation management regimes play a key role.

In the late 1970s, the introduction and cultivation of various saline tolerant crops with brackish water started in the Negev Desert of Israel [7]. Today farmers in the Negev region grow olives using DI or sub-surface drip irrigation (SDI) with brackish ground water (EC ~4.5 dS m^{-1}) from the local aquifer, as they have no alternative for other economical irrigation water source [8]. Olive trees are generally tolerant of drought and salinity [9,10]. However, salinity tolerance in olives is a cultivar specific trait. The main active root zone distribution in olives trees is at a depth of 30 to 60 cm [11,12] and various studies have reported that the upper critical limit of soil EC for normal olive development is 4 to 6 dS m^{-1} [12–15]. In olives trees, the maximum root growth rate can be achieved under fresh water irrigation and the high root mortality rate and root growth restriction occurs under moderately saline irrigation (4.2 dS m^{-1}) [16–19]. Irrigation water salinity of 4 dS m^{-1} limits significant production of the potential yield possible with good quality water [15] and there is a gradual buildup of soil salinity over the years in the root zone [16]. Therefore, an appropriate management of irrigation regime and salinity in root zone is necessary to optimize yield and oil quality in olive orchards irrigated with saline water [15,20].

In the long term, the commitment to utilizing marginal irrigation water sources, such as brackish water, may be fundamentally unsustainable, in particular, in arid lands where precipitation is too low to leach the accumulated salts from the active root zone [21]. There is a higher risk of soil salinization if rainfall is lower than 250 mm and the salts are not leached from the upper 60 cm depth [22–24]. The Negev region has an arid climate with high rates of evapotranspiration (about 2600 mm $year^{-1}$) and low rainfall (70 to 125 mm/year) [8,25]. When SDI was employed it reduced evaporation and improved irrigation water-use efficiency with olive yield similar to DI irrigation [26,27]. However, in SDI systems, salt accumulation above the dripper is high and does not offer an advantage over DI in regard to soil salt distribution under conditions of high evaporative demand [28,29]. In arid and semiarid areas, using SDI placed at shallow depths (about 20 cm) resulted in large amounts of salt accumulation near the soil surface [30], specifically located above the dripline [31,32]. When salts accumulate in soil surface layers, sprinkler irrigation is commonly used in SDI plots to leach salts below the drip tapes, but, in the long term it affects the economic sustainability of SDI [30]. Nevertheless, it was recently demonstrated [33,34] that a sequential practice of sprinkler irrigation for potato germination, followed by low discharge shallow SDI with brackish irrigation water, can result in similar potato yields to traditional methods that utilize sprinkler irrigation with fresh water.

There is high transient salinity and sodicity risk associated with saline water SDI in orchards [35] and they change with the amount and quality of infiltrated water, evapotranspiration rates, and rainfall [36]. When water quality of EC >2.5 dS m^{-1} and SAR >4 was used in olive and other orchards with SDI, there was a significant increase in soil salinity and sodicity values at 0–60 cm soil depths [37–40]. Most studies which examined the salinity and/or sodicity effect on olive growth and yield are short term (<8 years) studies [19,24,41,42] and, consequently, a severe accumulation of salts in the soil profile was not reported.

As mentioned, the introduction and cultivation of salt-tolerant crops in the arid regions in conjunction with brackish irrigation water for the past few decades has resulted in increasing soil salinity. In the current study, we quantify the salinity and sodicity spatial distribution in an olive orchard following twenty years of irrigation with brackish water. The motivation for this study stems from recent reports on continues decrease in yields (Figure 1) and the eventual uprooting of some olive orchards due to unprofitability. Therefore, it is necessary to understand the sustainability of olive cultivation under saline brackish water with SDI, so that secondary salinization is prevented and the soil can be reclaimed for agriculture in future years. The main objective of this study is to fill the knowledge gap regarding the spatial distribution of salinity and sodicity in long term sub-surface drip irrigated soils with brackish irrigation water. Given the relatively high distance (1 m) between drippers, we hypothesized that a high level of salinity and sodicity will be established between nearby drippers.

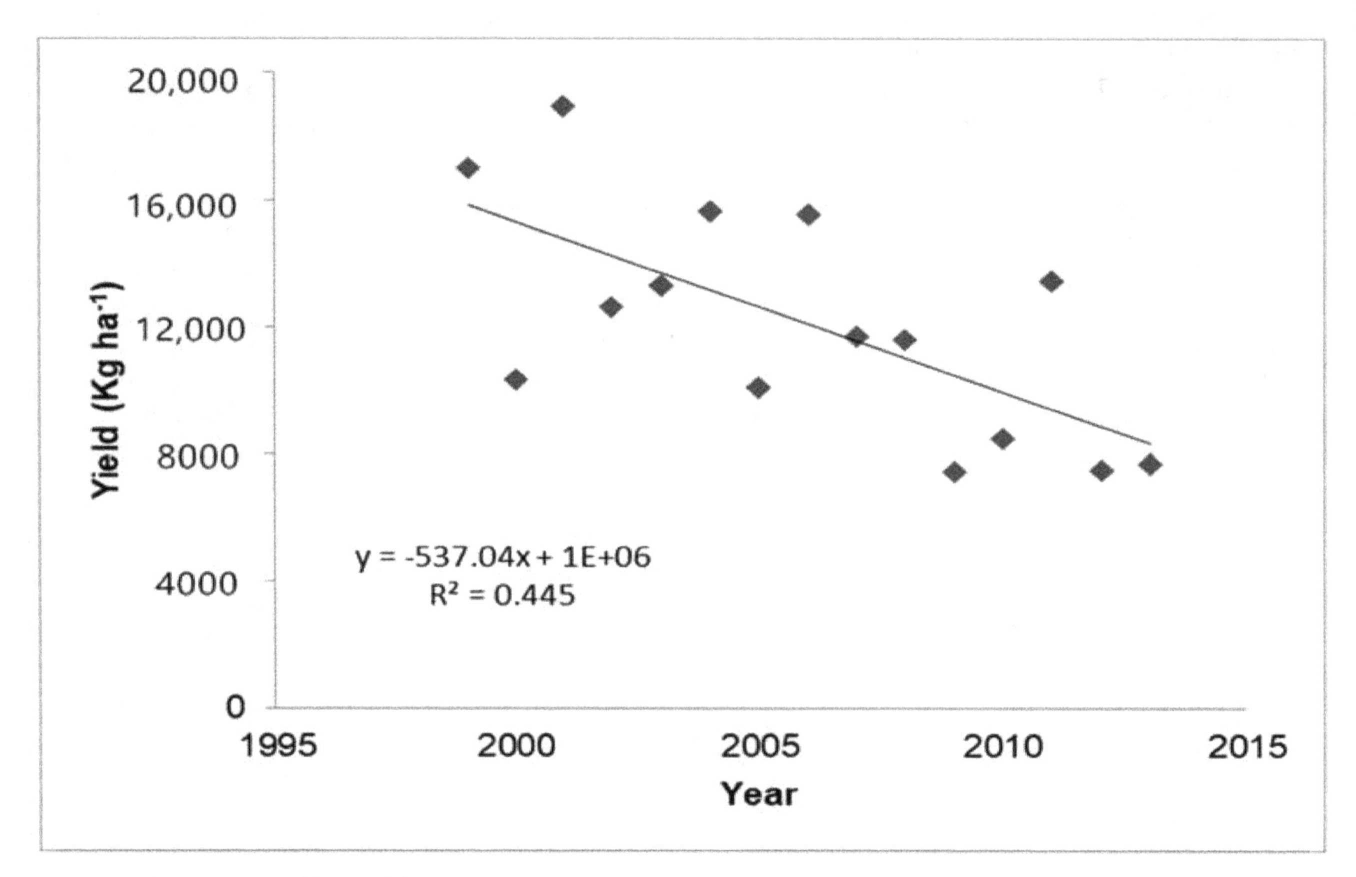

Figure 1. Yield trend for 15 years of the Barnea olive variety grown in the Revivim orchard.

2. Materials and Methods

2.1. Site Description

A field investigation was carried out in the olive (*Olea europaea*) cultivar 'Barnea' orchard of Kibbutz Revivim (31.0436° N, 34.7212° E), located in the central Negev Desert, Israel. The climatic conditions of the location are of the typical arid type, with cooler night temperatures and hot, dry summers (Figure 2). The mean annual rainfall ranges from 75 to 125 mm [25]. During the 2014/2015 season, the total precipitation was 105 mm and cumulative potential evapotranspiration was 2500 mm (Figure 3). The olive orchard was planted in 1995 and has been irrigated with brackish groundwater (EC = 4.4 dS m^{-1}) since then, for approximately 20 years. Water quality parameters of the brackish irrigation water are presented in Table 1.

During the first 15 years, DI was used but later converted to SDI by placing drip laterals at 20 cm soil depth and about 1 m distance from the tree line with an emitter flow-rate of 4 L h^{-1} and 1 m distance between nearby emitters. An initial tree spacing of about 3 × 7 m was first established and after ten years, each alternate tree within a row was uprooted, giving the current spacing of 6 × 7 m. Irrigation was scheduled according to class evaporation pan located nearby the orchard. Specifically, in average, a factor of 35% to 60 % was used to calculate the irrigation amounts from the predetermined cumulative pan evaporation (class A pan) [8]. Accordantly, irrigation intervals were scheduled every 3 days during summer and every 7 days during winter. Approximately, 800 mm plus an excess of 100 mm, as the leaching requirement of irrigation water, was applied annually.

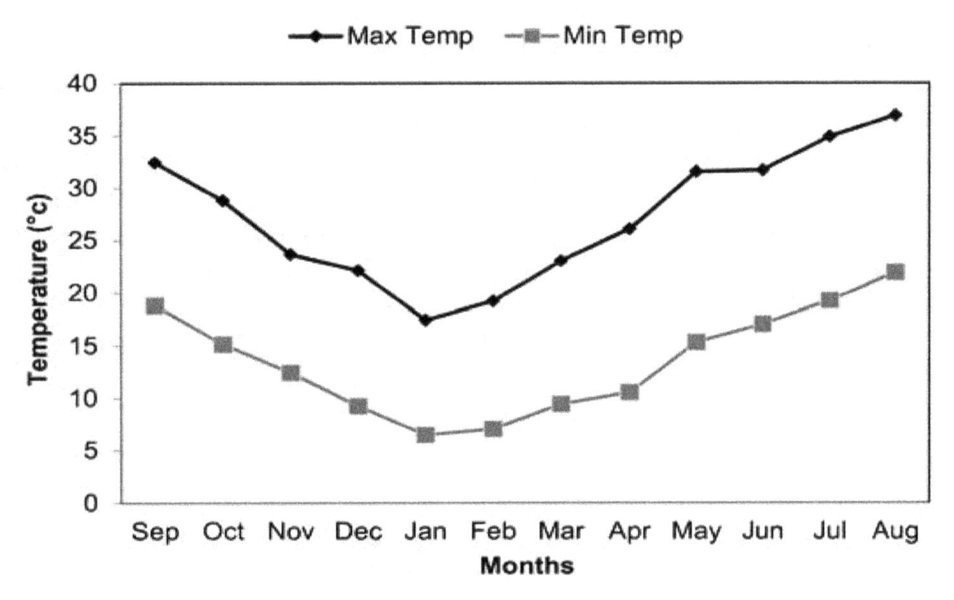

Figure 2. Minimum and maximum temperature in Revivim during 2014–2015.

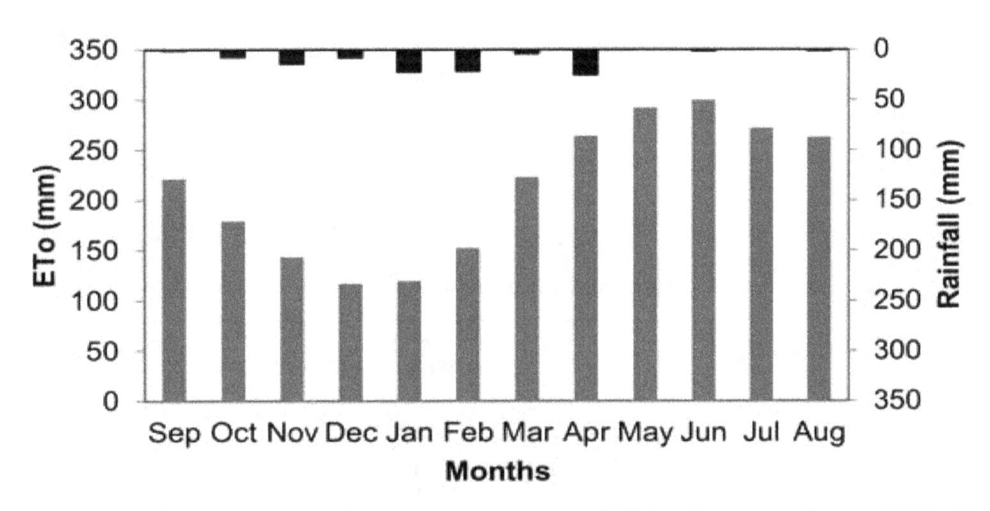

Figure 3. Rainfall and potential evapotranspiration (ETo) in Revivim during 2014–2015.

Table 1. Irrigation water quality parameters (Mekerot Water Company).

Parameter	Value
Boron (mg L^{-1})	1.200
Calcium (mg L^{-1})	171.000
Chloride (mg L^{-1})	1120.000
Electrical Conductivity dS m^{-1}	4.400
$CaCO_3$ (mg L^{-1})	748
HCO_3 (mg L^{-1})	301
Potassium (mg L^{-1})	1900
Magnesium (mg L^{-1})	78.000
Sodium (mg L^{-1})	684.000
pH	7.000
Total organic carbon (mg L^{-1})	<0.200
Total dissolved matter (mg L^{-1})	2697.000
SAR (meq $L^{-1})^{0.5}$	10.900

2.2. Soil Sampling and Analysis

Comprehensive soil sampling was carried out to explore the spatial distribution of salinity and sodicity along and perpendicular to the drip-line, representing a total area of 2 m^2 (Figure 4). Soil samples were collected from 41 locations along the drip-line between three nearby emitters that were perpendicular and diagonal to the central emitter. At each sampling location, disturbed soil samples (n = 246) were taken from six depths: 0–5, 5–10, 10–15, 15–30, 30–45, and 45–60 cm. In addition, representative intake soil samples were taken near the sampling locations mentioned above, from which the bulk density of each layer was calculated (Table 2). The gravimetric water content (WC) was measured shortly after the sampling event from the differences in weight before and after drying at 105 °C for 24 h. The rest of the soil samples were air-dried and thereafter passed through a 2-mm sieve.

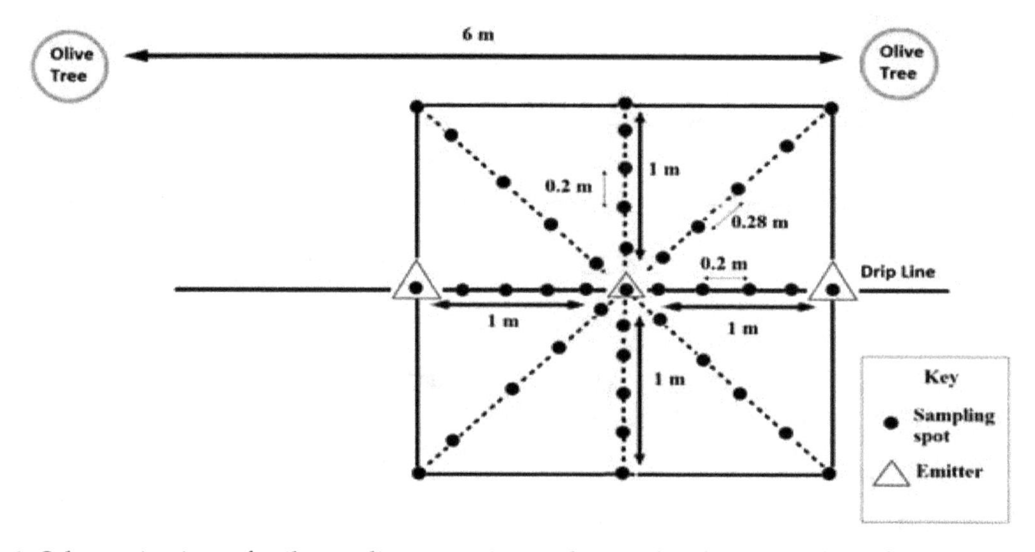

Figure 4. Schematic view of soil sampling spots in an olive orchard under sub surface drip irrigation.

Table 2. Revivim olive orchard soil properties.

Depth	Texture	Sand	Silt	Clay	Organic Matter	Bulk Density	CEC	SSA*	$CaCO_3$
(cm)		(%)	(%)	(%)	(%)	(g/cm^3)	(meq/100g)	(g/m^2)	(%)
0–5	Loamy Sand	80.83 ± 5.77	11.67 ± 5.77	7.50 0.00	10.22 ± 0.05	0.96 ± 0.017	8.54 ± 1.13	28.29 0.00	8.0 ± 3.5
5–10	Sandy Loam	70.83 ± 1.44	16.67 ± 2.89	12.50 ± 2.5	5.41 ± 3.28	1.09 ± 0.006	9.27 ± 0.77	57.19 ± 14.45	8.9 ± 1.5
10–15	Sandy Loam	67.50 0.00	19.17 ± 2.89	13.33 ± 2.89	1.62 ± 0.17	1.46 ± 0.006	8.61 ± 0.22	62.00 ± 16.69	8.5 ± 2.3
15–30	Sandy Loam	65.83 ± 5.20	19.17 ± 5.20	15.00 0.00	1.60 ± 0.20	1.48 ± 0.006	8.59 ± 0.63	71.64 0.00	9.5 ± 1.6
30–45	Loam	50.83 ± 3.82	28.33 ± 1.44	20.83 ± 2.89	1.53 ± 0.21	1.45 ± 0.006	9.19 ± 0.94	105.35 ± 16.69	8.6 ± 1.2
45–60	Loam	54.17 ± 3.82	27.50 ± 2.50	18.33 ± 1.44	1.38 ± 0.11	1.47 ± 0.006	9.03 ± 0.19	90.90 ± 8.34	10.9 ± 1.9

± Standard deviation, SSA*—specific surface area (calculated according to [43]).

The concentration of the main cations (Na, K, Mg, and Ca) in the soil solution was obtained from the extraction of the soil to distilled water ratio of 1:1. Samples were shaken on an end-over shaker and then centrifuged at 4000 rpm for 10 min. The supernatant was analyzed for soluble cations, bicarbonate, and chloride concentration. The cation exchange capacity (CEC) was measured by the sodium acetate method [44] and the exchangeable cations concentrations (xNa, xK, xMg, and xCa) from the sodium acetate extraction [45]. The cations concentration was measured by atomic adsorption spectrophotometer (Analyst 400, ParkinElmer) and Chloride (Cl^{-1}) concentration by Chloride Analyzer (926, Sherwood).

The sodium adsorption ratio (SAR) and the exchangeable sodium ratio (ESR) were calculated according to the Gapon equation.

$$\frac{xNa}{xCa + xMg} = K_G \cdot \frac{Na}{\sqrt{0.5 \cdot (Ca + Mg)}} \equiv ESR = K_G \cdot SAR \tag{1}$$

where, the concentrations of the soluble and exchangeable cation are in meq L^{-1} and meq Kg^{-1}, respectively. The K_G is the Gapon selectivity coefficient.

Contours map of the spatial distribution of the WC, EC, Cl^{-1}, SAR, and ESP in the soil profile were established with Surfer software (version 8, Golden Software, Colorado, USA) using the Kriging regression.

3. Results and Discussion

From the soil properties (Table 2) it can be seen that the texture in the examined soil layers changes from loamy-sand in the top soil layer (0–5cm), sandy-loam in the middle ones (5–30 cm), and loam in the deeper layers (30–60 cm). A distinct difference in organic matter (OM) percentage could be observed from 10.2% (0–5 cm), 5.4% (5–10 cm) and similar values ranging from 1.62% (10–15 cm) to 1.38% (45–60 cm). The bulk density exhibited an inverse linear correlation to OM content ($BD = 1.54 - 0.06 \times OM$, $R^2 = 0.93$) rather than any of size fractions; ranging from about 1 g cm^{-3} in the top soil layer (0–10 cm) and exhibited similar values of about 1.45 g cm^{-3} for the rest of the soil profile. The above observation may imply a higher water holding capacity in the top soil layers, due to water adsorption and/or structures formation induced by the level of soil OM.

In the followings, the spatial distribution obtained for water content, salinity, and sodicity quantities are presented for two 60 cm soil transect: (i) along the drip line and (ii) perpendicular to the drip line (crossing the middle dripper), (Figure 4). In addition, a three dimensional visualization is presented as a counter map calculated from all measured data points of the four transects for a given soil layer (Table 2).

3.1. Water Content Spatial Distribution

In Figure 5 the spatial distribution obtained for the WC is presented for the sampled transect along (Figure 5a) and perpendicular (Figure 5b) to the drip line. The WC distribution demonstrates that relatively higher WC can be found directly above and below the location of the emitters (i.e., 20 cm depth). A typical wetting bulb of relatively light-texture soil can be observed with the bulb radius; the horizontally wetted radius is less than the vertically wetted depth radius [46,47]. The near-saturation zone was located about 20 cm from the emitters from which a gradual reduction in WC can be observed up to 50 cm distance, which is located in the middle, between two nearby drippers. At this location, the WC above the emitters is the lowest one, suggesting that there is no significant overlap between nearby emitters. The relatively large distance between the emitters (i.e., 1 m) and the corresponding spatial distribution of the WC also affected the salinity and sodicity spatial distribution, as is demonstrated below. Regarding the perpendicular transect (Figure 5b), it should be noted that +100 cm on the x-axis is towards the tree-line and –100 cm is towards the road, i.e. away from the tree-line. Toward the tree-line, there is a gradual reduction in WC which is likely due to root water uptake. Away from the tree-line, the reduction in WC may stem from higher evaporation rates, due to less shading from the tree.

The three dimensional visualization (Figure 6) shows an entire 2 m^2 view for the spatial WC distribution at six individual depths (Table 2). It is clearly illustrated that down to 30 cm (the three top layers), the dryer zone prevails toward the tree line compared to the corresponding locations, away from the tree line. The dryer WC zone may indicate water uptake by the active root zone [11,12]. The relatively low overlap between the wetting fronts of the nearby emitters is also illustrated, suggesting that in the long-term, the solute fluxes, due to convection, dispersion, and diffusion might have reached the wetting front of individual emitter and accumulated at this location. Consequently, in the long-term, higher salinity can be expected between emitters and perpendicular to the emitter.

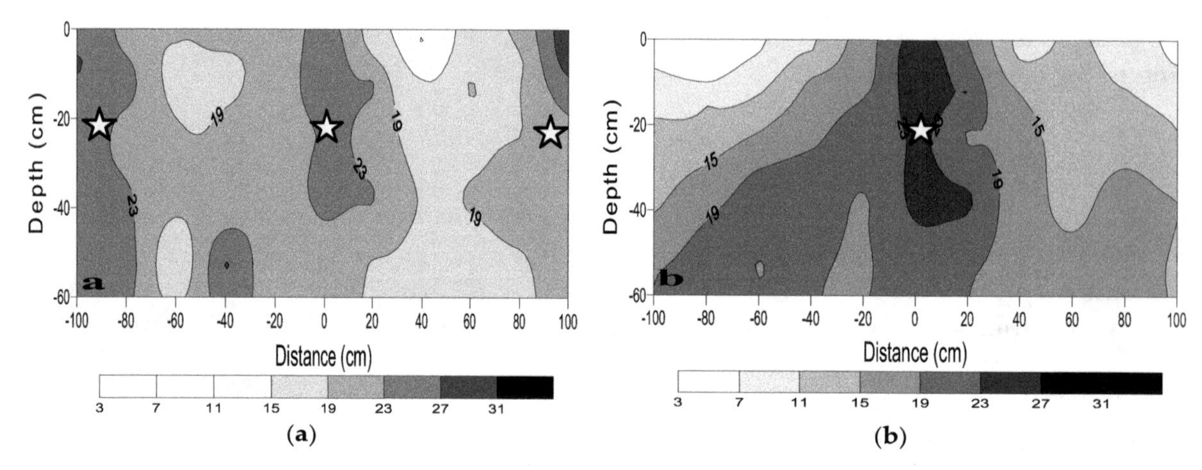

Figure 5. Gravimetric water content (%) distribution (**a**) along the drip line and (**b**) perpendicular to the drip line. The black and white stars indicate the location of the drippers.

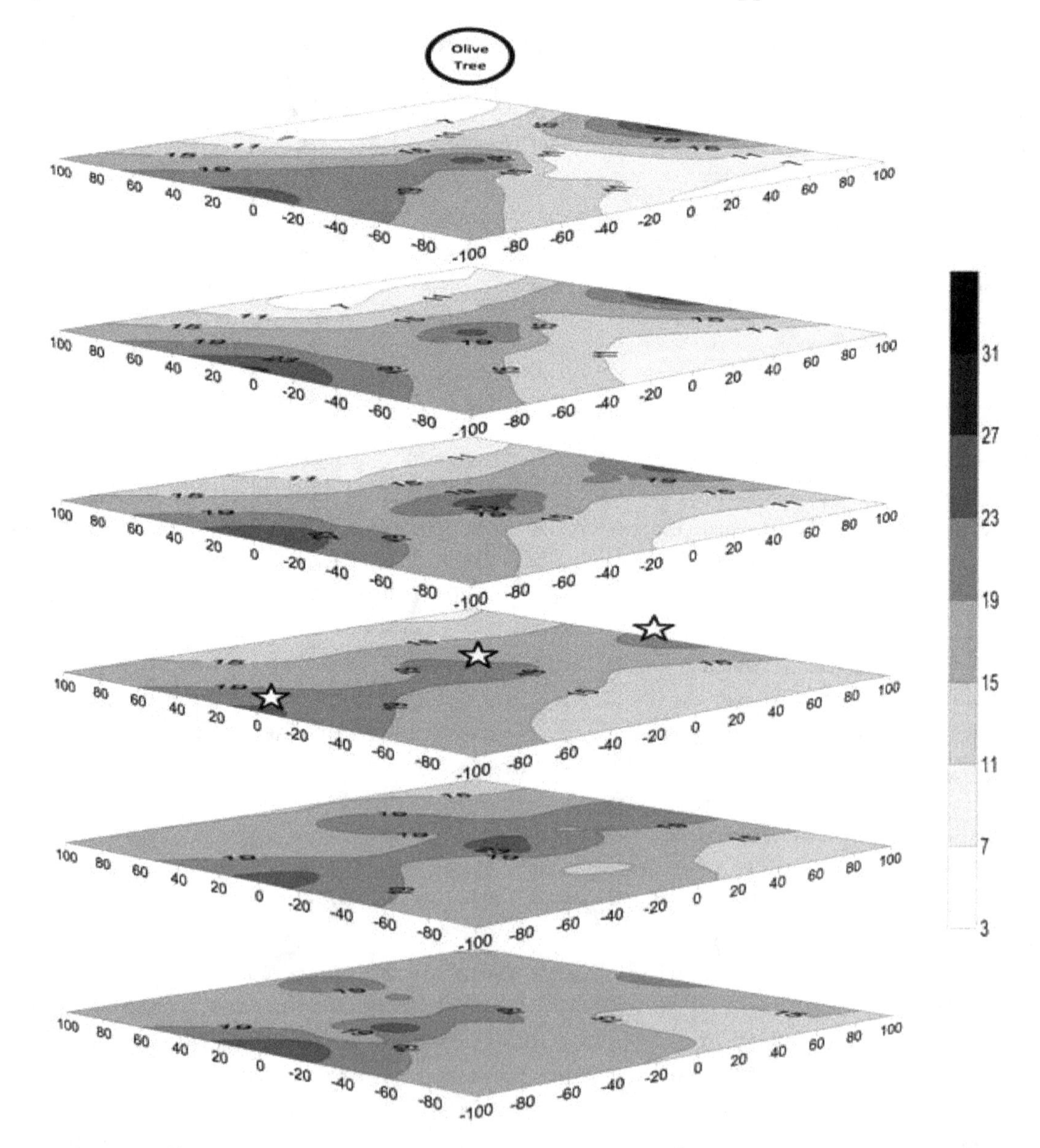

Figure 6. An overall gravimetric water content (%) distribution at all six depths, 0–5, 5–10, 10–15, 15–30, 30–45, and 45–60 cm. The black and white stars indicate the location of the drippers.

3.2. *Salinity Spatial Distribution*

Two quantities were used to describe the long-term accumulation of salinity: (i) electrical conductivity (EC)—representing the total salinity, and (ii) chloride concentration—as a soil native conservative tracer. The EC distribution is shown for the transect along (Figure 7a) and perpendicular (Figure 7b) to the drip line. The chloride distribution is shown in Figure 7c and d for transects along and perpendicular to the drip line, respectively. For both transects, salinity and chloride distribution exhibited similar patterns. Specifically, both quantities demonstrated a leaching zone above and below the emitters and salt accumulations zone between nearby emitters. For the transect along the drip-line, the highest salinity prevailed above the drip line in the middle of two nearby emitters. Nevertheless, the salinity values below the drip-line are also very high and may reduce water uptake by the olive trees' roots, due to the high osmotic pressure, even if a high water content is maintained. Regarding the perpendicular transect, a distinct, uneven distribution could be observed. Specifically, the salt accumulation away from the tree (−100 to 0 cm) is significantly higher than the one obtained toward the tree line (0 to 100 cm). The lowest salinity obtained near the tree line may be explained by a reduced evaporation and capillary rise toward the soil surface, due to the surface shading by the olive trees. However, the entire zone exhibited very-high to extreme values of EC, which indicates the salinization of the olive plantation, as clearly illustrated from the three dimensional visualization of the entire 2 m^2 view of the spatial EC distribution at six individual depths (Figure 8). The representation of the entire domain emphasizes the extreme values of salinity above the drip-line and away from the tree-line. A clear pattern could not be observed, due to the large salinity spectrum that was considered in this counter map. Nevertheless, the leaching zones above and below the emitter is clearly demonstrated, indicating moderate to high salinity levels.

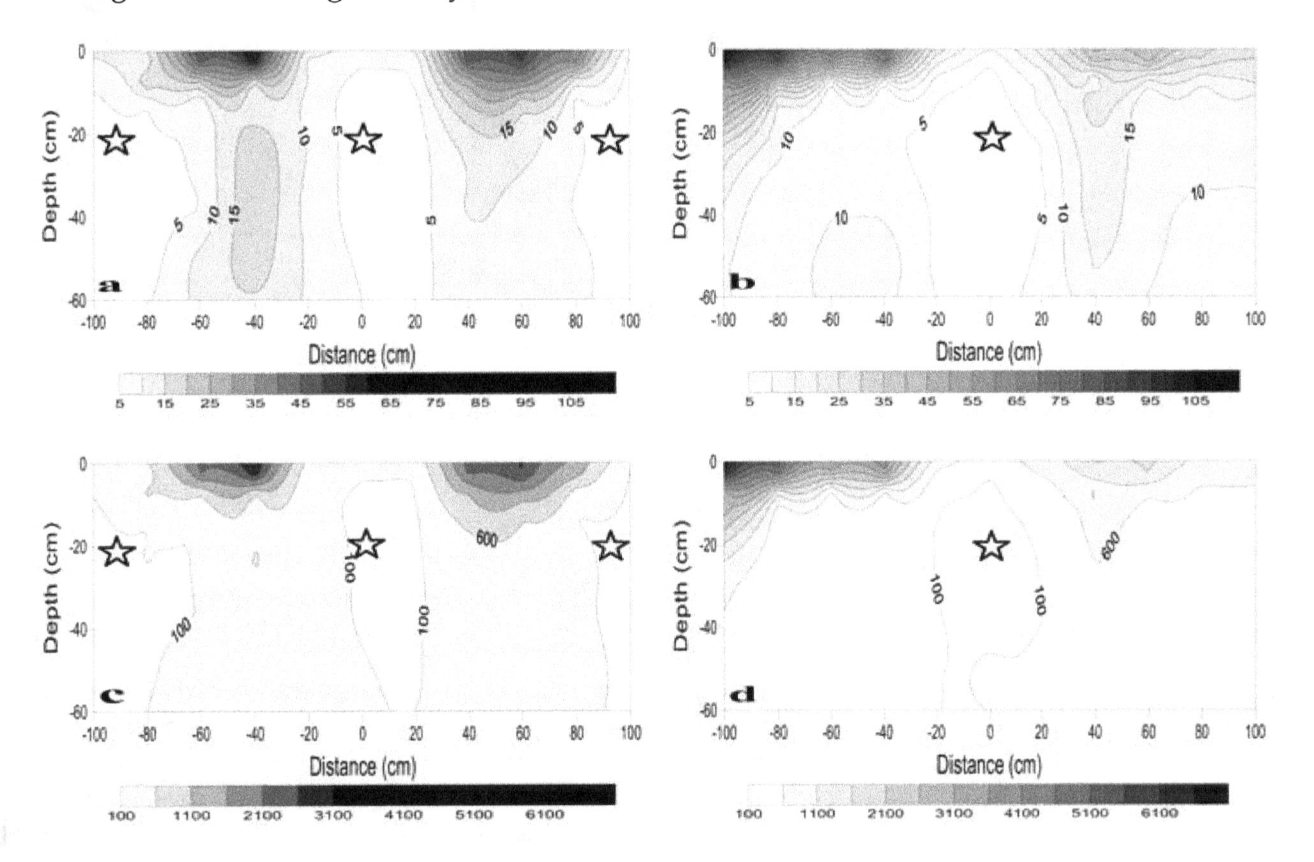

Figure 7. Electrical conductivity (dS m^{-1}) distribution (**a**) along the drip line, (**b**) perpendicular to the drip line, and soil chloride (mg L^{-1}) distribution (**c**) along the drip line and (**d**) perpendicular to the drip line. The black and white stars indicate the location of the drippers.

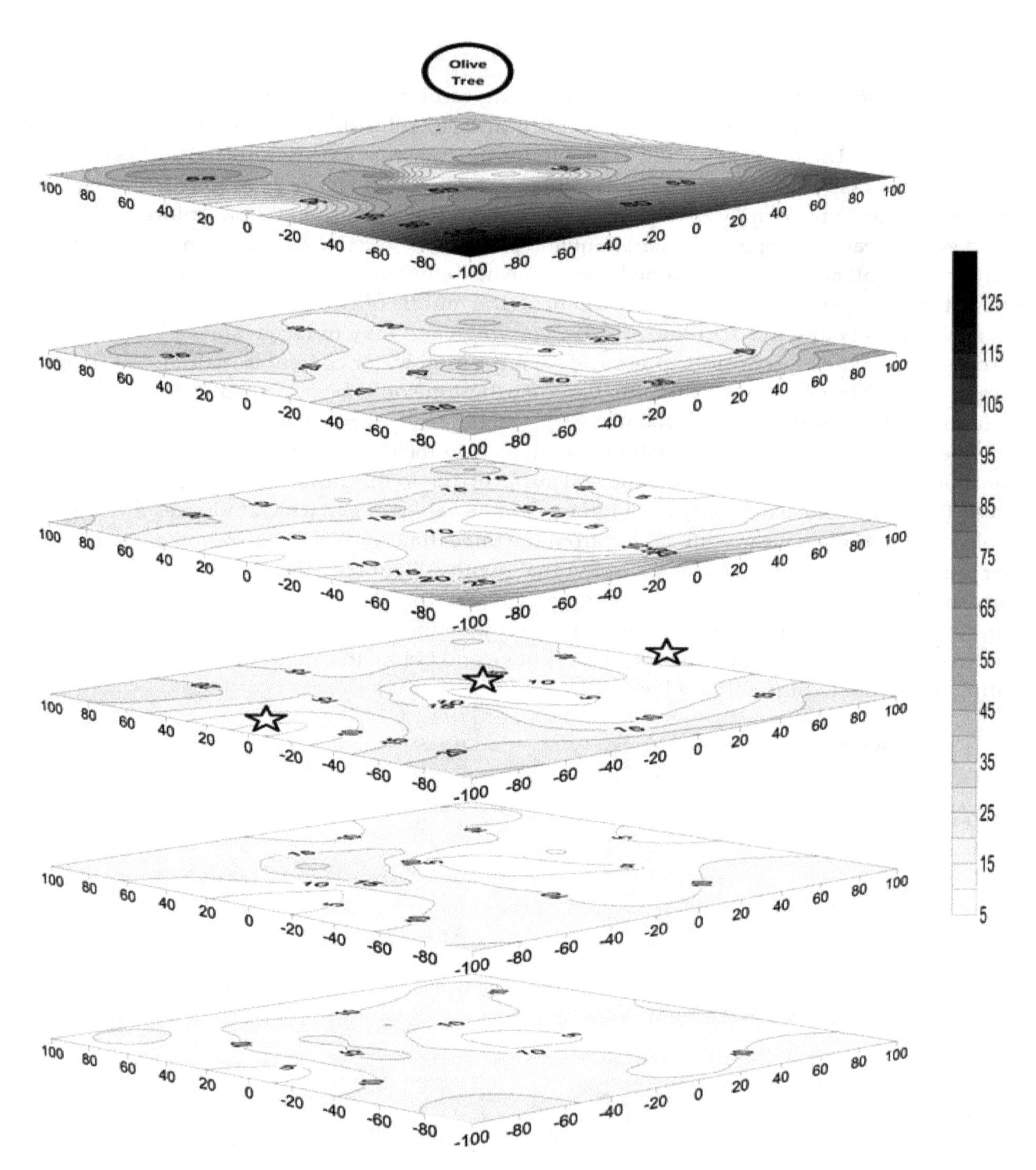

Figure 8. An overall distribution of the electrical conductivity (dS m^{-1}) at all six depths, 0–5, 5–10, 10–15, 15–30, 30–45, and 45–60 cm. The black and white stars indicate the location of the drippers.

As mentioned, the mean annual rainfall in this region is below 125 mm and may be distributed over ten low-rain events (Figure 3). Under these conditions, there is insufficient rain to leach the accumulated salts from the soil surface below the active root zone. This minimal rainfall can also exacerbate the salinity problem by bringing surface salts (0–15 cm) to the root zone (30–60 cm) after one or several rainfall events. The salinity observed in the orchard soils is far above the normal threshold salinity level for olive growth, i.e., a soil EC value of 4 to 6 dS m^{-1} is the accepted critical limit for normal olive growth [12–15]. To leach the excess salts from the root zone, high rainfall events >600 mm are required [48,49] or sprinkler irrigation has to be implemented in order to leach salts, but the long-term economic sustainability of this system is questionable [30].

3.3. Sodicity Spatial Distribution

The outcome of the long-term sodification is described by the calculated values of the SAR as a measure for the sodicity of the liquid phase and by the sodium adsorption percentage (ESP = 100 xNa/CEC) as a measure the sodicity of the solid phase. The spatial SAR distribution is shown for the transect along (Figure 9a) and perpendicular (Figure 9b) to the drip line. The spatial ESP distribution is shown in Figure 9c,d for transects along and perpendicular to the drip line, respectively. In general terms, the spatial distribution patterns obtained for the SAR and ESP resemble the one obtained for the salinity (Figure 7), demonstrating that a higher salinity in the soil solution resulted in higher SAR and consequently higher ESP. The SAR values obtained between nearby emitters, above a below the dripline, exhibited values >15% and reached values even higher than 30% at the top soil layers (Figure 10). Therefore, sodicity hazardous of soil structure degradation which can negatively affects soil hydraulic properties should be considered. Nevertheless, since high sodicity levels were accompanied by high salinity levels, the latter, may offset the negative sodicity effect on the stability of soil structure. The fact that sodicity levels increased with salinity implied a chemical equilibrium between the soil-solution and solid phase. In support of this argument is the linear correlation obtained from all samples between the ESR and SAR (Figure 11) with a slope of 0.0134, which is close to the commonly accepted value of the Gapon constant, $K_G = 0.015$, e.g., [50]. In addition, a positive linear correlation was obtained (data is not shown) between ESP and SAR (ESP = 0.77SAR + 3.34, $R^2 = 0.73$). It is well established [51] that if cation exchange reactions have reached equilibrium, the ESP values are similar to the SAR at the range of 0 to 40.

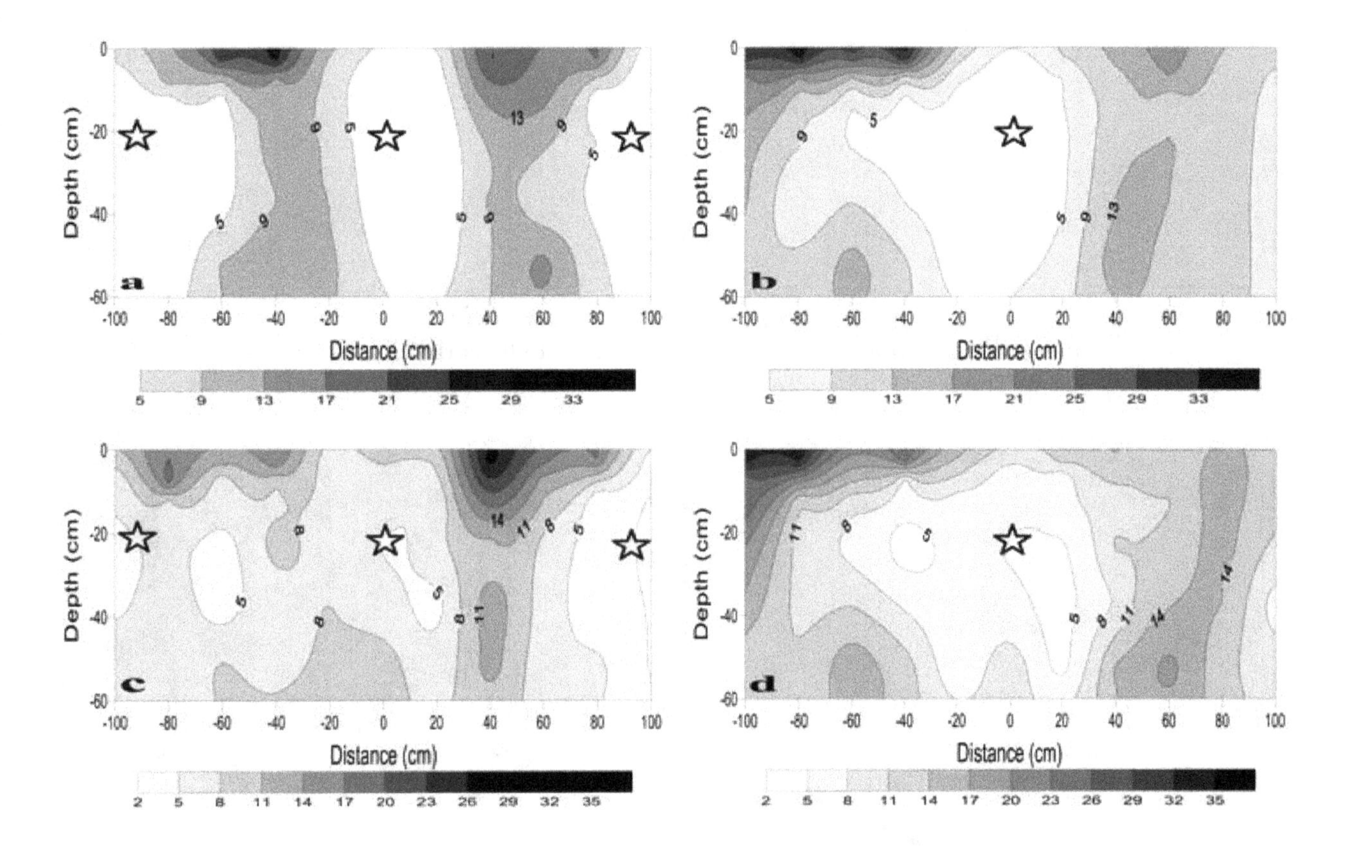

Figure 9. SAR (meq L^{-1})$^{0.5}$ distribution (**a**) along the drip line, (**b**) perpendicular to the drip line and, ESP (%) distribution, (**c**) along the drip line, and (d) perpendicular to the drip line. The black and white stars indicate the location of the drippers.

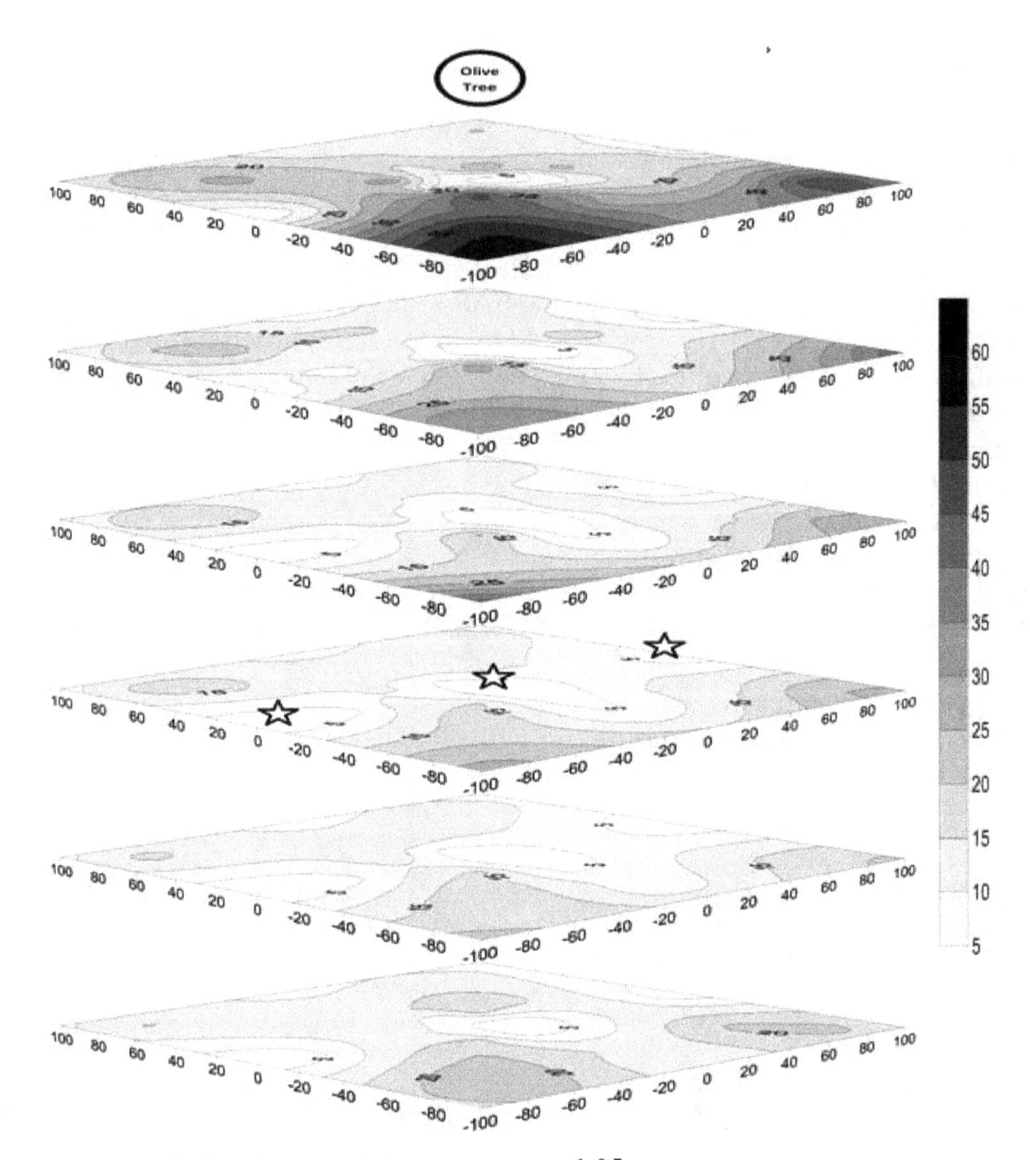

Figure 10. An overall distribution of the SAR (meq $L^{-1})^{0.5}$ at all six depths, 0–5, 5–10, 10–15, 15–30, 30–45, and 45–60 cm. The black and white stars indicate the location of the drippers.

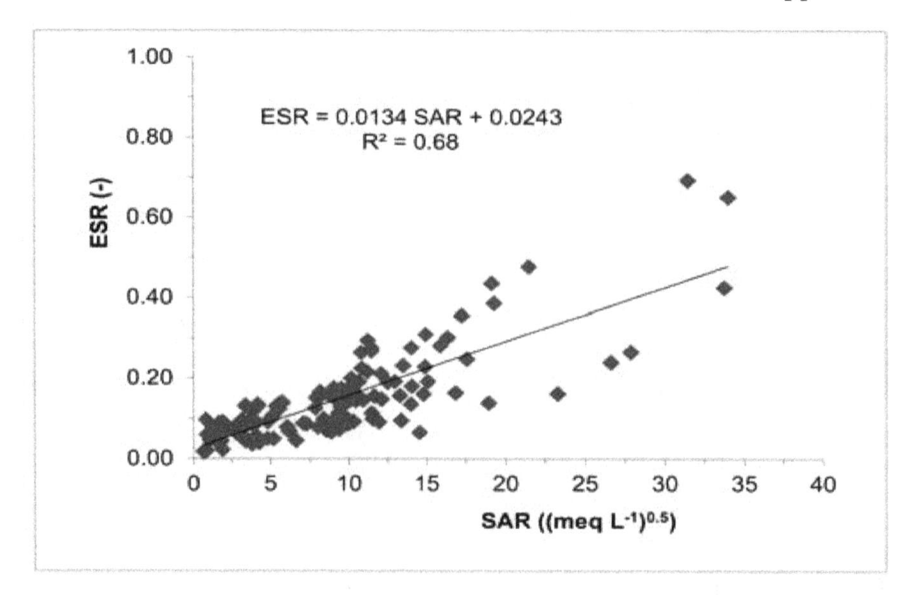

Figure 11. Exchangeable sodium ratio (ESR) as a function of sodium adsorption ratio (SAR).

4. Summary and Conclusions

The main goal of this study was to quantify the long term development of salinity and sodicity in an olive orchard grown in an arid region and irrigated with brackish water for two decades using DI and, subsequently, SDI. The study was motivated by reports on olive orchard uprooting in the Negev Desert, due to the continual reduction in olive yields. We assumed that under the climate conditions that prevail in this arid region, long term salinization and sodification at the active root zone is inevitable, in particular, under SDI with brackish irrigation water. The results of this study clearly demonstrate that following twenty years of irrigation with brackish irrigation water, salinization and sodification took place in the examined soil profile (0–60 cm), which represents the active root zone of the olive trees. The relatively large distance (1 m) between nearby drippers resulted in no significant overlaps between the wetting fronts of two nearby emitters. Consequently, a relatively small area of salt leaching could be observed below the emitters, with EC values close to the ones in the brackish irrigation water. However, moderate salt buildup took place above the emitters. The salinity buildup between nearby emitters were above the salinity threshold level for olive trees with extreme EC values above the drip line and high ones below it. The spatial distribution of the sodicity levels resembled the ones obtained for salinity, corresponding to high sodicity levels (in terms of SAR and ESP) where salinization took place. The linear correlation obtained between the sodicity quantities (i.e., ESR vs. SAR and ESP vs. SAR) implies that chemical equilibrium has been reached between the brackish irrigation water, soil solution, and the solid phase.

The results of this study show that in arid regions, the benefits of water saving, attributed to SDI, are masked by soil salinization and sodification that was induced by this irrigation method. The quantification of the long term suitability of brackish water irrigation with SDI may assist in improving the irrigation system design, for example, by significantly reducing the distance of nearby emitters and increasing the allocated leaching fraction. Finally, this study emphasizes the current necessity for salinity and sodicity reclamation in the studied region. Any alternative practices of replacing olives trees and further introduction of even higher salinity tolerant plants (e.g., jojoba) in this region will intensify the salt buildup without leaving any option for soil reclamation in the future.

Author Contributions: The study was conceived and designed by G.A., J.R.K. and P.T. performed the field experiment. All authors took a part in the data analysis, interpretation, and writing the paper. All authors have read and approved the final manuscript.

Acknowledgments: We thank E.N. and his team from Kibbutz Revivim for providing the agricultural facilities. We also appreciate the technical assistance and support by Y.M.

References

1. Ladeiro, B. Saline agriculture in the 21st century: Using salt contaminated resources to cope food requirements. *J. Bot.* **2012**, *2012*, 310705. [CrossRef]
2. Siegel, S.M. *Let There be Water: Israel's Solution for a Water-Starved World*; Macmillan: New York, NY, USA, 2015.
3. Fita, A.; Rodríguez-Burruezo, A.; Boscaiu, M.; Prohens, J.; Vicente, O. Breeding and domesticating crops adapted to drought and salinity: A new paradigm for increasing food production. *Front. Plant Sci.* **2015**, *6*, 978. [CrossRef] [PubMed]
4. De Oliveira, A.B.; Alencar, N.L.M.; Gomes-Filho, E. Comparison between the water and salt stress effects on plant growth and development. In *Responses of Organisms to Water Stress*; Akinci, S., Ed.; IntechOpen: London, UK, 2013.
5. Uddin, M.N.; Hossain, M.A.; Burritt, D.J. Salinity and drought stress: Similarities and differences in oxidative responses and cellular redox regulation. *Water Stress Crop Plants Sustain. Approach* **2016**, *1*, 86–101.
6. Muscolo, A.; Sidari, M.; Panuccio, M.R.; Santonoceto, C.; Orsini, F.; De Pascale, S. Plant responses in saline and arid environments: An overview. *Eur. J. Plant Sci. Biotechnol.* **2011**, *5*, 1–11.

7. Pasternak, D.; Aronson, J.A.; Ben-Dov, J.; Forti, M.; Mendlinger, S.; Nerd, A.; Sitton, D. Development of new arid zone crops for the Negev Desert of Israel. *J. Arid Environ.* **1986**, *11*, 37–59. [CrossRef]
8. Dag, A.; Tugendhaft, Y.; Yogev, U.; Shatzkin, N. Commercial cultivation of olive (*Olea europaea* L.) with saline water under extreme desert conditions. In Proceedings of the Fifth International Symposium on Olive Growing, Izmir, Turkey, 27 September 2004; pp. 279–284.
9. Gucci, R.; Tattini, M. Salinity tolerance in olive. *Hortic. Rev.* **1997**, *21*, 177–214.
10. Gucci, R.; Caruso, G. Environmental stresses and sustainable olive growing. In Proceedings of the XXVIII International Horticultural Congress on Science and Horticulture for People (IHC2010): Olive Trends Symposium, Lisbon, Portugal, 22 August 2010; pp. 19–30.
11. Fernández, J.E.; Moreno, F.; Cabrera, F.; Arrue, J.L.; Martín-Aranda, J. Drip irrigation, soil characteristics and the root distribution and root activity of olive trees. *Plant Soil* **1991**, *133*, 239–251. [CrossRef]
12. Weissbein, S.; Wiesman, Z.; Ephrath, Y.; Silberbush, M. Vegetative and reproductive response of olive cultivars to moderate saline water irrigation. *HortScience* **2008**, *43*, 320–327. [CrossRef]
13. Ayers, R.S.; Westcot, D.W. *Water Quality for Agriculture. Irrigation and Drainage Paper No. 29*; Food and Agriculture Organization of the United Nations: Rome, Italy, 1985.
14. Aragüés, R.; Puy, J.; Royo, A.; Espada, J.L. Three-year field response of young olive trees (*Olea europaea* L., cv. Arbequina) to soil salinity: Trunk growth and leaf ion accumulation. *Plant Soil* **2005**, *271*, 265–273. [CrossRef]
15. Ben-Gal, A. Salinity and olive: From physiological responses to orchard management. *Isr. J. Plant Sci.* **2011**, *59*, 15–28. [CrossRef]
16. Weissbein, S. *Characterization of New Olive (Olea Europea L.) Varieties Response to Irrigation with Saline Water in the Ramat Negev Area*; Ben Gurion University: Beersheba, Israel, 2006.
17. Rewald, B.; Rachmilevitch, S.; Ephrath, J.E. Salt stress effects on root systems of two mature olive cultivars. *Acta Hortic.* **2011**, *888*, 109–118. [CrossRef]
18. Hill, A.; Rewald, B.; Rachmilevitch, S. Belowground dynamics in two olive varieties as affected by saline irrigation. *Sci. Hortic.* **2013**, *162*, 313–319. [CrossRef]
19. Soda, N.; Ephrath, J.E.; Dag, A.; Beiersdorf, I.; Presnov, E.; Yermiyahu, U.; Ben-Gal, A. Root growth dynamics of olive (*Olea europaea* L.) affected by irrigation induced salinity. *Plant Soil* **2017**, *411*, 305–318. [CrossRef]
20. Wiesman, Z.; Itzhak, D.; Dom, N.B. Optimization of saline water level for sustainable Barnea olive and oil production in desert conditions. *Sci. Hortic.* **2004**, *100*, 257–266. [CrossRef]
21. Tal, A. Rethinking the sustainability of Israel's irrigation practices in the Drylands. *Water Res.* **2016**, *90*, 387–394. [CrossRef]
22. Keller, J.; Bliesner, R.D. *Sprinkle and Trickle Irrigation*; Springer: New York, NY, USA, 1990.
23. Metochis, C. Irrigation of 'Koroneiki' olives with saline water. *Olivae* **1999**, *76*, 22–24.
24. Melgar, J.C.; Mohamed, Y.; Serrano, N.; García-Galavís, P.A.; Navarro, C.; Parra, M.A.; Benlloch, M.; Fernández-Escobar, R. Long term responses of olive trees to salinity. *Agric. Water Manag.* **2009**, *96*, 1105–1113. [CrossRef]
25. Bruins, H.J. Ancient desert agriculture in the Negev and climate-zone boundary changes during average, wet and drought years. *J. Arid Environ.* **2012**, *86*, 28–42. [CrossRef]
26. Lamm, F.R.; Bordovsky, J.P.; Schwankl, L.J.; Grabow, G.L.; Enciso-Medina, J.; Peters, R.T.; Colaizzi, P.D.; Trooien, T.P.; Porter, D.O. Subsurface drip irrigation: Status of the technology in 2010. *Trans. ASABE* **2012**, *55*, 483–491. [CrossRef]
27. Martínez, J.; Reca, J. Water use efficiency of surface drip irrigation versus an alternative subsurface drip irrigation method. *J. Irrig. Drain. Eng.* **2014**, *140*. [CrossRef]
28. Dorta-Santos, M.; Tejedor, M.; Jiménez, C.; Hernández-Moreno, J.M.; Palacios-Díaz, M.P.; Díaz, F.J. Evaluating the sustainability of subsurface drip irrigation using recycled wastewater for a bioenergy crop on abandoned arid agricultural land. *Ecol. Eng.* **2015**, *79*, 60–68. [CrossRef]
29. Dorta-Santos, M.; Tejedor, M.; Jiménez, C.; Hernández-Moreno, J.M.; Díaz, F.J. Using marginal quality water for an energy crop in arid regions: Effect of salinity and boron distribution patterns. *Agric. Water Manag.* **2016**, *171*, 142–152. [CrossRef]
30. Roberts, T.L.; White, S.A.; Warrick, A.W.; Thompson, T.L. Tape depth and germination method influence patterns of salt accumulation with subsurface drip irrigation. *Agric. Water Manag.* **2008**, *95*, 669–677. [CrossRef]
31. Oron, G.; De Malach, Y.; Gillerman, L.; David, I.; Rao, V.P. Improved saline-water use under subsurface drip irrigation. *Agric. Water Manag.* **1999**, *39*, 19–33. [CrossRef]

32. Thompson, T.L.; Pang, H.C.; Li, Y.Y. The potential contribution of subsurface drip irrigation to water-saving agriculture in the western USA. *Agric. Sci. China* **2009**, *8*, 850–854.
33. Trifonov, P.; Lazarovitch, N.; Arye, G. Increasing water productivity in arid regions using low-discharge drip irrigation: A case study on potato growth. *Irrig. Sci.* **2017**, *35*, 287–295. [CrossRef]
34. Trifonov, P.; Lazarovitch, N.; Arye, G. Water and Nitrogen Productivity of Potato Growth in Desert Areas under Low-Discharge Drip Irrigation. *Water* **2018**, *10*, 970. [CrossRef]
35. Mounzer, O.; Pedrero-Salcedo, F.; Nortes, P.A.; Bayona, J.M.; Nicolás-Nicolás, E.; Alarcón, J.J. Transient soil salinity under the combined effect of reclaimed water and regulated deficit drip irrigation of Mandarin trees. *Agric. Water Manag.* **2013**, *120*, 23–29. [CrossRef]
36. Oster, J.D.; Shainberg, I. Soil responses to sodicity and salinity: Challenges and opportunities. *Soil Res.* **2001**, *39*, 1219–1224. [CrossRef]
37. Levy, G.J.; Fine, P.; Goldstein, D.; Azenkot, A.; Zilberman, A.; Chazan, A.; Grinhut, T. Long term irrigation with treated wastewater (TWW) and soil sodification. *Biosyst. Eng.* **2014**, *128*, 4–10. [CrossRef]
38. Bedbabis, S.; Trigui, D.; Ahmed, C.B.; Clodoveo, M.L.; Camposeo, S.; Vivaldi, G.A.; Rouina, B.B. Long-terms effects of irrigation with treated municipal wastewater on soil, yield and olive oil quality. *Agric. Water Manag.* **2015**, *160*, 14–21. [CrossRef]
39. Ayoub, S.; Al-Shdiefat, S.; Rawashdeh, H.; Bashabsheh, I. Utilization of reclaimed wastewater for olive irrigation: Effect on soil properties, tree growth, yield and oil content. *Agric. Water Manag.* **2016**, *176*, 163–169. [CrossRef]
40. Erel, R.; Eppel, A.; Yermiyahu, U.; Ben-Gal, A.; Levy, G.; Zipori, I.; Schaumann, G.E.; Mayer, E.; Dag, A. Long-term irrigation with reclaimed wastewater: Implications on nutrient management, soil chemistry and olive (*Olea europaea* L.) performance. *Agric. Water Manag.* **2019**, *213*, 324–335. [CrossRef]
41. Bader, B.; Aissaoui, F.; Kmicha, I.; Salem, A.B.; Chehab, H.; Gargouri, K.; Boujnah, D.; Chaieb, M. Effects of salinity stress on water desalination, olive tree (*Olea europaea* L. cvs 'Picholine','Meski'and 'Ascolana') growth and ion accumulation. *Desalination* **2015**, *364*, 46–52. [CrossRef]
42. Ben-Gal, A.; Beiersdorf, I.; Yermiyahu, U.; Soda, N.; Presnov, E.; Zipori, I.; Crisostomo, R.R.; Dag, A. Response of young bearing olive trees to irrigation-induced salinity. *Irrig. Sci.* **2017**, *35*, 99–109. [CrossRef]
43. Banin, A.; Amiel, A. A correlative study of the chemical and physical properties of a group of natural soils of Israel. *Geoderma* **1970**, *3*, 185–198. [CrossRef]
44. Rhoades, J.D. Cation exchange capacity. In *Agronomy Monograph No. 9 Methods of Soil Analysis: Part 2. Chemical and Microbiological Properties*, 2nd ed.; Page, A.L., Miller, R.H., Kearney, D.R., Eds.; ASA, SSSA: Madison, WI, USA, 1986; pp. 149–157.
45. Thomas, G.W. Exchangeable cations. In *Agronomy Monograph No 9 Methods of Soil Analysis. Part 2. Chemical and Microbiological Properties*, 2nd ed.; Page, A.L., Miller, R.H., Keerney, D.R., Eds.; ASA, SSSA: Madison, WI, USA, 1986; pp. 159–164.
46. Fares, A.; Parsons, L.R.; Wheaton, T.A.; Morgan, K.T.; Simunek, J.; Van Genuchten, M.T. Simulated drip irrigation with different soil types. *Proc. Fla. State Hortic. Soc.* **2001**, *114*, 22–24.
47. Sevostianova, E.; Leinauer, B.; Sallenave, R.; Karcher, D.; Maier, B. Soil salinity and quality of sprinkler and drip irrigated warm-season turfgrasses. *Agron. J.* **2011**, *103*, 1773–1784. [CrossRef]
48. Ben-Hur, M.; Li, F.H.; Keren, R.; Ravina, I.; Shalit, G. Water and salt distribution in a field irrigated with marginal water under high water table conditions. *Soil Sci. Soc. Am. J.* **2001**, *65*, 191–198. [CrossRef]
49. Lado, M.; Bar-Tal, A.; Azenkot, A.; Assouline, S.; Ravina, I.; Erner, Y.; Fine, P.; Dasberg, S.; Ben-Hur, M. Changes in chemical properties of semiarid soils under long-term secondary treated wastewater irrigation. *Soil Sci. Soc. Am. J.* **2012**, *76*, 1358–1369. [CrossRef]
50. Levy, R.; Hillel, D. Thermodynamic Equilibrium Constants of Sodium-Calcium Exchange in Some Israel SOILS1. *Soil Sci.* **1968**, *106*, 393–398. [CrossRef]
51. United States Department of Agriculture. Diagnosis and improvement of saline and alkali soils. In *Agriculture Handbook*; US Government Printing Office: Washington, DC, USA, 1954; Volume 60, pp. 83–100.

5

Potential of Deficit and Supplemental Irrigation under Climate Variability in Northern Togo, West Africa

Agossou Gadédjisso-Tossou [1,2,*], Tamara Avellán [1] and Niels Schütze [2]

1 United Nations University Institute for Integrated Management of Material Fluxes and of Resources (UNU-FLORES), Ammonstrasse 74, 01067 Dresden, Germany; avellan@unu.edu
2 Institute of Hydrology and Meteorology, Technische Universität Dresden, 01069 Dresden, Germany; niels.schuetze@tu-dresden.de
* Correspondence: Agossou.Gadedjisso-Tossou@tu-dresden.de

Abstract: In the context of a growing population in West Africa and frequent yield losses due to erratic rainfall, it is necessary to improve stability and productivity of agricultural production systems, e.g., by introducing and assessing the potential of alternative irrigation strategies which may be applicable in this region. For this purpose, five irrigation management strategies, ranging from no irrigation (NI) to controlled deficit irrigation (CDI) and full irrigation (FI), were evaluated concerning their impact on the inter-seasonal variability of the expected yields and improvements of the yield potential. The study was conducted on a maize crop (*Zea mays* L.) at a representative site in northern Togo with a hot semi-arid climate and pronounced dry and wet rainfall seasons. The OCCASION (Optimal Climate Change Adaption Strategies in Irrigation) framework was adapted and applied. It consists of: (i) a weather generator for simulating long climate time series; (ii) the AquaCrop model, which was used to simulate the irrigation system during the growing season and the yield response of maize to the considered irrigation management strategies; and (iii) a problem-specific algorithm for optimal irrigation scheduling with limited water supply. We found high variability in rainfall during the wet season which leads to considerable variability in the expected yield for rainfed conditions (NI). This variability was significantly reduced when supplemental irrigation management strategies (CDI or FI) requiring a reasonably low water demand of about 150 mm were introduced. For the dry season, it was shown that both irrigation management strategies (CDI and FI) would increase yield potential for the local variety TZEE-W up to 4.84 Mg/ha and decrease the variability of the expected yield at the same time. However, even with CDI management, more than 400 mm of water is required if irrigation would be introduced during the dry season in northern Togo. Substantial rainwater harvesting and irrigation infrastructures would be needed to achieve that.

Keywords: AquaCrop model; maize; deficit irrigation; crop-water production function; West Africa

---

1. Introduction

The present world population of 7.3 billion will increase to 9.7 billion by 2050 [1]. Similarly, the medium variant of the UN Population Division [2] predictions disclose that the total population of the West African region would increase from 350 million in 2015 to 450 million in 2030, and nearly 800 million in 2050. FAO [3] estimates that agricultural production will have to rise by 60% by 2050 to meet the world's projected demands for food and feed. In West Africa, Liniger et al. [4] reported that food production should increase by 70% by 2050 to meet the necessary caloric requirements. However, a lack of available water for agricultural production, the energy sector, and other forms of anthropogenic water consumption is already harming several parts of the world. This lack of water is projected to become more severe with the growing population, rising temperatures, and altering precipitation patterns [5]. The variation of the food diets in many developing countries compound this problem and lead to the demand for more processed food and animal proteins by consumers [6].

The World Bank [7] reports that the rate of increase in food demand is projected to be higher in developing than in developed countries. These are also the regions that are subject to a wide yield gap. The world demand (billion tons) of cereals was 1.20 in 1974, 1.84 in 1997 and is expected to be 2.50 in 2020 [8]. In addition, van Ittersum et al. [9] pointed out that Sub-Saharan Africa (SSA) is the region with lowest food security because by 2050 its demand for cereals will almost triple, whereas current levels of cereal consumption already rely on considerable importations.

Lobell and Gourdji [10] pointed out that, in the past several decades, air temperatures have been increasing in most of the main cereal cropping areas around the world. They added that the changes in temperature and the intensity and seasonal volume of rainfall are impacting soil moisture. In turn, soil moisture is of high importance for crop production. In developing countries, particularly, the changes in these climatic variables over time are likely to have a damaging impact on water accessibility, which in turn affects crop yield. Kotir [11] and Druyan [12] stressed the fact that researchers have described Sub-Saharan Africa as the most sensitive region to the impacts of climate variabilities and change because of its dependence on rainfed agriculture and low capacity for adaptation. Moreover, Sarr [13] contended that the West African region has faced decades of severe drought, which have affected agricultural production substantially. The observations already show the late onset and early cessation dates of rainfall and the reduction of length of growing period.

According to the Togolese Ministry of the Environment and Forestry (MERF) [14], in the dry savannah of northern Togo, a West African country, the wet season, which spanned six months in the 1970s, has reduced to five or four months nowadays. Consequently, on the one hand, a substantial amount of rainwater falls within a short period causing flooding, while, on the other hand, frequent dry spells in the wet season lead to crop failure [15]. In addition, there is no rainfed agricultural activity during the dry season in northern Togo because of a lack of rainfall [16].

Researchers and practitioners are putting more focus on producing more with limited resources in agriculture to meet the food demand and at the same time address the adverse effects of climate change [17–19]. Agriculture, which accounts for 38% of Togo's gross domestic product, provides over 20% of export earnings and employs 70% of the active population. Togolese agriculture is predominantly rainfed [20,21]. According to the International Commission on Irrigation and Drainage (ICID) [22], rainfed agriculture is "agriculture without application of irrigation. It may be without, or with a drainage system." A promising practice to overcome water shortage in rainfed cropping systems is supplemental irrigation (SI). The ICID [22] defines SI as: "the addition of small amounts of water to essentially rainfed crops during times when rainfall fails to provide sufficient moisture for normal plant growth, in order to improve and stabilize yields." SI practice increases yields and water productivity in rainfed cropping systems [23]. In addition, conventional irrigation systems can be used to improve crop productivity. The ICID [22] defines conventional irrigation as: "the replenishment of soil water storage in plant root zone through methods other than natural precipitation".

Irrigation scheduling is the procedure of deciding when, where, and how much water to apply [24] for irrigation. Farmers can apply the total crop-water requirements or more in the right period if water is available. This practice is called full irrigation (FI). When water provisions are limited, or irrigation expenses are great, FI may be substituted by deficit irrigation (DI) [25]. This is limited irrigation scheduling in agriculture [26]. DI can be controlled or otherwise. Uncontrolled DI is equivalent to rainfed agriculture. English [27] and English and Raja [28] defined controlled deficit irrigation (CDI) as the concept of intentionally and systematically under-irrigating a crop. English [27] developed an analytical framework to evaluate the profit when optimizing water use. Thus, he included implicitly economic aspects in the definition. Later, Lecler [29] provided a more explicit definition: "CDI is an optimization strategy by which net returns are maximized by lessening the volume of irrigation water applied to a crop to a level that results in some yield loss caused by water stress". Recently, Fereres and Soriano [30] defined CDI as the application of water below full crop-water requirements or evapotranspiration. The objective of applying limited water is to cope with scarce water supplies and improve productivity. Kögler and Söffker [31] reported that CDI practice contributes to saving up to 20–40% irrigation water at yield reductions under 10%. It can contribute to increasing farmers' net income where water is scarce [27]. Thus, CDI is an irrigation management practice that contributes to enhancing food security.

Many studies that applied the simulation-based approach to assess deficit irrigation strategies failed to consider the variability of relevant climate factors—such as precipitation and temperature—and soil properties [32,33]. Semenov [34] and Brumbelow and Georgakakos [35], among others, analyzed possible impacts of climate variability and climate change on agriculture using process-based simulation models. Most of these studies only look at rainfed or non-irrigated sites or assumed full irrigation. Few researchers, including Schütze and Schmitz [36] and Brumbelow and Georgakakos [35], assessed limited irrigation systems and the impact of climate variability on crop-water production functions (CWPF). Brumbelow and Georgakakos [35] derived probability distribution functions of CWPF (CWPF-PDs) using climate change scenarios data of the Intergovernmental Panel on Climate Change (IPCC). Schütze and Schmitz [36] delved into the CWPF concept and suggested a stochastic framework in the form of a decision support tool for Optimal Climate Change Adaption Strategies in Irrigation (OCCASION) for deriving site-specific stochastic CWPFs (SCWPFs). To perform such analyses, one needs to utilize crop models to simulate the potential or expected crop yield for a given soil, climate, and management practice condition.

Several crop simulation models such as DSSAT [37], AquaCrop [38–40], DAISY [41], CropWat [42], APSIM [43], and PILOTE [44] are available in the literature to simulate yield response to water. It is important to recognize that most of these models show substantial complexities and require several data to run. Most of these models require many parameters to run, and many are not readily available in the field and need to be determined experimentally [45]. Exceptionally, the AquaCrop model uses relatively few explicit and mostly intuitive parameters and input variables, requiring simple methods for their derivation [46]. For instance, unlike AquaCrop, the DSSAT model requires input data about crop genetics and pest management [37], while APSIM requires NO_3 and NH_4 content of the soil layers [43].

Few studies have investigated irrigation management strategies on crops in the dry savannah area of northern Togo [20]. Therefore, this study assessed the potential of deficit and supplemental irrigation in northern Togo. Specifically, the study aimed at: (i) characterizing the climate of a water-scarce site in northern Togo, West African region; and (ii) evaluating five irrigation management strategies, ranging from no irrigation (NI) to CDI and FI for a maize crop (*Zea mays* L.) at a representative site in northern Togo with pronounced dry and wet rainfall seasons.

2. Materials and Methods

2.1. Study Area

Togo is a small West African francophone country. It is bordered by the Bight of Benin and Burkina Faso in the south and north, respectively. Togo is bound in the west by Ghana and in the east by Benin. Geographically, it lies between latitudes 6° N and 11° N, and longitudes 0° E and 2° E. It covers a surface of 56,600 km^2 and has a long, narrow profile, stretching more than 550 km from north to south but not exceeding 160 km in width [47]. Its population is estimated to be 6,191,155 [48].

We conducted this study in the Dapaong district, northern Togo (Figure 1). Dapaong belongs to the Southern-Guinea-Savannah agro-ecological zone [49]. The principal rainfed crops grown include maize (*Zea mays*), sorghum (*Sorghum bicolor*), and pearl millet (*Pennisetum glaucum*), mainly for subsistence, while cash crops such as cotton (*Gossypium hirsutum*) are also cultivated. Some vegetables and legumes such as okra (*Abelmoschus esculentus*), cowpea (*Vigna unguiculata*), and soybean (*Glycine max*) are grown in association with the cereals mentioned above. The vegetation type is a woody savannah, with noticeable agricultural farms. The primary tree species are *Parkia biglobosa, Butyrospermum parkii,* and *Acacia sieberiana* [50]. The Togolese Institute of Agricultural Research (ITRA) [51] and Didjeira et al. [52] identified maize crop as the staple food in Togo, and it represents 60% of the cereals consumed by the population. On the farms close to the houses, the main cropping system is intercropping (cereal–legume mixtures), while on the farms far from the houses, farmers practice monoculture [53]. Since cotton is grown with a high level of pesticides, intercropping is not possible on cotton farms. Hoes and cutlasses are the primary tools of cultivation.

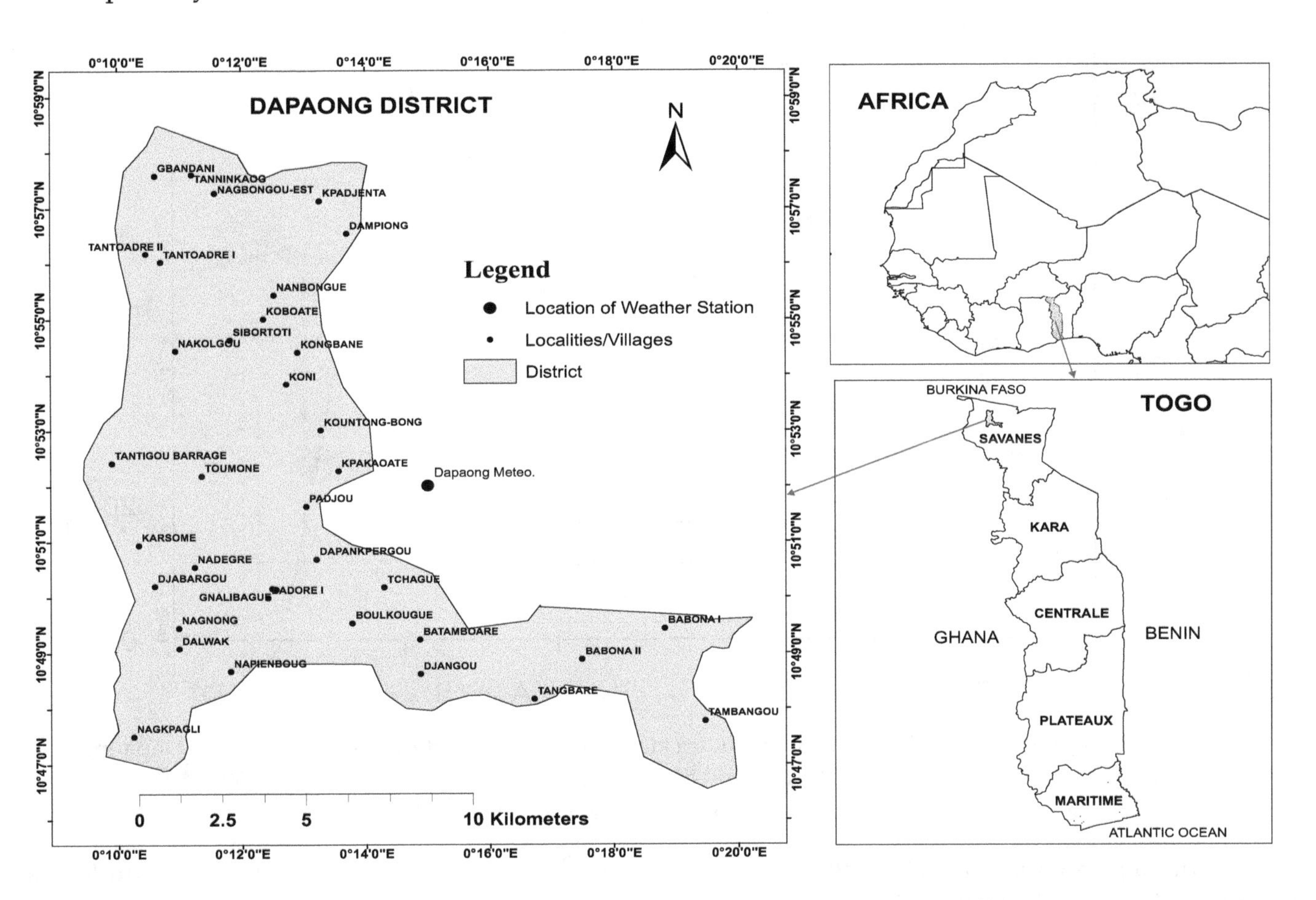

Figure 1. Map of northern Togo indicating the study area (Dapaong district).

According to Köppen–Geiger's climatic classification, the climate of Dapaong district is hot semi-arid (BSh) [54]. The period from mid-April to mid-October is humid, while in the other months dry conditions predominate in Dapaong. The months from June to September show high rainfall (Figure 2). These high annual values of rainfall are sufficient for rainfed cereal crops in northern Togo. The annual rainfall is, however, very unequally dispersed. From November to March (or sometimes April), there is practically no rainfall in the area. From May to October, a substantial amount of rainfall is recorded. Consequently, northern Togo is characterized by a single wet season in a year. This explains why farmers adopt intercropping to obtain the range of crops they need. Introducing irrigated crops in the dry season may help farmers to sustain their production. The mean annual temperature is 28.1 °C, and the annual total precipitation is 1050 mm. The mean daily maximum temperature of the driest month is around 37 °C, whereas the mean daily minimum temperature of the wettest month is 20 °C (Figure 2). In January and February, a robust dusty wind named harmattan, blowing in the northeast direction from the Sahara Desert, increases the dryness of the weather in the area [16].

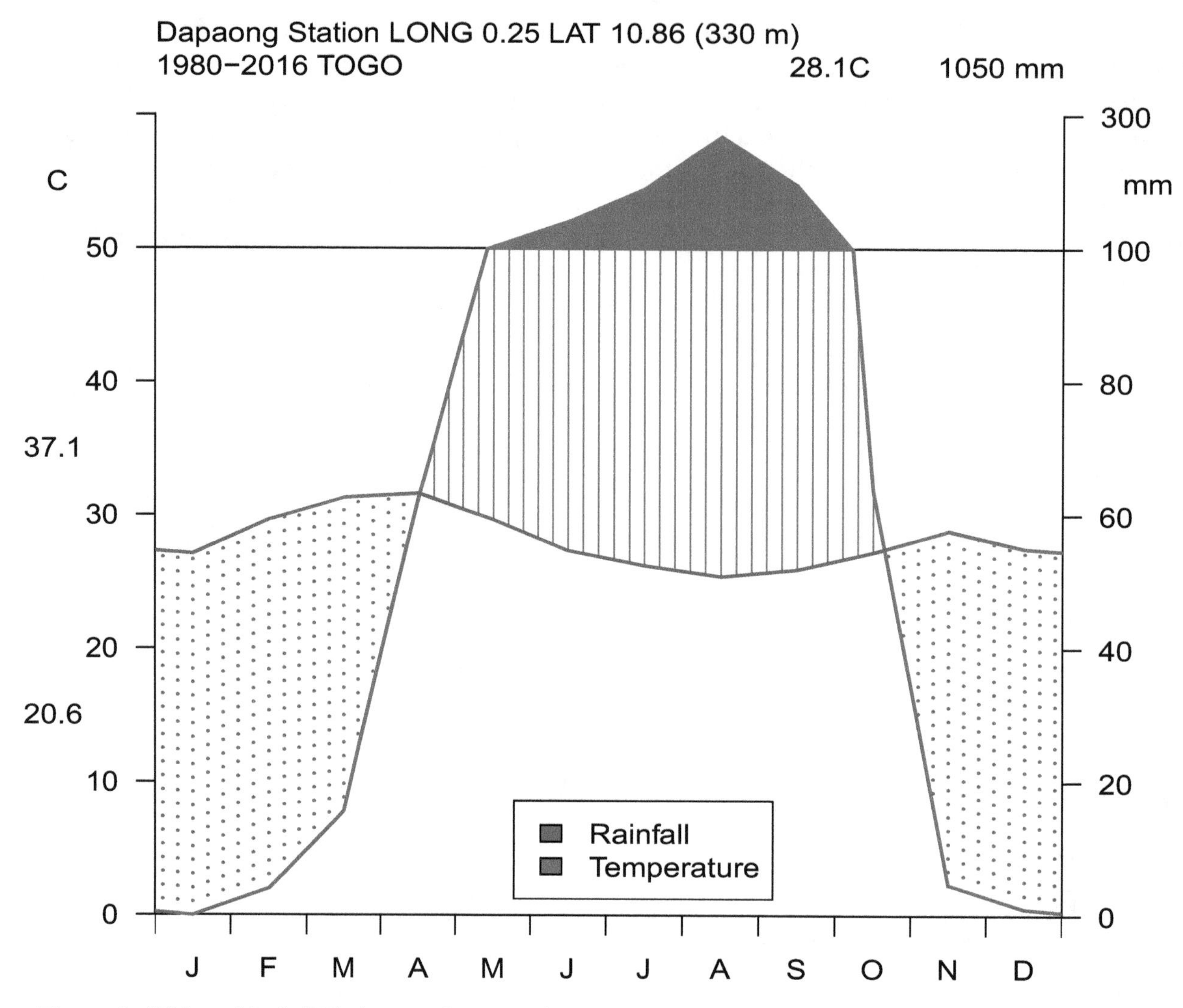

Figure 2. Walter–Lieth [55] climate diagram for northern Togo based on data collected at Dapaong Meteorological Station (Latitude: 10°51′44.10″ N, Longitude: 0°12′27.43″ E, Altitude: 330 m above sea level). Rainfall and temperature data were measured between 1980 and 2016.

With a population density of 96 inhabitants per km^2, over 88% of the population live under the poverty line (US$ 2/day) [56,57]. Complicated communal land tenure favors men, and encourages farm fragmentation. Women access only marginal lands characterized by reduced soil fertility. Most farmers are smallholders with less than 1.5 ha of land under cultivation [53]. Crop yields are generally low due to erratic rainfall, low soil fertility, low-quality seeds, and inappropriate land preparation tools, among others. Farmers' livelihood depends on small-scale farms with low input, and mixed crop–livestock agriculture. Regarding poultry, most farmers have local hens, cocks, and guinea fowls in their houses. Some families raise local dwarf goats and pigs [53].

2.2. Methods

2.2.1. Adapted Framework for the Evaluation of Irrigation Management Alternatives

In this study, we investigated five irrigation management strategies. These are NI, CDI for supplemental irrigation, CDI for conventional irrigation, FI for supplemental irrigation, and FI for conventional irrigation. The NI is equivalent to the rainfed system, the type of agriculture most farmers are practicing in Dapaong. When rainfall is unevenly distributed throughout the wet season, farmers have the option to apply an optimal amount of irrigation water to supplement the shortage (CDI for SI) or use the fully required amount (FI for SI). On the other hand, in the dry season, farmers can deliberately apply an optimal amount of irrigation water (CDI for conventional irrigation) or fully irrigate the plants (FI for conventional irrigation). When combining these strategies with dry and wet seasons, we obtain the following: (i) NI for the wet season (WS-NI); (ii) CDI for supplemental irrigation system in the wet season (WS-CDI); (iii) full irrigation for supplemental irrigation system in the wet season (WS-FI); (iv) CDI for conventional irrigation system in the dry season (DS-CDI); and (v) full irrigation for conventional irrigation system in the dry season (DS-FI). In this study, one should bear in mind that we only dealt with the physiological and agronomical aspects of DI—crop response to different irrigation regimes—without any economic evaluation. The summary can be seen in Table 1.

Table 1. Irrigation management strategies investigated.

Type of Irrigation System	Irrigation Management Strategies			Application Scenarios	
	Limited Supply		Full Supply		
	Uncontrolled	Controlled	Controlled	Wet Season (WS)	Dry Season (DS)
No irrigation	NI	–	–	x	–
Supplemental irrigation	–	CDI	FI	x	–
Conventional irrigation	–	CDI	FI	–	x

CDI, controlled deficit irrigation; FI, full irrigation; NI, no irrigation.

The OCCASION framework was adapted and used to assess the five irrigation management strategies mentioned above (Figure 3). The adapted framework consists of: (i) a weather generator for simulating long climate time series; (ii) the AquaCrop model, which was used to simulate the irrigation system during the growing season and the yield response of maize to the considered irrigation management strategies (Figure 3, Loop 1); and (iii) a problem-specific algorithm for optimal irrigation scheduling with limited water supply (Figure 3, Loop 2). The latter is named Global Evolutionary Technique for OPTimal Irrigation Scheduling (GET-OPTIS) (For more details, see [33]). A range of given maximum volumes of water is then assigned; a complete CWPF can be derived. The produced CWPF characterizes the maximum yields that can be attained with a given amount of water and is designated the potential CWPF. Then, the crop simulation model was run for a long-term climate time series data yielding a necessary amount of CWPFs. Also, optimized irrigation schedules are obtained. Subsequently, the resulting CWPFs were analyzed, and the SCWPFs obtained through parameters of descriptive statistics such as mean, median, and probability of exceedance, among others. SCWPFs are empirical probability functions where, for every volume of applied irrigation water, the marginal distribution function of the yield related to it can be derived. The probability of exceedance represents the reliability that a specific yield can be achieved [32].

2.2.2. Processing of Climate Data and Set-Up of the LARS Weather Generator

Historical weather observations, including daily maximum temperature, daily minimum temperature, daily rainfall, daily wind speed, daily minimum humidity, and daily maximum humidity were obtained from the nearest meteorological station to the study site—courtesy of the National Weather Service of Togo. These daily weather data available at the station range from 1983 to 2011. In addition, the observed monthly rainfall and maximum and minimum temperatures data from 1980 to 2016 were provided. These monthly data were utilized to characterize the climate of northern Togo with the climate diagram of Walter and Lieth [55]. The Dapaong meteorological station is located at latitude 10°51′44.10″ N, longitude 0°12′27.43″ E, and altitude 330 m above sea level (Figure 1). The solar radiation data, as well as sunshine hours data, were not available at Dapaong weather station. As a substitute, the uncorrected gridded incident solar radiation from the Prediction of Worldwide Energy Resource dataset from the National Aeronautics and Space Administration project NASA-POWER [58] was utilized. Van Wart et al. [59] showed that NASA-POWER is a good source of climate data for crop yields simulation studies. It is publicly accessible, shows acceptable general agreement with ground data for incident solar radiation, and has been used by similar previous studies (See Section 2.2.4).

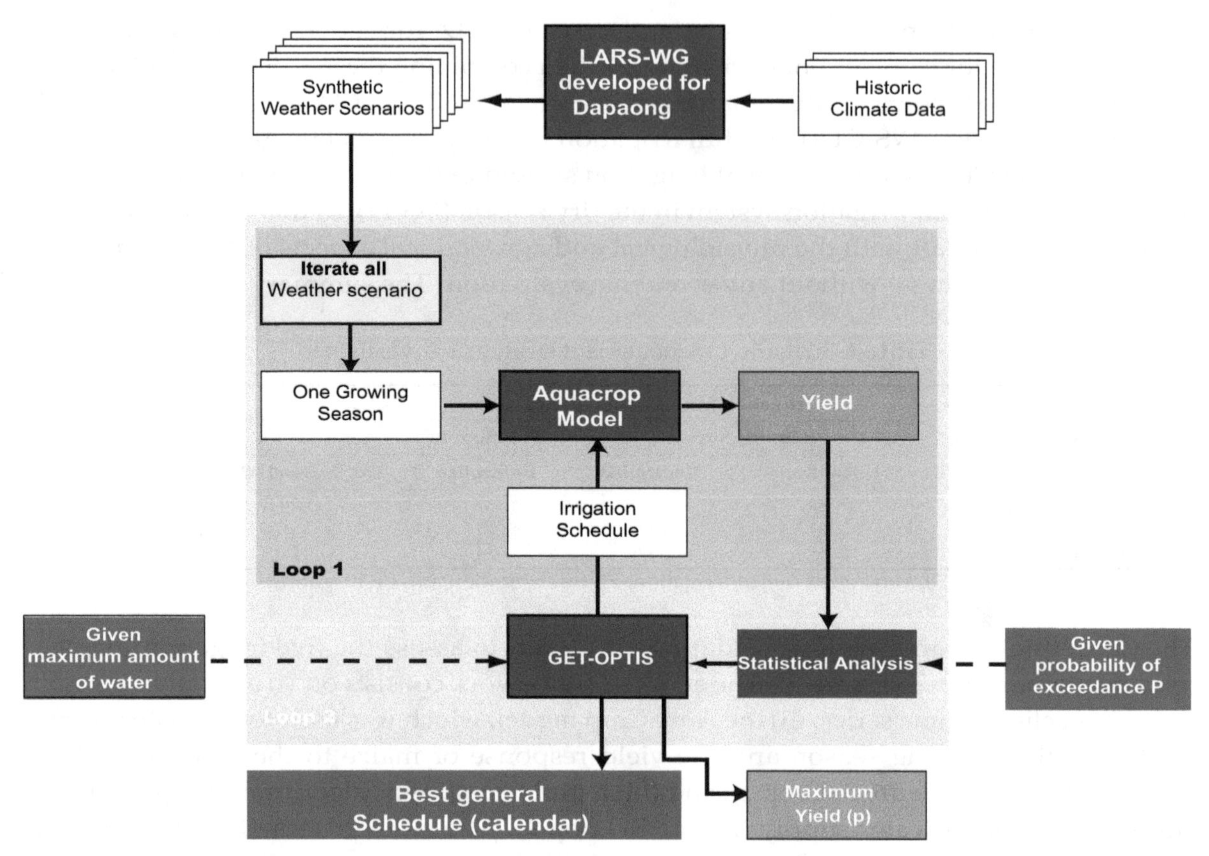

Figure 3. General framework for generating stochastic crop water production functions (adapted from Schütze and Schmitz [36]).

Since the 29-year period (1983–2011) of weather data is not long enough to be used in the assessment of climate variability effect on crop yield, the Long Ashton Research Station Weather Generator version 4.0 (LARS-WG)—a stochastic weather generator—was used to generate a 100-year period of near future climate data. In this study, out of the existing weather generators, LARS-WG was used for two reasons. Firstly, it uses more complex distributions for weather variables and has been tested for diverse climates and found to be better than some other weather generators such as WGEN [60] (Appendix A). Secondly, Semenov [61] recently tested LARS-WG at different locations across the world and revealed its ability to model rainfall extremes with acceptable performance. Similarly, Mehan et al. [62] provided insights into the suitability of LARS-WG for use with water resource applications. Guo et al. [63] suggested performing more than a single realization when generating weather data using LARS-WG for hydrologic and environmental applications. We assessed the performance of the LARS-WG in simulating weather data of Dapaong by comparing the observed and the simulated data with the Kolmogorov–Smirnov test (KS-test). We used the KS-test for the comparison of the probability distributions for each month. The KS-test is a non-parametric and distribution-free test that tries to determine if two datasets are extensively different and come from different distributions. It is an alternative to the Chi-square goodness of fit test. The KS-test compares the two empirical distribution functions as in Equation (1) [64].

$$D = |E_1(i) - E_2(i)| \quad (1)$$

where E_1 and E_2 represent the empirical distribution functions of the two distributions, and D is the absolute difference between them.

The KS-test examines for changes in distributions coming from the generated and observed weather. The KS-test calculates a test statistic and an equivalent p-value [65]. It shows how likely it is that the generated and observed data originate from the same distribution. If the p-value is very low and below the significance level, set to 0.01 or 0.05, the simulated climate is unlikely to be the same as the "true" climate. Although a p-value of 0.05 is the standard significance level employed in most statistics, the authors of the LARS-WG model recommended that a p-value of 0.01 should be considered as the satisfactory significance level.

The calibrated LARS-WG for Dapaong was then used to forecast the 100-year daily rainfall and temperature data mentioned above for a near future. For this, the outputs of the General Circulation Models (GCMs) HADCM3 (Hadley Centre Coupled Model version 3) of the IPCC Special Report on Emission Scenarios (SRES) A2 were inputted into LARS-WG. The HADCM3 is the product of the UK Meteorological Office, gridded as $2.5^\circ \times 3.75^\circ$. These long-term data were used to run the AquaCrop model to assess the five irrigation management strategies.

2.2.3. Description and Set-Up of the Crop Simulation Model

AquaCrop, a water-driven crop simulation model, was developed in 2009 by the Food and Agriculture Organization (FAO) of the United Nations [38–40]. The development of the AquaCrop model is based on the algorithm of yield response to water in FAO Irrigation and Drainage Paper No. 33 [66]. AquaCrop evolves from the previous Doorenbos and Kassam [66] K_y approach (Equation (2)), where relative evapotranspiration (ET) is pivotal in calculating yield.

$$\frac{(Y_x - Y_a)}{Y_x} = K_y \left[\frac{(ET_x - ET_a}{ET_x} \right] \quad (2)$$

where Y_x and Y_a are the maximum and actual yield, respectively; ET_x and ET_a are the maximum and actual evapotranspirations, respectively; and K_y is the proportionality factor between relative yield loss and relative reduction in evapotranspiration.

AquaCrop simulates crop yield in four steps: crop development, crop transpiration, biomass formation, and yield formation [40]. Four water stress response coefficients are considered in the model. These are related to canopy expansion, stomatal conductance, canopy senescence, and harvest index [67].

2.2.4. Soil Data and Calibration of the Crop Simulation Model

We retrieved the physical characteristics data of soils in Dapaong from Poss [68]. These measured soil physical characteristics were used as input into the Soil Water Hydraulic Properties Calculator (http://hydrolab.arsusda.gov/soilwater/Index.htm) to compute various soil hydraulic parameters required to run AquaCrop. We used this soil water hydraulic properties calculator because it has been employed in previous studies in the West African region (e.g., Akumaga et al. [69]). These include volumetric soil water content at field capacity, permanent wilting point, saturation, and saturated hydraulic conductivity (Table 2). Poss [68] classified the soil of Dapaong as sandy loam. According to the World Reference Base for Soil Resources, the soil in northern Togo is characterized Dystric-Ferric Luvisols [70,71].

Table 2. The soil description and properties of Dapaong (See Poss [68]).

Soil Depth (cm)	Texture			OM (%)	dB (g/cm)	SAT (Vol.%)	FC (Vol.%)	PWP (Vol.%)	Ksat (mm/da)	Textural Class
	Sand (%)	Silt (%)	Clay (%)							
0–20	72.5	20.5	7.0	1.5	1.5	42.7	13.3	5.3	1252.6	Sandy Loam
20–50	72.0	19.0	9.0	0.9	1.6	40.8	13.5	5.9	503.0	Sandy Loam
50–110	66.5	18.0	15.5	0.7	1.6	39.9	18.3	10.0	239.5	Sandy Loam

FC, field capacity; PWP, permanent wilting point; SAT, saturation (SAT); Ksat, saturated hydraulic conductivity; dB, soil bulk density; OM, organic matter content in the soil.

Regarding the crop parameters, some of them were assumed to be conservative. The values of conservative parameters used in our study are the same as values proposed by FAO [72] (not presented here). The others, non-conservative or crop-specific, were estimated using measured data retrieved from the ITRA [51], Didjeira et al. [52], and Worou and Saragoni [73] studies conducted in northern Togo (Table 3). These data were used to fine-tune the maize parameters to the local agronomic and management conditions of the study area before running the simulations in AquaCrop. These parameters include information about sowing, canopy cover, canopy senescence, flowering, rooting depth, harvest index, soil management, and the maize cultivar used. Regarding the calibration of the canopy cover, we used the options in AquaCrop to estimate the initial canopy cover (CCo) from sowing rate, seed weight, seed number and estimated germination rate. Subsequently, the canopy expansion rates were automatically estimated by AquaCrop after we entered the phenological dates such as dates of emergence, maximum canopy cover, senescence and maturity. The AquaCrop model simulations were run in growing degree day (GDD) calculated from temperature data used as climate input. Geerts et al. [74], Salemi et al. [75], and Silvestro et al. [76] reported on the most sensitive parameters in AquaCrop obtained through sensitivity analysis testing. The essential crop-specific parameters used to calibrate the AquaCrop model for simulating maize growth and productivity for the study area are presented in Table 3. It should be noted that the calibration of AquaCrop model in this study is preliminary, thus the conclusions that emanated from the simulations are qualitative. The main idea was to compare the irrigation management strategies assessed in this study qualitatively.

Table 3. Non-conservative parameters adjusted and agronomic information for Dapaong, Togo.

Parameter Description	Value	Units or Meaning
Time from sowing to emergence	7 (135)	DAP(GDD)
Time to maximum canopy cover	60 (1109)	DAP(GDD)
Time from sowing to maximum rooting depth	67 (1257)	DAP(GDD)
Time from sowing to start of canopy senescence	76 (1408)	DAP(GDD)
Time from sowing to maturity	100 (1898)	DAP(GDD)
Time from sowing to flowering	54 (1018)	DAP(GDD)
Duration of flowering	10 (183)	DAP(GDD)
Length of building up HI	42 (778)	DAP(GDD)
Maximum effective rooting depth, Z	1	meter
Minimum effective rooting depth, Zn	0.3	meter
Reference harvest index, HI	50	%
Cultivar (TZEE-W)	–	TZEE-W
Planting method	–	Direct sowing
Planting density	62,500	Plants/ha
Soil fertility	65	Moderate (%)
Surface mulches	0	%
Curve number, CN	66	–
Readily Evaporable water, REW	2	mm

DAP, days after planting; GDD, growing degree days; HI, harvest index.

Table 4 summarizes the potential and selected sources of the input data used in this study and reasons for selecting these specific sources.

Table 4. Input data sources.

Type of Data	Possible Sources	Selected Sources for the Study	Reasons of Selecting Specific Sources for the Study
Temperature, rainfall, wind speed, and humidity	-Local meteorological station -Observed data online (NOAA, etc.) -Satellite data (NASA, etc.)	Local meteorological station	Observed data with no missing values
Solar radiation and sunshine hours	-Observed data online (NOAA, etc.) -Satellite data (NASA, etc.)	Satellite data (NASA-POWER project)	Publicly accessible, shows acceptable general agreement with ground data
Soil data	-Poss [68] -National soil survey -FAO Harmonized World Soil Database -ISRIC Soil Geographic Databases	Poss [68]	Publicly accessible and with good resolution (field)
Crop data: conservative parameters	AquaCrop manual	AquaCrop manual	In line with AquaCrop model
Crop data: non-conservative parameters	-AquaCrop manual -ITRA [51], Didjeira et al. [52], and Worou and Saragoni [73]	ITRA [51], Didjeira et al. [52], and Worou and Saragoni [73]	Specific to the maize variety used in the study

2.2.5. Optimal Irrigation Scheduling with Limited Water Supply

Matlab, AquaCrop interface, and Plugin-ACsaV40 (version 4; http://www.fao.org/aquacrop/en/) were used to simulate multiple projects for successive years. The soil and crop phenological data described in Tables 2 and 3, respectively, were used to calibrate AquaCrop. First, AquaCrop was run for a given amount of irrigation water for the maize crop under a specific climate scenario during the dry season of the Dapaong area. GET-OPTIS was employed as irrigation scheduling optimizer and crop yield maximizer. Then, we iterated over a range of given water volumes. As a result, a complete crop-water production function (CWPF) was derived. The 100-year maize crop simulations were run for the wet season as well as the dry season to assess the irrigation management strategies described above, in northern Togo.

3. Results and Discussion

3.1. Traits of the Climate in Dapaong

The temperature is high during the dry season reaching 37 °C and 26 °C maximum and minimum temperatures, respectively, while, in the wet season, the maximum temperature is 30 °C and the minimum temperature is close to 26 °C (Figure 4a). Due to these high temperatures, especially in the dry season, it is likely that the evapotranspiration is relatively high in the area. This argument is corroborated by Djaman and Ganyo [77] who found that the potential annual reference evapotranspiration—computed using the FAO-56 Penman–Monteith method—in northern Togo is higher than 1800 mm on average.

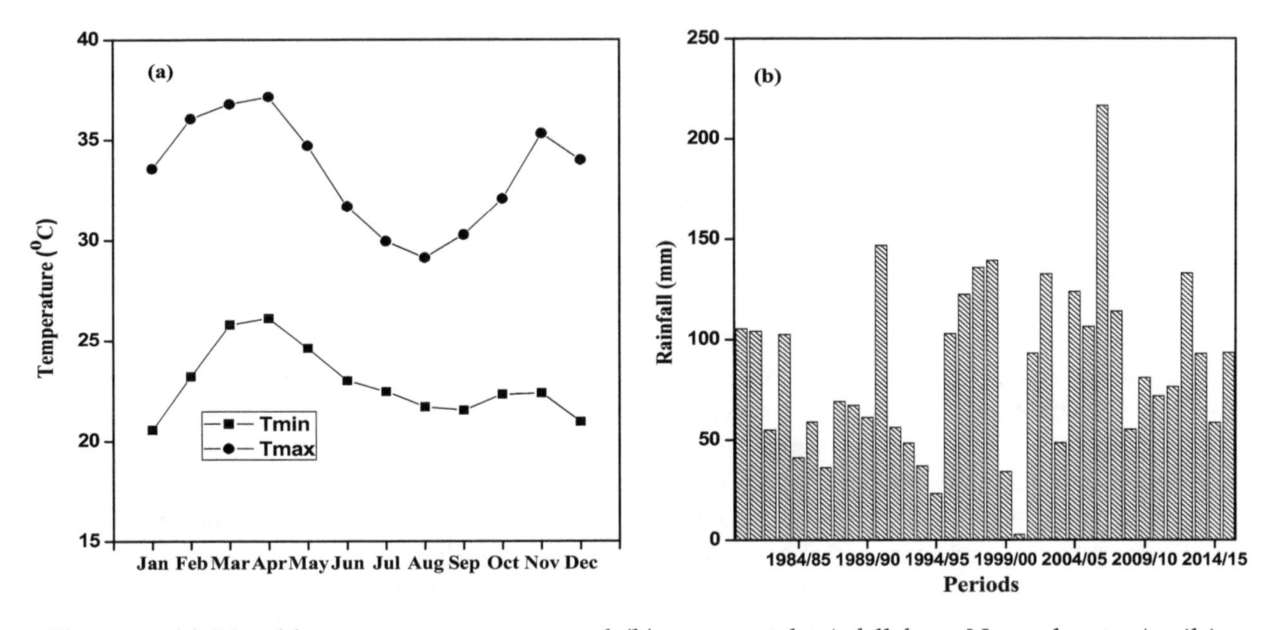

Figure 4. (**a**) Monthly mean temperature; and (**b**) mean total rainfall from November to April in Dapaong, Togo (1980–2016).

Figure 4b depicts the mean total rainfall during the dry season (November–April) in Dapaong district. The rainfall recorded during the dry season varies significantly from year to year. On average, the total rain that falls within this period is lower than 85 mm. In some years, the volume of rain which falls in the same period is up to 100 mm. The highest amount was reached in 2006/2007 (216 mm). Globally, this rainfall occurs on an average of five days only. Thus, none of the main cereals grown in the area such as maize, millet, and sorghum can survive under the dry season climatic conditions without an additional water supply. These findings prove again the fact that farmers only grow crops during the wet season. Overall, the climate of Dapaong in northern Togo is unfavorable to agricultural activities throughout the year because of its vagaries and uncertainties compromising crop yield. These results are in agreement with studies by Ogounde and Abotchi [16].

3.2. Validation and Application of the LARS Weather Generator

The LARS-WG model showed robust compliance between observed and simulated data for the maximum as well as minimum temperatures (Table 5). These findings showed no significant differences between the observed and simulated temperatures for all months. All *p*-values were close to one. It means that the observed and simulated data were from the same distribution. Therefore, based on these results, we conclude that the performance of the LARS-WG model in simulating the climatic variables such as minimum and maximum temperatures of Dapaong district is satisfactory. Similar results were obtained by Semenov et al. [60] at 18 sites in the USA and Europe. However, the standard deviations of the monthly mean simulated values are less than half of the standard deviations of observed values for all months. This means that the extreme temperature values in the minimum and maximum temperatures simulated are smaller than in the observed data.

The observed and simulated rainfall values for most of the months do not correlate significantly (Table 5). This result agrees with studies by Osman et al. [78] in Iraq. However, there are significant differences between December and January, when LARS-WG was incapable of reproducing the observed rainfall, partly because these periods are the driest during the dry season. The standard deviations of the monthly mean rainfall of observed and predicted values are similar for January, February, and April (Table 5). These results imply that there are fewer extreme rainfall values in the dry months, which are of our interest in this study. Overall, the performance of LARS-WG in predicting the rainfall of the Dapaong area is at an acceptable level. It means that the quality of the long-term data that were generated based on these calibration results is not affected.

3.3. Evaluation of Irrigation Management Strategies

3.3.1. Wet Season—Rainfed and Supplemental Irrigation Systems

➢ *Maize Crop under Rainfed Conditions (WS-NI)*

While Figure 5a shows the results of the expected maize crop yields that can be achieved during the rainfed cropping system, Figure 5b portrays the rainfall statistics within the same period. The volume of rainwater that falls within the cropping period of the wet season in Dapaong ranges from 450 mm to 1100 mm approximately. The frequency of the rainfall is high, between 600 mm and 900 mm (Figure 5b). The distribution of the expected rainfed yields is moderately skewed left with a higher coefficient in absolute values (1.91) (Figure 5a). The standard deviation of the expected yields obtained under rainfed conditions is higher than in the case of irrigated maize, regardless of the volume of water used, in northern Togo (See Section 3.3.2). These results show that the variability, as well as the uncertainty, in the yields, are higher under the rainfed conditions (WS-NI) than under the dry season CDI and FI. The high variability under rainfed conditions is likely due to inadequate rainfall distribution and dry spells in the wet season [79]. On average, the expected maize crop yield achieved in the wet season is 3.5 Mg/ha (Figure 5a). These results agree with the findings by Didjeira et al. [52] who indicated the range of 3.5–5 Mg/ha as the expected yield for the maize variety used in this study. Similarly, these results are in line with that of Fosu-Mensah [80] who reported that, in sub-humid Ghana under projected climate change (2030–2050) for scenario A1B of IPCC, the rainfed maize grain yield varies from 3.16 Mg/ha to 4.09 Mg/ha. Therefore, the calibrated AquaCrop model in this study performs well. These results can be improved if data on more site-specific parameters are made available. Akumaga et al. [69] suggested that the AquaCrop model can be utilized as a tool in the study and modeling of maize productivity in West African region.

Table 5. Kolmogorov–Smirnov test statistics for rainfall, maximum and minimum temperatures in Dapaong.

Month	RAINFALL				MAXIMUM TEMPERATURE				MINIMUM TEMPERATURE			
	SD of Observed Data	SD of Simulated Data	K-S	*p*-Value	SD of Observed Data	SD of Simulated Data	K-S	*p*-Value	SD of Observed Data	SD of Simulated Data	K-S	*p*-Value
January	0.11	0.17	0.57	0.00	1.34	0.46	0.11	1.00	1.60	0.52	0.05	1.00
February	14.89	13.89	0.17	0.84	1.24	0.41	0.16	0.91	1.66	0.48	0.11	1.00
March	19.53	31.90	0.15	0.94	0.73	0.28	0.11	1.00	1.04	0.36	0.16	0.91
April	44.99	43.27	0.11	1.00	0.95	0.43	0.11	1.00	0.84	0.43	0.11	1.00
May	38.39	44.09	0.05	1.00	1.25	0.40	0.11	1.00	0.83	0.38	0.11	1.00
June	54.58	53.28	0.03	1.00	0.89	0.32	0.05	1.00	0.72	0.30	0.05	1.00
July	69.68	85.13	0.05	1.00	0.72	0.37	0.05	1.00	0.57	0.26	0.05	1.00
August	85.44	99.96	0.06	1.00	0.60	0.30	0.05	1.00	0.57	0.26	0.11	1.00
September	61.77	68.39	0.08	1.00	0.58	0.37	0.05	1.00	0.58	0.25	0.05	1.00
October	43.20	57.00	0.01	1.00	1.04	0.40	0.11	1.00	0.83	0.27	0.05	1.00
November	12.00	15.73	0.13	0.98	0.83	0.26	0.11	1.00	1.34	0.34	0.11	1.00
December	5.66	10.98	0.26	0.36	1.04	0.43	0.05	1.00	1.38	0.44	0.05	1.00

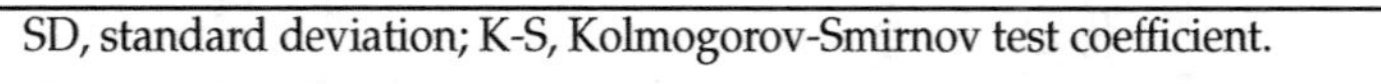
SD, standard deviation; K-S, Kolmogorov-Smirnov test coefficient.

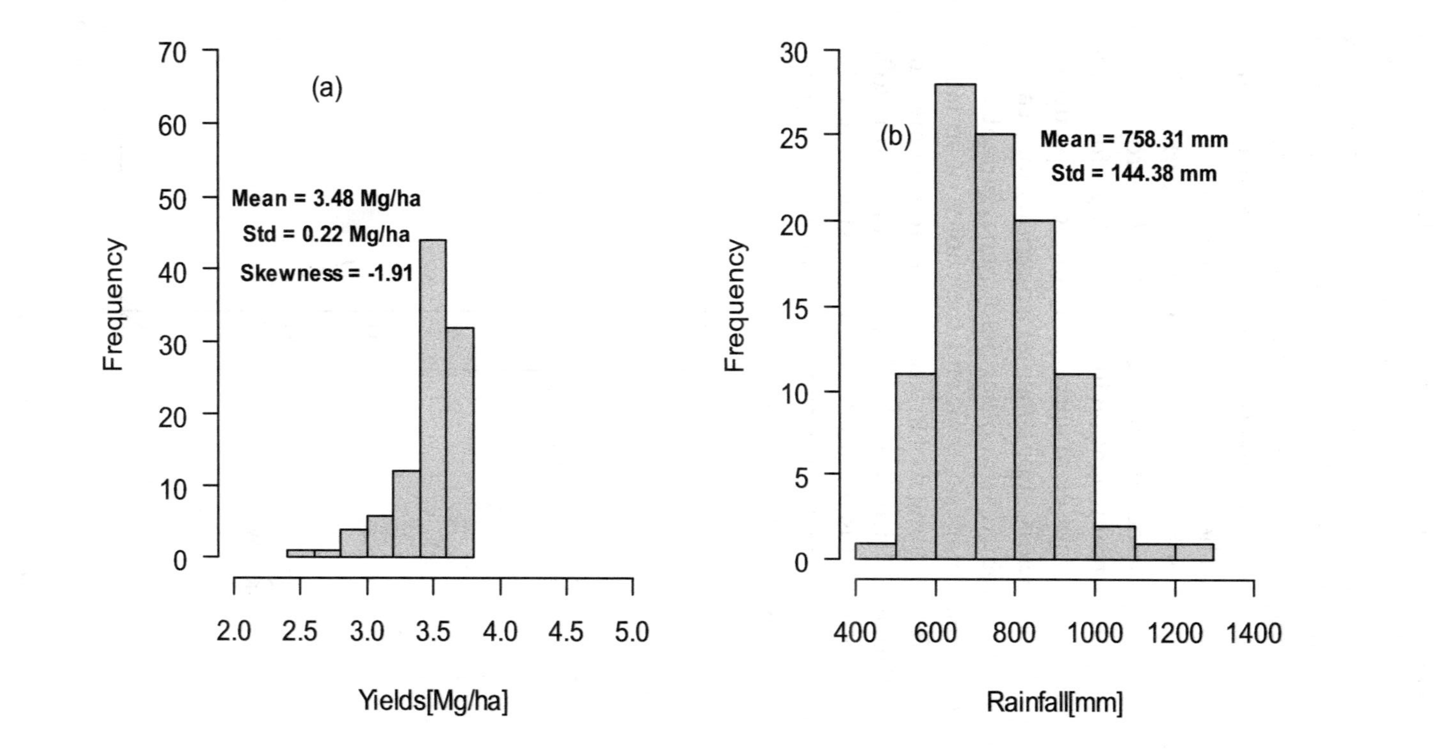

Figure 5. Histogram of distributions of: (**a**) expected yield of maize grown in a rainfed system (WS-NI); and (**b**) the rainfall during the wet season in Dapaong.

➢ *Maize under Supplemental Irrigation (WS-CDI and WS-FI)*

To improve yield while reducing its variability at the same time, one may apply supplemental irrigation during the rainfed cropping system whenever the crops are experiencing severe water stress, and rainfall is not occurring. The stochastic crop-water production functions for supplemental irrigation conditions are shown in Figure 6a. It can be hypothesized that, when more than 150 mm supplemental irrigation water is applied, the variation in the resulting expected crop yield is likely due to the variation of temperature and radiation in the area. These assumptions are supported by the nearly symmetric distributions of the corresponding expected crop yields (Figure 6a). Besides, at volumes of supplemental water lower than 150 mm, the variation in the expected crop yield can result from the combined effects of the uneven distribution of rainfall and the climate parameters mentioned above. The 90% of SCWPF exceedance probability of yield achievement seems to be the best option for enhancing food security in northern Togo. This might be because it is the only option which helps to achieve the highest level of crop yield improvement (15% or more) (Figure 6a). Applying supplemental irrigation in northern Togo for maize crop cultivation will not only contribute to improving crop grain yield and enhancing food security [81–83] but also help to improve farmers' livelihood. Nevertheless, supplemental irrigation alone cannot improve the rainfed yields significantly; it needs to be combined with other field management aspects such as soil preparation and fertility, pests and diseases management, and the choice of suitable crop varieties. It can be concluded that CWPF is a useful planning tool to assess water requirement for crops, especially in water-scarce regions. Heng et al. [84] and Stricevic et al. [85] reported that, due to its sufficient degree of simulation accuracy, the AquaCrop model is a valuable tool for estimating crop productivity under rainfed conditions, deficit and supplemental irrigation, and on-farm water management strategies for improving the efficiency of water use in agriculture.

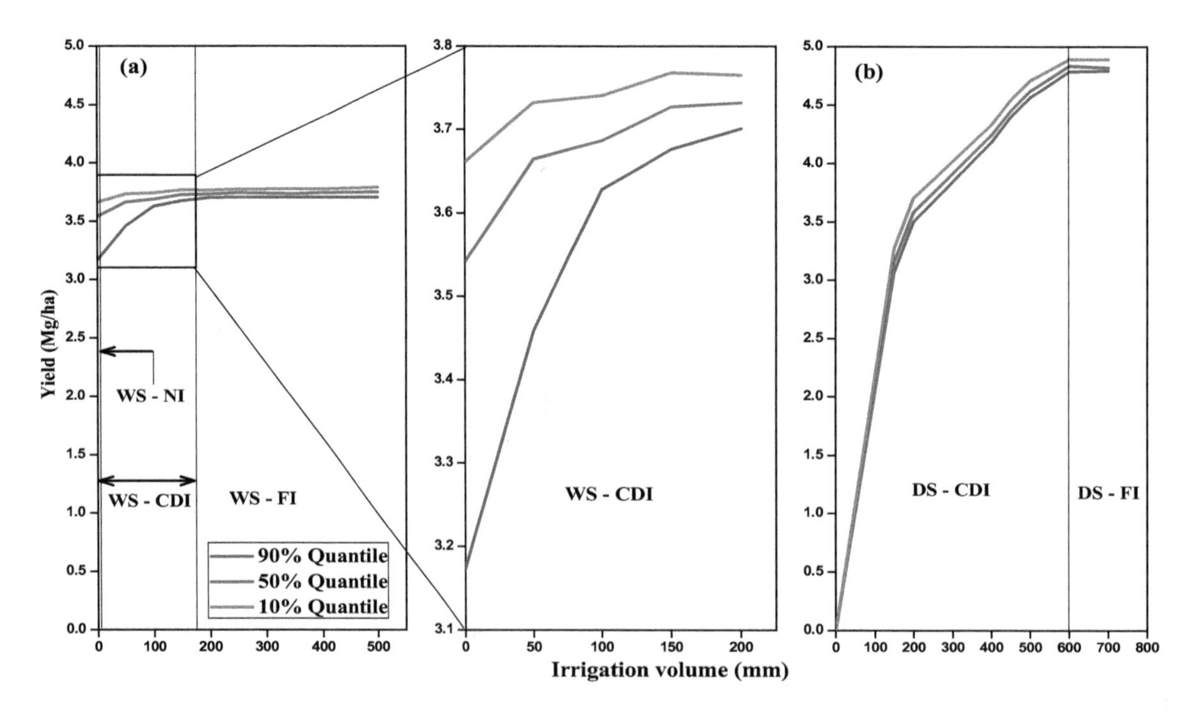

Figure 6. Stochastic crop-water production function for: (**a**) rainfed and supplemental irrigated systems in the wet season; and (**b**) optimized conventional irrigation system in the dry season for maize in Dapaong.

Figure 7 shows the detailed results of the expected yields at various amounts of supplemental irrigation water. With supplemental irrigation (WS-CDI), the rainfed yield increased from 3.48 Mg/ha to 3.74 Mg/ha. The yield becomes constant when the volume of water applied is equal to or greater than 150 mm. Then, the variability in the yields as well as the skewness decreases in absolute value.

These results imply that supplemental irrigation is beneficial up to 150 mm. Above this value, the advantages of supplemental irrigation (WS-FI) become insignificant. Therefore, rainfed maize crop yields may be improved in northern Togo by applying supplemental irrigation assuming that water is available.

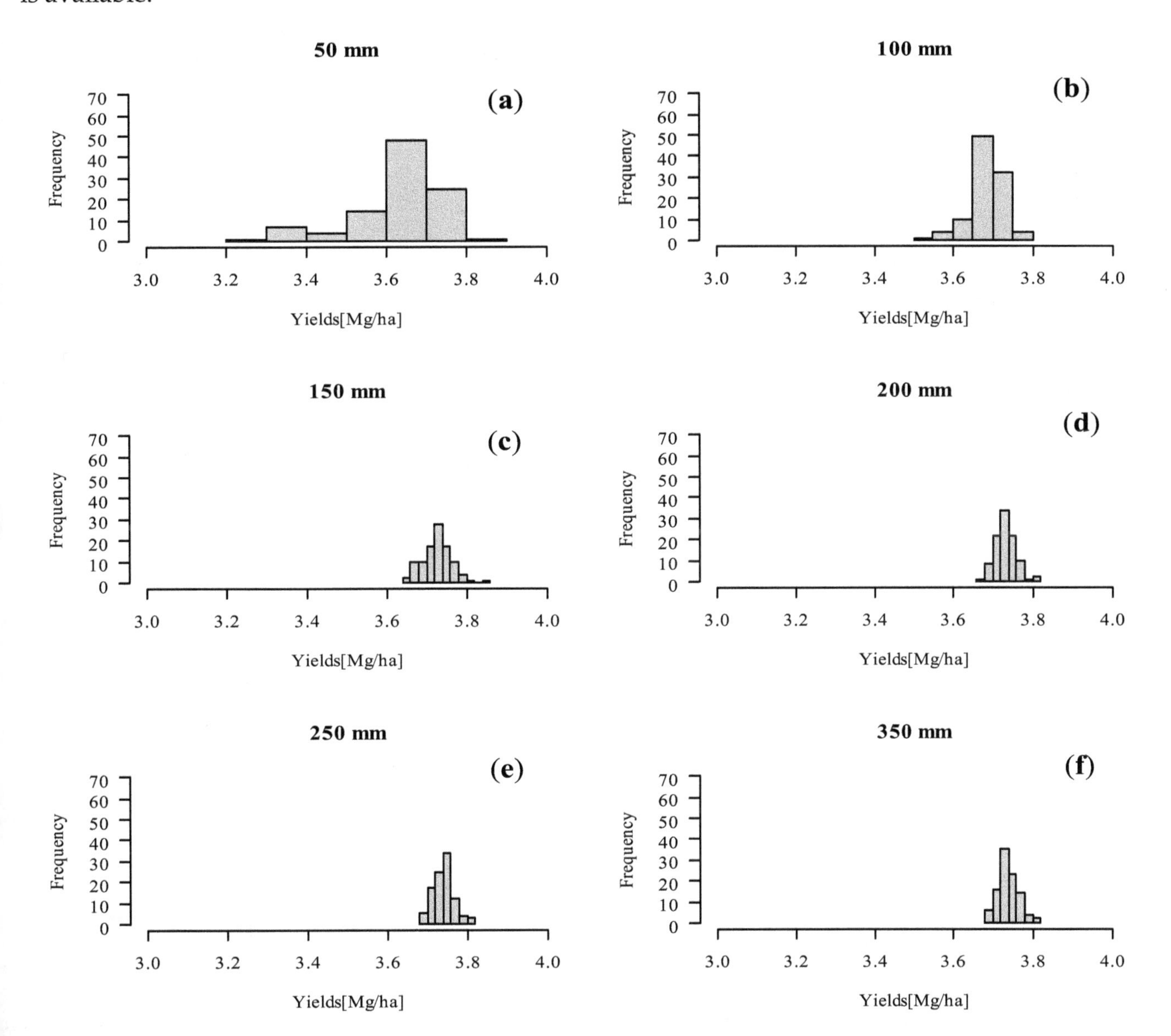

Figure 7. Histogram of distributions of expected yield using water for supplemental irrigation of maize in the wet season in Dapaong: (**a**) 50 mm; (**b**) 100 mm; and (**c**) 150 mm (WS-CDI); and (**d**) 200 mm; (**e**) 250 mm; and (**f**) 350 mm (WS-FI).

3.3.2. Dry Season—Conventional Irrigation System (DS-CDI and DS-FI)

Figure 6b shows the stochastic crop-water production functions (SCWPF) for optimized irrigated maize crop in the dry season in northern Togo. The quantile percentage represents the probability of exceedance. Since rainfall can be ruled out, it is believed that, when the optimal full irrigation conditions are met, the variation of temperature and radiation can explain the variability in the expected crop yield. These assumptions are corroborated by the nearly symmetric distributions of the expected crop yields at full irrigation (Figure 8). These findings are supported by the results presented by Schütze and Schmitz [36]. These two parameters are part of the yield defining factors, as highlighted in the papers explaining the principles of ecology production [86]. In addition, for volumes of water lower than full irrigation, the variation in the expected crop yield can result from the combined effects of drought stress on crops and the climate parameters mentioned above. The maximum expected yields were 4.79 Mg/ha (90% quantile) and 4.89 Mg/ha (10% quantile) at near full irrigation (600 mm) (Figure 6b). The controlled deficit irrigation ranges from 0 to 600 mm for maize in northern Togo. The DS-CDI strategy seems to save water with an insignificant reduction in the grain yield relative to full irrigation [87–92]. Overall, growing maize crop in the dry season in northern Togo may be feasible under CDI if water is available. Irrigation is vital for improving crop yield and stabilizing crop production [93] amidst the threats of climate change [94].

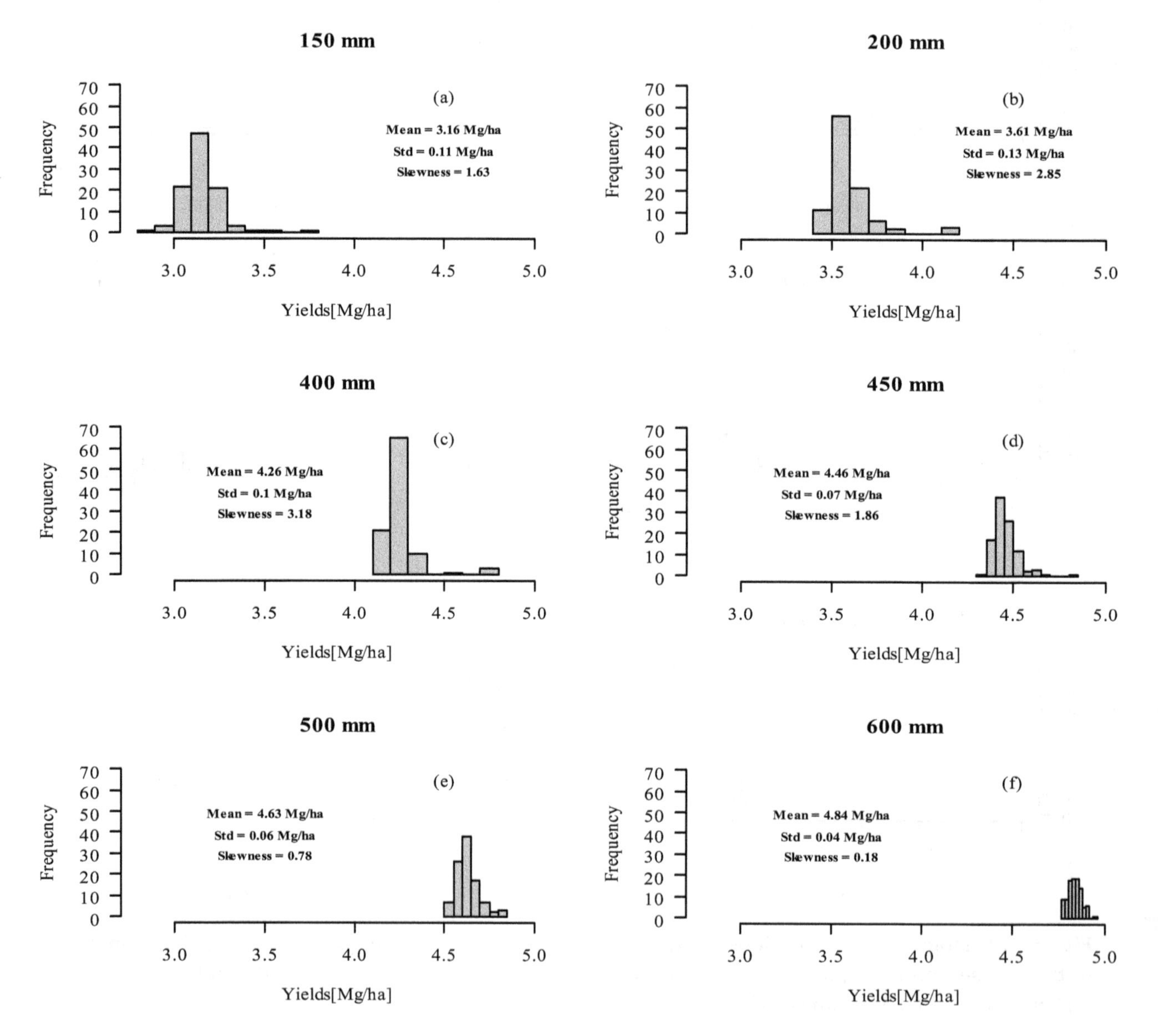

Figure 8. Histogram of distributions of expected yield using water for irrigation of maize in the dry season in Dapaong: (**a**) 150 mm; (**b**) 200 mm; (**c**) 400 mm; (**d**) 450 mm; and (**e**) 500 mm (DS-CDI); and (**f**) 600 mm (DS-FI).

In Figure 8, detailed results of the expected yields at various amounts of irrigation water are given. There is a change in the histogram distribution among the various volumes of irrigation water. The average expected yields concerning the amount of irrigation water used range from 3.16 Mg/ha to 4.84 Mg/ha at 150 mm and 600 mm, respectively. With the increasing application of irrigation water (DS-CDI), the yield increases to a level at which additional water supply fails to raise the crop yield any further (around 600 mm). Thus, the latter volume of water is assumed to be near full irrigation. The frequency distribution shows a positive sign for all the histograms. The coefficients of skewness of the expected yields for 150 mm, 200 mm, and 400 mm water volumes are 1.63, 2.85, and 3.18, respectively. On the contrary, at 600 mm volume of water (DS-FI), the distributions of the expected yields are symmetrical. In addition, the standard deviation is relatively low for the yields at these volumes of water. Abedinpour et al. [95] reported that the AquaCrop model can predict maize yield with acceptable accuracy under variable irrigation in a semi-arid environment.

3.4. Summary of the Discussion

The variability in rainfall during the wet season (WS-NI) was high, inducing a considerable variability in the expected yield for rainfed conditions. The variability in the expected yield would decrease significantly if supplemental irrigation (WS-CDI or WS-FI) were applied. At the same time, supplemental irrigation would improve the expected yields and contribute to avoiding crop failure. The dry season irrigation management strategies (DS-CDI and DS-FI) would increase yield potential and decrease the variability of expected yield at the same time. Thus, the application of supplemental or dry season irrigation management strategies investigated in this study would help to enhance food availability in the West African region.

There are a few caveats that readers should keep in mind when interpreting the results of this study: The AquaCrop model in this study was calibrated with crop and soil data retrieved from previous studies conducted in the area. Thus, the conclusions derived from the outputs of the model simulation are qualitative—ranking of the irrigation management strategies assessed in the study. There are several uncertainties in the general circulation model outputs as well as crop model simulations. The uncertainties related to crop yield exist because AquaCrop assumes a disease- and pest-free environment and considers no effect of weed or extreme climate events such as flooding. Another point worth considering is that, by concluding that there is potential for deficit and supplemental irrigation for maize crop in northern Togo, we assumed that a proper soil fertility management is guaranteed, and water is available for irrigation management. Finally, it is important to note that substantial investments in irrigation infrastructure, as well as extension services to farmers, would be necessary to enhance food security in northern Togo. The calibrated crop model needs to be validated with experimental data to improve the accuracy of the resulting simulations.

4. Conclusions

The AquaCrop model was used to assess the potential of deficit and supplemental irrigation in the dry savannah area of northern Togo under climate variability. For this, the climate of the study area was characterized. The performance of the weather generator used to produce the long-term time series climate data for the crop simulation was also evaluated. In summary, the climate of northern Togo is unimodal with the dry season ranging from November to April. According to Köppen–Geiger's classification, the climate is hot semi-arid in northern Togo. During the dry season the mean maximum and minimum temperatures are 35 °C and 25 °C, respectively, and the mean total rainfall is 85 mm. In short, the performance of the LARS Weather Generator in predicting the climate of northern Togo was found satisfactory. Overall, we found that the deficit irrigation water requirement ranges from 0 to 600 mm. The maximum expected maize grain yield that can be reached under irrigated conditions is 4.84 Mg/ha with TZEE-W local variety. The rainfed yield can be improved from 3.48 to 3.74 Mg/ha with 150 mm of supplemental irrigation water. At the same time, the variability in the yield was significantly reduced. Irrigation practice in agriculture helps to lower crop yield variability as well as crop failure.

Thus, growing maize crop in the dry season in northern Togo may be feasible. In general, irrigation can help to alleviate food insecurity, while supplemental irrigation is a climate-related management practice for crop yield improvement. The latter also contributes to improving farmers' livelihood. Further maize crop genetic improvements would be needed to fine-tune the seeds to the dry season climate. Irrigation infrastructures would be needed to implement in northern Togo the irrigation management strategies investigated in this study. In addition, realistic irrigation water pricing and cost recovery policies should be enforced and followed by all stakeholders to maintain the irrigation infrastructures and ensure the viability of the system. Institutional reforms relevant to the development and management of irrigation systems should be made. The complicated land tenure issue in northern Togo needs to be addressed to incentivize investment in, and management of, irrigation systems. Moreover, the institutional arrangement—market and connectivity among farmers and other agents—should be improved.

To develop regional water management strategies, the adapted framework used in this study may be applied to other sites in the West African region. Field experiments are needed to validate the results of this study before the implementation of its recommendations. In addition, the framework can be extended by adding a soil variability dimension to it. The analysis can be made more comprehensive by considering farmers' socioeconomic characteristics.

Author Contributions: A.G.-T. and N.S. developed the concept and design of the numerical experiment. A.G.-T. carried out the simulations, analyzed the data, and wrote the manuscript; T.A. and N.S. critically reviewed the manuscript. All authors revised and approved the final manuscript.

Acknowledgments: This research received logistical assistance from the United Nations University Institute for Integrated Management of Material Fluxes and Resources (UNU-FLORES) and Technische Universität Dresden (TU Dresden), Germany. We extend our thanks to the administration of the national meteorological service of Togo for providing us with the climate data. Our gratitude goes to the editor and anonymous reviewers whose comments and suggestions expressively contributed to the improvement of this paper. Our thanks also go to Atiqah Fairuz Salleh for her editorial input to the manuscript.

Appendix A

List of Abbreviations

APSI	Agricultural Production Systems Simulator
MCDI	Controlled Deficit Irrigation
CWPF	Crop-Water Production Functions
DAISY	Danish simulation model for transformation and transport of energy and matter in the soil plant atmosphere system
DAP	Days After Planting
DI	Deficit Irrigation
DS	Dry Season
DSSAT	Decision Support System for Agrotechnology Transfer
FI	Full Irrigation
GET-OPTIS	Global Evolutionary Technique for OPTimal Irrigation Scheduling
GDD	Growing Degree Days
HI	Harvest index
ICID	International Commission on Irrigation and Drainage
IPCC	Intergovernmental Panel on Climate Change
ISRIC	International Soil Reference and Information Centre
ITRA	Togolese Institute of Agricultural Research
LARS-WG	Long Ashton Research Station Weather Generator
MERF	Togolese Ministry of the Environment and Forestry
NI	No Irrigation
OCCASION	Optimal Climate Change Adaption Strategies in Irrigation
PILOTE	An operative crop model for soil water balance and yieldestimations under conventional tillage
REW	Readily Evaporable Water
SCWPF	Stochastic Crop-Water Production Functions
SI	Supplemental Irrigation
SSA	Sub-Saharan Africa
WGEN	Weather Generator
WS	Wet Season

References

1. UN DESA. *World Population Prospects: The 2015 Revision, Key Findings and Advance Tables*; UN DESA: New York, NY, USA, 2015.

2. UN DESA. *World Population Prospects: The 2017 Revision, Key Findings and Advance Tables*; UN DESA: New York, NY, USA, 2017.
3. FAO. *Climate-Smart Agriculture Sourcebook*; FAO: Rome, Italy, 2013; ISBN 978-92-5-107720-7.
4. Liniger, H.; Mekdaschi Studer, R.; Hauert, C.; Gurtner, M. *Sustainable Land Management in Practice: Guidelines and Best Practices for Sub-Saharan Africa*; TerrAfrica, World Overview of Conservation Approaches and Technologies (WOCAT) and Food and Agriculture Organization of the United Nations (FAO): Rome, Italy, 2011; ISBN 9789250000000.
5. Elliott, J.; Deryng, D.; Müller, C.; Frieler, K.; Konzmann, M.; Gerten, D.; Glotter, M.; Flörke, M.; Wada, Y.; Best, N.; et al. Constraints and potentials of future irrigation water availability on agricultural production under climate change. *Proc. Natl. Acad. Sci. USA* **2014**, *111*, 3239–3244. [CrossRef] [PubMed]
6. Edgerton, M.D. Increasing Crop Productivity to Meet Global Needs for Feed, Food, and Fuel. *Plant Physiol.* **2009**, *149*, 7–13. [CrossRef] [PubMed]
7. World Bank. *World Development Report 2008: Agriculture for Development*; World Bank: Washington, DC, USA, 2008.
8. Rosegrant, M.W.; Paisner, M.S.; Siet, M.; Witcover, J. *2020 Global Food Outlook*; International Food Policy Research Institution: Washington, DC, USA, 2001; pp. 1–24.
9. van Ittersum, M.K.; van Bussel, L.G.J.; Wolf, J.; Grassini, P.; van Wart, J.; Guilpart, N.; Claessens, L.; de Groot, H.; Wiebe, K.; Mason-D'Croz, D.; et al. Can sub-Saharan Africa feed itself? *Proc. Natl. Acad. Sci. USA* **2016**, *113*, 14964–14969. [CrossRef] [PubMed]
10. Lobell, D.B.; Gourdji, S.M. The Influence of Climate Change on Global Crop Productivity. *Plant Physiol.* **2012**, *160*, 1686–1697. [CrossRef] [PubMed]
11. Kotir, J.H. Climate change and variability in Sub-Saharan Africa: A review of current and future trends and impacts on agriculture and food security. *Environ. Dev. Sustain.* **2011**, *13*, 587–605. [CrossRef]
12. Druyan, L.M. Studies of 21st-century precipitation trends over West Africa. *Int. J. Climatol.* **2011**, *31*, 1415–1424. [CrossRef]
13. Sarr, B. Present and future climate change in the semi-arid region of West Africa: A crucial input for practical adaptation in agriculture. *Atmos. Sci. Lett.* **2012**, *13*, 108–112. [CrossRef]
14. Ministère de l'Environnement et des Ressources Forestières (MERF). *Plan d'Action National d'Adaptation aux Changements Climatiques (PANA)*; MERF: Lome, Togo, 2009. (In French)
15. Mcsweeney, C.; New, M.; Lizcano, G. *UNDP Climate Change Country Profiles, Togo*; School of Geography and Environment, Oxford University: Oxford, UK, 2009.
16. Ogounde, L.; Abotchi, T. *Quelques contraintes à la croissance Agricole dans la région des Savanes du Nord-Togo. Bulletin de la société Neuchâteloise de Geographie*; Société Neuchâteloise de Geographie: Neuchâtel, Switzerland, 2003. (In French)
17. Dobermann, A.; Nelson, R.; Beever, D.; Bergvinson, D.; Crowley, E.; Denning, G.; Griller, K.; d'Arros Hughes, J.; Jahn, M.; Lynam, J.; et al. *Solutions for Sustainable Agriculture and Food Systems—Technical Report for the Post-2015 Development Agenda*; The United Nations Sustainable Development Solutions Network (UNSDSN): New York, NY, USA, 2013.
18. Rockström, J.; Williams, J.; Daily, G.; Noble, A.; Matthews, N.; Gordon, L.; Wetterstrand, H.; De Clerck, F.; Shah, M.; Steduto, P.; et al. Sustainable intensification of agriculture for human prosperity and global sustainability. *Ambio* **2017**, *46*, 4–17. [CrossRef]
19. Godfray, H.C.J.; Garnett, T. Food security and sustainable intensification. *Philos. Trans. R. Soc. Lond. B. Biol. Sci.* **2014**, *369*, 1–10. [CrossRef]
20. Bolor, J.K. Analyse de l'état actuel de développement de l'irrigation au Togo. In *Irrigation in West Africa: Current Status and a View to the Future*; Namara, R.E., Sally, H., Eds.; International Water Management Institute (IWMI), Colombo, Sri Lanka: Ouagadougou, Burkina Faso, 2010; pp. 305–312.
21. Jalloh, A.; Nelson, G.C.; Thomas, T.S.; Zougmoré, R.; Roy-Macauley, H. *West African Agriculture and Climate Change: A Comprehensive Analysis*; IFPRI Research Monograph; International Food Policy Research: Washington, DC, USA, 2013.
22. International Commission on Irrigation and Drainage (ICID). Basic Introduction: Irrigation. Available online: http://www.icid.org/res_irrigation.html (accessed on 10 September 2018).
23. Rockström, J.; Hatibu, N.; Oweis, T.; Wani, S.; Barron, J.; Bruggeman, A.; Qiang, Z.; Farahani, J.; Karlberg, L. Managing Water in Rainfed Agriculture. In *Water for Food, Water for Life: A Comprehensive Assessment of Water Management in Agriculture*; Molden, D., Ed.; Earthscan: London, UK, 2007; pp. 315–352.

24. Pereira, L.S. Higher performance through combined improvements in irrigation methods and scheduling: A discussion. *Agric. Water Manag.* **1999**, *40*, 153–169. [CrossRef]
25. English, M.J.; Nuss, G.S. Designing for Deficit Irrigation. *J. Irrig. Drain. Div.* **1982**, *108*, 91–106.
26. Djaman, K.; Irmak, S.; Rathje, W.R.; Martin, D.L.; Eisenhauer, D.E. Maize evapotranspiration, yield production functions, biomass, grain yield, harvest index, and yield response factors under full and limited irrigation. *Am. Soc. Agric. Biol. Eng.* **2013**, *56*, 273–293.
27. English, M. Deficit Irrigation. I: Analytical Framework. *J. Irrig. Drain. Eng.* **1990**, *116*, 399–412. [CrossRef]
28. English, M.; Raja, S.N. Perspectives on deficit irrigation. *Agric. Water Manag.* **1996**, *32*, 1–14. [CrossRef]
29. Lecler, N.L. Integrated methods and models for deficit irrigation planning. In *Agricultural Systems Modeling and Simulation*; Lecler, N.L., Peart, R.M., Eds.; Marcel Dekker Inc.: New York, NY, USA, 1998; pp. 283–299.
30. Fereres, E.; Soriano, M.A. Deficit irrigation for reducing agricultural water use. *J. Exp. Bot.* **2006**, *58*, 147–159. [CrossRef] [PubMed]
31. Kögler, F.; Söffker, D. Water (stress) models and deficit irrigation: System-theoretical description and causality mapping. *Ecol. Model.* **2017**, *361*, 135–156. [CrossRef]
32. Kloss, S.; Pushpalatha, R.; Kamoyo, K.J.; Schütze, N. Evaluation of Crop Models for Simulating and Optimizing Deficit Irrigation Systems in Arid and Semi-arid Countries Under Climate Variability. *Water Resour. Manag.* **2012**, *26*, 997–1014. [CrossRef]
33. Schütze, N.; De Paly, M.; Shamir, U. Novel simulation-based algorithms for optimal open-loop and closed-loop scheduling of deficit irrigation systems. *J. Hydroinformatics* **2012**, *14*, 136–151. [CrossRef]
34. Semenov, M.A. Development of high-resolution UKCIP02-based climate change scenarios in the UK. *Agric. For. Meteorol.* **2007**, *144*, 127–138. [CrossRef]
35. Brumbelow, K.; Georgakakos, A. Consideration of Climate Variability and Change in Agricultural Water Resources Planning. *J. Water Resour. Plan. Manag.* **2007**, *133*, 275–285. [CrossRef]
36. Schütze, N.; Schmitz, G.H. OCCASION: New Planning Tool for Optimal Climate Change Adaption Strategies in Irrigation. *J. Irrig. Drain. Eng.* **2010**, *136*, 836–846. [CrossRef]
37. Jones, J.W.; Hoogenboom, G.; Porter, C.H.; Boote, K.J.; Batchelor, W.D.; Hunt, L.A.; Wilkens, P.W.; Singh, U.; Gijsman, A.J.; Ritchie, J.T. The DSSAT cropping system model J.W. *Eur. J. Agron.* **2003**, *18*, 235–263. [CrossRef]
38. Hsiao, T.C.; Heng, L.; Steduto, P.; Rojas-Lara, B.; Raes, D.; Fereres, E. Aquacrop-The FAO crop model to simulate yield response to water: III. Parameterization and testing for maize. *Agron. J.* **2009**, *101*, 448–459. [CrossRef]
39. Raes, D.; Steduto, P.; Hsiao, T.C.; Fereres, E. Aquacrop-The FAO crop model to simulate yield response to water: II. main algorithms and software description. *Agron. J.* **2009**, *101*, 438–447. [CrossRef]
40. Steduto, P.; Hsiao, T.C.; Raes, D.; Fereres, E. Aquacrop-the FAO crop model to simulate yield response to water: I. concepts and underlying principles. *Agron. J.* **2009**, *101*, 426–437. [CrossRef]
41. Hansen, S.; Jensen, H.E.; Nielsen, N.E.; Svendsen, H. *DAISY: A Soil Plant System Model. Danish simulation Model for Transformation and Transport of Energy and Matter in the Soil Plant Atmosphere System*; The National Agency for Environmental Protection: Copenhagen, Denmark, 1990.
42. Smith, M. *CROPWAT: A Computer Program for Irrigation Planning and Management*; Food and Agriculture Organization of the United Nations, Ed.; FAO irrigation and drainage paper 46; ISBN1 9251031061. Food and Agriculture Organization of the United Nations: Rome, Italy, 1992; ISBN2 9251031061.
43. Keating, B.A.; Carberry, P.S.; Hammer, G.L.; Probert, M.E.; Robertson, M.J.; Holzworth, D.; Huth, N.I.; Hargreaves, J.N.G.; Meinke, H.; Hochman, Z.; et al. An overview of APSIM, a model designed for farming systems simulation. *Eur. J. Agron.* **2003**, *18*, 267–288. [CrossRef]
44. Mailhol, J.C.; Olufayo, A.A.; Ruelle, P. Sorghum and sunflower evapotranspiration and yield from simulated leaf area index. *Agric. Water Manag.* **1997**, *35*, 167–182. [CrossRef]
45. Iqbal, M.A.; Shen, Y.; Stricevic, R.; Pei, H.; Sun, H.; Amiri, E.; Penas, A.; del Rio, S. Evaluation of the FAO AquaCrop model for winter wheat on the North China Plain under deficit irrigation from field experiment to regional yield simulation. *Agric. Water Manag.* **2014**, *135*, 61–72. [CrossRef]
46. Vanuytrecht, E.; Raes, D.; Steduto, P.; Hsiao, T.C.; Fereres, E.; Heng, L.K.; Garcia Vila, M.; Mejias Moreno, P. AquaCrop: FAO's crop water productivity and yield response model. *Environ. Model. Softw.* **2014**, *62*, 351–360. [CrossRef]
47. Department of Immigration and Citizenship (DIC). *Togolese Community Profile*; Department of Immigration and Citizenship, Commonwealth of Australia: Lomé, Togo, 2007.

48. RGPH. *Recensement Générale de la population et de l'habitat. Direction Générale de la Statistique et de la Comptabilité Nationale*; RGPH: Lomé, Togo, 2010.
49. Ali, E. A review of agricultural policies in independent Togo. *Int. J. Agric. Policy Res.* **2017**, *5*, 104–116. [CrossRef]
50. Poch, R.M.; Ubalde, J.M. Diagnostic of degradation processes of soils from northern Togo (West Africa) as a tool for soil and water management. In *Proceedings of the Workshop for Alumni of the M.Sc. Programmes in Soil Science, Eremology and Physical Land Resources*; Langouche, D., Van Ranst, E., Eds.; Workshop IC-PLR: Ghent, Belgium, 2006; pp. 187–194.
51. Institut Togolais de Recherche Agronomique (ITRA). *Bien cultiver et conserver le maïs. Collection Brochures et Fiches Techniques*; ITRA: Lomé, Togo, 2008. (In French)
52. Didjeira, A.; Adourahim, A.A.; Sedzro, K. *Situation de référence sur les principales céréales cultivées au Togo: Maïs, Riz, Sorgho, Mil*; ITRA: Lomé, Togo, 2007.
53. Desplat, A.; Rouillon, A. *Diagnostic agraire dans la région des Savanes au Togo: Cantons de Nioukpourma, Naki-Ouest et Tami*; Master de Recherche, Institut des Sciences et Industries du Vivant et de L'environnement, AgroParisTech: Paris, France, 2011.
54. Kottek, M.; Grieser, J.; Beck, C.; Rudolf, B.; Rubel, F. World Map of the Köppen-Geiger climate classification updated. *Meteorol. Z.* **2006**, *15*, 259–263. [CrossRef]
55. Walter, H.; Lieth, H.H.F. *Klimadiagramm-Weltatlas*; G. Fischer Verlag: Jena, Germany, 1967.
56. Institut National de la Statistique et des Etudes Economiques et Démographiques (INSEED). *Profil de pauvreté: Togo*; INSEED: Lomé, Togo, 2016.
57. Institut National de la Statistique et des Etudes Economiques et Démographiques (INSEED). Statistiques Nationales: Togo 2015. Available online: http://togo.opendataforafrica.org/# (accessed on 13 September 2018).
58. NASA POWER Project-Agroclimatology Data. Available online: https://power.larc.nasa.gov/data-access-viewer/ (accessed on 10 September 2017).
59. Van Wart, J.; Grassini, P.; Yang, H.; Claessens, L.; Jarvis, A.; Cassman, K.G. Creating long-term weather data from thin air for crop simulation modeling. *Agric. For. Meteorol.* **2015**, *209–210*, 49–58. [CrossRef]
60. Semenov, M.A.; Brooks, R.J.; Barrow, E.M.; Richardson, C.W. Comparison of the WGEN and LARS-WG stochastic weather generators for diverse climates. *Clim. Res.* **1998**, *10*, 95–107. [CrossRef]
61. Semenov, M.A. Simulation of extreme weather events by a stochastic weather generator. *Clim. Res.* **2008**, *35*, 203–212. [CrossRef]
62. Mehan, S.; Guo, T.; Gitau, M.; Flanagan, D.C.; Mehan, S.; Guo, T.; Gitau, M.W.; Flanagan, D.C. Comparative Study of Different Stochastic Weather Generators for Long-Term Climate Data Simulation. *Climate* **2017**, *5*, 26. [CrossRef]
63. Guo, T.; Mehan, S.; Gitau, M.W.; Wang, Q.; Kuczek, T.; Flanagan, D.C. Impact of number of realizations on the suitability of simulated weather data for hydrologic and environmental applications. *Stoch. Environ. Res. Risk Assess.* **2018**, *32*, 2405–2421. [CrossRef]
64. Chakravarti, I.M.; Laha, R.G.; Roy, J. *Handbook of Methods of Applied Statistics*; Wiley: New York, NY, USA, 1967.
65. Semenov, M.A.; Barrow, E.M. Use of a stochastic weather generator in the development of climate change scenarios. *Clim. Chang.* **1997**, *35*, 397–414. [CrossRef]
66. Doorenbos, J.; Kassam, A.H. *Yield Response to Water*; FAO Irrigation and Drainage Paper No. 33; FAO: Rome, Italy, 1979.
67. Greaves, G.E.; Wang, Y.-M. Assessment of FAO AquaCrop Model for Simulating Maize Growth and Productivity under Deficit Irrigation in a Tropical Environment. *Water* **2016**, *8*, 557. [CrossRef]
68. Poss, R. *Etude Morphopédologique du Nord du Togo à [au] 1/500,000*; Institut français de recherche scientifique pour le développement en coopération (ORSTOM): Lomé, Togo, 1996.
69. Akumaga, U.; Tarhule, A.; Yusuf, A.A. Validation and testing of the FAO AquaCrop model under different levels of nitrogen fertilizer on rainfed maize in Nigeria, West Africa. *Agric. For. Meteorol.* **2017**, *232*, 225–234. [CrossRef]
70. IUSS Working Group WRB. *World Reference Base for Soil Resources 2014, Update 2015 International Soil Classification System for Naming Soils and Creating Legends for Soil Maps*; FAO: Rome, Italy, 2015.
71. Worou, K.S. *Sols Dominants du Togo—Corrélation avec la Base de Référence Mondiale. Quatorzième Réunion du Sous-Comité ouest et Centre Africain de Corrélation des sols. Rapport sur les Ressources en Sols du Monde 98*; Food and Agriculture Organization of the United Nations (FAO): Rome, Italy, 2002. (In French)

72. Raes, D.; Steduto, P.; Hsiao, T.C.; Fereres, E. Chapter 2—Users guide. In *Reference Manual: AquaCrop, Version 4.0*; FAO, Land and Water Division: Rome, Italy, 2012; pp. 1–164.
73. Worou, S.; Saragoni, H. *La Culture du Maïs de Contre Saison Est-elle Possible au Togo Meridional? Premières Conclusions d'une Experimentation sur la Station de Recherche Agronomique d'ativémé*; Institut français de recherche scientifique pour le développement en coopération (ORSTOM): Lomé, Togo, 1988.
74. Geerts, S.; Raes, D.; Garcia, M.; Miranda, R.; Cusicanqui, J.A.; Taboada, C.; Mendoza, J.; Huanca, R.; Mamani, A.; Condori, O.; et al. Simulating Yield Response of Quinoa to Water Availability with AquaCrop. *Agron. J.* **2009**, *101*, 499–508. [CrossRef]
75. Salemi, H.; Amin, M.; Soom, M.; Lee, T.S.; Farhad Mousavi, S.; Ganji, A.; Kamilyusoff, M. Application of AquaCrop model in deficit irrigation management of Winter wheat in arid region. *Afr. J. Agric. Res.* **2011**, *610*, 2204–2215. [CrossRef]
76. Silvestro, P.C.; Pignatti, S.; Yang, H.; Yang, G.; Pascucci, S.; Castaldi, F.; Casa, R. Sensitivity analysis of the Aquacrop and SAFYE crop models for the assessment of water limited winter wheat yield in regional scale applications. *PLoS ONE* **2017**, *12*, 1–30. [CrossRef] [PubMed]
77. Djaman, K.; Ganyo, K. Trend analysis in reference evapotranspiration and aridity index in the context of climate change in Togo. *J. Water Clim. Chang.* **2015**, *6*, 848–864. [CrossRef]
78. Osman, Y.; Abdellatif, M.; Al-Ansari, N.; Knutsson, S.; Jawad, S. Climate Change and Future Precipitation in Arid Environment of Middle East: Case study of Iraq. *J. Environ. Hydrol.* **2017**, *25*, 1–18.
79. Assefa, S.; Biazin, B.; Muluneh, A.; Yimer, F.; Haileslassie, A. Rainwater harvesting for supplemental irrigation of onions in the southern dry lands of Ethiopia. *Agric. Water Manag.* **2016**, *178*, 325–334. [CrossRef]
80. Fosu-Mensah, B.Y. *Modelling the Impact of Climate Change on Maize (Zea mays L.) YieLd under Rainfed Conditions in Sub-Humid Ghana*; United Nations University–Institute for Natural Resources in Africa (UNU-INRA): Accra, Ghana, 2013.
81. Chauhan, C.P.S.; Singh, R.B.; Gupta, S.K. Supplemental irrigation of wheat with saline water. *Agric. Water Manag.* **2008**, *95*, 253–258. [CrossRef]
82. Fox, P.; Rockström, J. Supplemental irrigation for dry-spell mitigation of rainfed agriculture in the Sahel. *Agric. Water Manag.* **2003**, *61*, 29–50. [CrossRef]
83. Wakchaure, G.C.; Minhas, P.S.; Ratnakumar, P.; Choudhary, R.L. Optimising supplemental irrigation for wheat (*Triticum aestivum* L.) and the impact of plant bio-regulators in a semi-arid region of Deccan Plateau in India. *Agric. Water Manag.* **2016**, *172*, 9–17. [CrossRef]
84. Heng, L.K.; Hsiao, T.; Evett, S.; Howell, T.; Steduto, P. Validating the FAO AquaCrop Model for Irrigated and Water Deficient Field Maize. *Agron. J.* **2009**, *101*, 488–498. [CrossRef]
85. Stricevic, R.; Cosic, M.; Djurovic, N.; Pejic, B.; Maksimovic, L. Assessment of the FAO AquaCrop model in the simulation of rainfed and supplementally irrigated maize, sugar beet and sunflower. *Agric. Water Manag.* **2011**, *98*, 1615–1621. [CrossRef]
86. Tittonell, P.; Giller, K.E. When yield gaps are poverty traps: The paradigm of ecological intensification in African smallholder agriculture. *Field Crops Res.* **2013**, *143*, 76–90. [CrossRef]
87. Bell, J.M.; Schwartz, R.; McInnes, K.J.; Howell, T.; Morgan, C.L.S. Deficit irrigation effects on yield and yield components of grain sorghum. *Agric. Water Manag.* **2018**, *203*, 289–296. [CrossRef]
88. Greaves, G.E.; Wang, Y.-M. Effect of regulated deficit irrigation scheduling on water use of corn in southern Taiwan tropical environment. *Agric. Water Manag.* **2017**, *188*, 115–125. [CrossRef]
89. Hergert, G.W.; Margheim, J.F.; Pavlista, A.D.; Martin, D.L.; Isbell, T.A.; Supalla, R.J. Irrigation response and water productivity of deficit to fully irrigated spring camelina. *Agric. Water Manag.* **2016**, *177*, 46–53. [CrossRef]
90. Kifle, M.; Gebretsadikan, T.G. Yield and water use efficiency of furrow irrigated potato under regulated deficit irrigation, Atsibi-Wemberta, North Ethiopia. *Agric. Water Manag.* **2016**, *170*, 133–139. [CrossRef]
91. Li, X.; Kang, S.; Zhang, X.; Li, F.; Lu, H. Deficit irrigation provokes more pronounced responses of maize photosynthesis and water productivity to elevated CO_2. *Agric. Water Manag.* **2018**, *195*, 71–83. [CrossRef]
92. Mustafa, S.M.T.; Vanuytrecht, E.; Huysmans, M. Combined deficit irrigation and soil fertility management on different soil textures to improve wheat yield in drought-prone Bangladesh. *Agric. Water Manag.* **2017**, *191*, 124–137. [CrossRef]
93. Lee, S.O.; Jung, Y. Efficiency of water use and its implications for a water-food nexus in the Aral Sea Basin. *Agric. Water Manag.* **2018**, *207*, 80–90. [CrossRef]

94. Gunn, K.M.; Baule, W.J.; Frankenberger, J.R.; Gamble, D.L.; Allred, B.J.; Andresen, J.A.; Brown, L.C. Modeled climate change impacts on subirrigated maize relative yield in northwest Ohio. *Agric. Water Manag.* **2018**, *206*, 56–66. [CrossRef]
95. Abedinpour, M.; Sarangi, A.; Rajput, T.B.S.; Singh, M.; Pathak, H.; Ahmad, T. Performance evaluation of AquaCrop model for maize crop in a semi-arid environment. *Agric. Water Manag.* **2012**, *110*, 55–66. [CrossRef]

6

Irrigation Water Quality—A Contemporary Perspective

Arindam Malakar [1], Daniel D. Snow [2,*] and Chittaranjan Ray [3]

[1] Nebraska Water Center, part of the Robert B. Daugherty Water for Food Global Institute, 109 Water Sciences Laboratory, University of Nebraska, Lincoln, NE 68583-0844, USA
[2] School of Natural Resources and Nebraska Water Center, part of the Robert B. Daugherty Water for Food Global Institute, 202 Water Sciences Laboratory, University of Nebraska, Lincoln, NE 68583-0844, USA
[3] Nebraska Water Center, part of the Robert B. Daugherty Water for Food Global Institute 2021 Transformation Drive, University of Nebraska, Lincoln, NE 68588-6204, USA
* Correspondence: dsnow1@unl.edu

Abstract: In the race to enhance agricultural productivity, irrigation will become more dependent on poorly characterized and virtually unmonitored sources of water. Increased use of irrigation water has led to impaired water and soil quality in many areas. Historically, soil salinization and reduced crop productivity have been the primary focus of irrigation water quality. Recently, there is increasing evidence for the occurrence of geogenic contaminants in water. The appearance of trace elements and an increase in the use of wastewater has highlighted the vulnerability and complexities of the composition of irrigation water and its role in ensuring proper crop growth, and long-term food quality. Analytical capabilities of measuring vanishingly small concentrations of biologically-active organic contaminants, including steroid hormones, plasticizers, pharmaceuticals, and personal care products, in a variety of irrigation water sources provide the means to evaluate uptake and occurrence in crops but do not resolve questions related to food safety or human health effects. Natural and synthetic nanoparticles are now known to occur in many water sources, potentially altering plant growth and food standard. The rapidly changing quality of irrigation water urgently needs closer attention to understand and predict long-term effects on soils and food crops in an increasingly fresh-water stressed world.

Keywords: crop uptake; food quality; geogenic; emerging contaminants; nanomaterials

---

1. Introduction

Irrigation is the controlled use of multiple water sources in a timely manner for increased or sustained crop production. Irrigation comprises of the water that is applied by an irrigation system during the growing season and also includes water applied during field preparation, pre-irrigation, weed control, harvesting, and for leaching salts from the root zone [1]. In 2015 it was estimated that in the United States irrigation alone accounted for 62% of water usage [1]. Globally, irrigation is the highest consumptive use of freshwater [2]. As the world's population grows, the risk increases that more people will be deprived of adequate food supplies in impoverished areas, particularly those subject to water scarcity [3]. Agricultural production of food needs to increase by an estimated 60% by 2050 to ensure global food security [3] and irrigation will increasingly be called upon to help meet this demand. In the race to enhance agricultural productivity, irrigation will become even more dependent on substandard sources of water. Therefore, it is of utmost importance to access our current state of knowledge and explore the effects of irrigation water quality on crops. This understanding will help ensure adequate crop production to meet increased demand as well as to maintain proper food and soil quality.

Groundwater exploitation (withdrawal for irrigation) can release naturally occurring geogenic contaminants, such as arsenic, from the solid phase to groundwater, while wastewater reuse can concentrate pesticides, pharmaceuticals and other emerging contaminants in irrigation water [4,5]. Use of untreated wastewater is becoming prevalent in developing countries where around 80–90% of wastewater remains untreated [6]. Polluted municipal, industrial or agricultural water used for irrigation significantly changes soil quality, increases the amount of trace elements in soil and plants, and acts as a source of various pathogens which affects food quality and safety [7,8]. Water of inadequate quality is a potential source of both direct and indirect contamination to food crops [9], and leads to increased contamination of soil and water [10,11]. In addition, the presence of synthetic and natural nanomaterials is beginning to be identified in crops [12–14]. In locations where excess irrigation is practiced, contaminants in soils are leached to the vadose zone, where they can contribute to geogenic contaminant mobilization and potentially increase contaminant levels in local groundwater [15]. Many aspects of water composition, such as hardness and iron content, also affect the suitability of a water source for newer, more efficient spray or drip irrigation techniques. Runoff, return flow, and leaching of irrigation water also contribute to local surface and groundwater contamination [16]. Increased usage of irrigation water has already led to impaired irrigation water and soil quality. Considering the presence of new contaminant types in different water sources (see Figure 1), it is essential to evaluate the impact of these contaminants within the context of modern agriculture. To date, very little research and regulatory attention has been paid to contaminants in irrigation water. Contamination of irrigation water supplies is likely to worsen unless additional efforts (research, guidelines, regulations, treatment methods) are brought to bear on this problem.

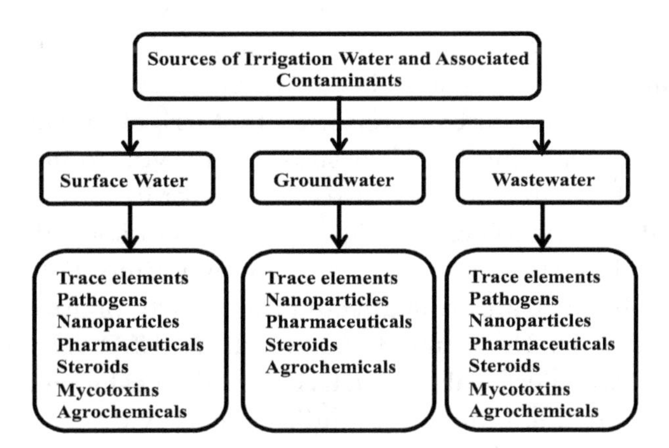

Figure 1. Main sources of irrigation water and different types of contaminants present in those sources impacting food, soil, and water quality. Note that surface water and wastewater are subject to similar types of contamination.

This review article looks at previous approaches to define irrigation water quality and compares to a current perspective with respect to impacts on human health. It evaluates the long-term effect and influence of the changing quality of water sources used in agricultural production. Although the article discusses traditional irrigation water quality concerns, such as salinization, it mainly emphasizes contemporary water quality issues like new or emerging contaminants, pathogens, geogenic trace elements and engineered nanomaterials. These contaminants are now widespread in various conventional and unconventional water sources used for modern-day irrigation. The article is organized as a short summary of conventional measures of irrigation water quality followed by a more detailed evaluation of the impact of contemporary irrigation water quality issues on soil and crop quality. Contemporary topics include emerging contaminants with separate sections on pharmaceuticals, antibiotics, steroid hormones, pesticides, cyanotoxins and mycotoxins, biological contaminants bacteria, virus and antibiotic resistance genes, modern inorganic contaminants, such as geogenic trace elements and nanomaterials. The review is summarized by considering the changing quality of water sources used for irrigation, and the need for additional work and improved regulation of irrigation water, especially for food production. The primary focus of this review is to recognize water quality issues that have a direct or indirect influence on surface soil contamination, crop uptake of contaminants and their potential to impact human health. This article shows a need for modern guidelines, regulations and research to understand the complex nature of irrigation water. Though it is a critically important topic from a human health standpoint, this review does not include an exhaustive discussion of contamination of irrigation water by human pathogens. While wastewater treatment technologies are constantly evolving and can address some of the issues presented here, a review of wastewater quality as a function of treatment technology is beyond the scope of this article. Moreover, treatment approaches are likely to be tailored to sources, and irrigation water sources are highly varied depending on climate, population, industry, crop, and livestock density.

2. Conventional Measures of Irrigation Water Quality

The effect of irrigation water composition on soil properties for crop production has been a focus for the past half century. Previous studies of water quality issues, and the suitability of freshwater sources for irrigation, have primarily been directed toward an understanding of potential problems to soil salinity, fertility and crop growth. For example, early work by the United States Geological Survey (USGS) [17] evaluated groundwater quality in Texas for irrigation and other potentially competing uses. A subsequent report by Schwennesen and Forbes characterized groundwater in San Simon Valley, Arizona and New Mexico, for domestic use and irrigation [18]. Clark reported on the chemical composition of groundwater in the Morgan Hill area of California [19], while Scofield and Headly [20] evaluated water composition with respect to irrigation potential. Most of these early works focused on understanding the impact of water quality on long-term viability of irrigation in arid regions of the United States.

Globally, irrigation water quality was described in Tanzania, Africa, with respect to pH and alkaline and alkaline-earth elements [21]. Taylor et al. reported that irrigation water pH was one of the main factors for wheat growth in Punjab, India [22]. A subsequent work reiterated that alkaline elements such as sodium play a crucial role in continued use of water for irrigation of cropland and quantified the maximum amount that may be tolerated [23]. The effects of soil salinization and trace element composition on crop growth have become more apparent over time. Eaton et al. reported that boron present in water around Hollister, California affected the growth of apricots and prunes [24]. In the subsequent years, the United States Department of Agriculture (USDA) conducted further studies and reported that sodium, boron and electrical conductivity are the best general measures for judging the suitability of water for irrigation [25]. From these studies, it was evident that continuous irrigation with water of marginal quality impacted soil and also affected crop growth [26–31]. In 1967, the American Society for Testing Materials (ASTM) developed a quantitative assessment of irrigation water quality, including new formulas for maximum permissible quantity of chloride and electrical conductivity based on infiltration rate, evapotranspiration rate, irrigation frequency and duration [32]. Traditionally, discussion on irrigation water quality has mainly focused on its effect on soil quality, and how soil quality was predicted to affect crop growth and yield. Color, turbidity, total dissolved solids (TDS), pH, specific conductance, odor and foam characterized the quality of water. Colorless, odorless, foamless water with minimum turbidity, TDS below 1000 mg L^{-1} at circumneutral pH and specific conductance below 1.5 mmhos/m is generally considered to be of good quality for irrigation purposes [33,34]. A higher TDS is not recommended for most crops as it can impact the salinity of soil and pore water will become highly concentrated when taken up by roots via osmosis. Excessive dissolved solids content, or salinity of irrigation water, has historically been the primary characteristic determining water suitability for irrigation. Salt accumulation in the crop root zone impedes water uptake and can eventually prevent plant growth altogether [34]. Excess salinity from sodium can affect soil structure and water infiltration. The proportion of sodium to calcium and magnesium is the primary factor controlling the hydraulic conductivity of water in soil [33–35]. Sodium is generally expressed as a sodium absorption ratio (SAR) [9]. Long-term irrigation of soils with elevated sodium concentrations relative to calcium and magnesium, bicarbonate, carbonate, and TDS will be limiting soil aggregate formation, which reduces infiltration and makes less water available to crops [34].

Seiler et al., under the National Irrigation Water Quality Program (NIWQP) of the U.S. Department of the Interior (DOI), studied the effect of irrigation-induced contamination of water, soil and biota in the western United States. NIWQP data from the 26 areas under study suggested that degradation of groundwater quality due to irrigation is a common occurrence [11,36]. The study indicated that selenium was the most common contaminant, followed by arsenic, uranium and molybdenum [11,37]. This study also suggested regular co-occurrence of these contaminants. For example, selenium was found to be elevated with uranium, and these contaminants were accumulating in the soils and affecting long-term suitability for crop production. This was one of the first reports to correlate trace element contamination in water sources used for irrigation to soil quality. These findings led to the appreciation of the intricate complexities of irrigation water quality and its role in ensuring proper crop growth and long-term food quality. These studies mainly focused on the impact of water quality on crop productivity and soil quality, while effects to food quality and safety were just beginning to be recognized.

3. Impact of Contemporary Irrigation Water Quality Issues on Soil and Crop Quality

Irrigation water quality has mainly been characterized with respect to effects on crop growth and yield, though an emerging and pressing issue relates to plant uptake and soil enrichment with inorganic and organic contaminants (Figure 2). These "new" issues with respect to irrigation water quality can lead to food quality and safety concerns, as well as affect crop growth and yield [38–41]. Wastewater reuse for irrigation contributes to increasing incidence of organic microcontaminants [42], such as pharmaceuticals and other synthetic organics in soils and crops. Increasing reliance on groundwater also contributes to the probability for elevated concentrations of natural geogenic contaminants such as arsenic and selenium in irrigation water and soils. Understanding the occurrence and fate of these new contaminants in irrigation water sources is paramount in limiting the effects to modern agricultural products [43]. Long term impacts to soil and crop quality (see Figure 2) need to be understood.

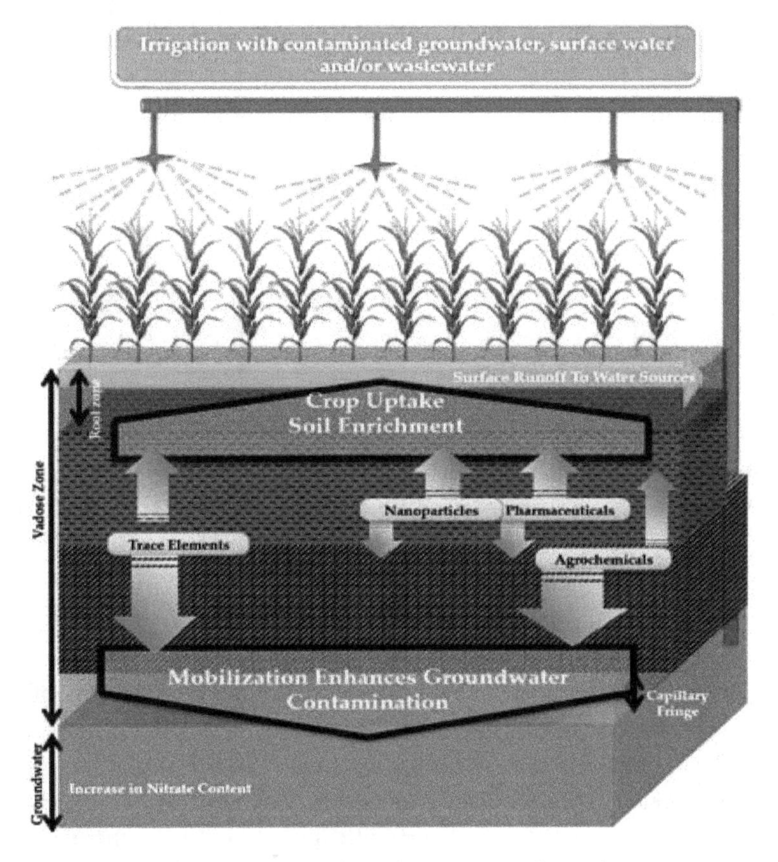

Figure 2. A conceptual model of the impact of inadequate quality of irrigation water sources on soil and crop quality.

3.1. Emerging Contaminants: Organic Pollutants

3.1.1. Pharmaceuticals

Traces of pharmaceuticals and personal care products have been identified in a variety of freshwater sources, including drinking water [44], groundwater [45], and surface water [45]. Pharmaceuticals can enter the water system from various sources, including direct disposal and human excretion into sewers leading to elevated concentrations of pharmaceuticals in wastewater [46]. Pharmaceuticals often detected in sewage sludge include non-steroidal anti-inflammatory drugs (NSAIDs), blood thinners, psychiatric drugs, antidiuretics and β-blockers [47–49]. Plant uptake of a wide variety of pharmaceutical groups like NSAIDs, antihistamine, β-blockers, calcium channel blockers, antiepileptics, steroid hormones, antidepressants, antineoplastic agents, anti-itch compounds, x-ray contrast agents, lipid-lowering agents, benzodiazepines, tranquilizers and veterinary drugs from soil and contaminated water has been observed and studied [50–52]. Wu et al. reported that a primary pathway for contamination by pharmaceuticals in food crops is through irrigation water [53,54]. For example, a recent study found traces of carbamazepine, caffeine, lamotrigine, gabapentin and acesulfame in a variety of vegetables grown with treated wastewater in Jordan [55]. Treated wastewater is well known to contain a large variety of pharmaceuticals and personal care products, many of which are known to accumulate in food crops [56,57]. The occurrence of these and other synthetic organic chemicals is likely to increase in water supplies, especially in areas with water scarcity, and irrigation with contaminated water will lead to soil contamination and plant uptake.

3.1.2. Antibiotics

Environmental contamination by antibiotic residues in food production systems is a growing problem worldwide, and the potential implications to proliferation of antibiotic resistance have been the subject of multiple reviews and opinion articles [58–60]. The occurrence of persistent antibiotic residues in various water sources [61,62] is well documented which not only includes municipal [63,64], agricultural [65,66] and hospital sewage [67,68] but also groundwater [69] and surface water [70–72]. The concentration range of antibiotics is generally measured at ng L^{-1} to a few μg L^{-1} in many water sources [73], though concentrated wastewater can have much higher levels [65]. Recent studies have demonstrated that plants can take up antibiotics (like amoxicillin, ketoconazole, lincomycin, oxytetracycline, sulfamethoxazole, sulfonamides, and tetracyclines) [74–76] and antibiotic contaminated irrigation water can play a significant role in the uptake [77]. The environmental fate and transport of antibiotics depend on various physical properties such as water solubility, lipophilicity, volatility and sorption potential [77].

The implications to human health due to the presence of antibiotics in food crops is not clear, but other potential adverse impacts include allergic reactions, disruption of digestive function and chronic toxic effects as a result of prolonged low-level exposure [78–81]. One of the major concerns for the increasing prevalence of antibiotics in the environment is the development and spread of antibiotic-resistant gene and bacteria [82], which is discussed in Section 3.2.2. Clearly, the absence of antibiotic residues in irrigation water cannot be assumed.

3.1.3. Steroids

Land application of livestock manure can contribute to accumulation of steroid hormones [83] and veterinary pharmaceuticals. Very low concentrations of natural steroid hormones, such as estrone, 17α-estradiol, 17β-estradiol, estriol, testosterone, androstenedione and progesterone that occur in animal waste and wastewater have been documented as accumulating in soil [84]. Traces of steroid hormones have also been reported in groundwater [85] and surface water [86], and often in treated municipal wastewater [87–91]. Wastewater treatment plants are known to discharge these hormones into river water and also other recipients [92]. Laboratory experiments have suggested that traces of steroid hormones and pharmaceuticals can be taken up in crops [93] and recent studies of food crops irrigated with treated municipal water have confirmed this can occur in the field. Further work is needed to understand the significance and impact of these chemicals in the environment and to human health [94], though at present, the reported concentrations are relatively low in comparison to other contaminants.

3.1.4. Agrochemicals

Regular use of pesticides in irrigated crops is also likely to lead to the occurrence of these residues in irrigation water and food crops, especially in regions where regulation and training in proper application of these substances is lacking. Application of large quantities of agrichemicals and improper management can create a substantial effect on the environment. The leaching and runoff of agrochemicals is a potential source of groundwater and surface water contamination [95,96]. The occurrence of agrichemical residues in vegetables has been documented [97] and their uptake by crops is well studied [98,99] and regulated. Leaching of nitrate from fertilizer over application to groundwater below is well reported, and accumulation of reactive nitrogen is also thought to initiate mobilization of other geogenic contaminants [15].

3.1.5. Cyanotoxins and Mycotoxins

Cyanotoxins, which comprise a large range of naturally produced organic compounds, are produced and released by cyanobacteria when they are present in large quantities (blooms) and especially when these organisms die off and decay in surface water. Cyanobacteria also referred to as blue-green algae, naturally occur in all freshwater ecosystems [100]. Warmer temperatures coupled with high nutrient concentrations are thought to favor conditions for algae blooms to form in surface water. Of many different groups of cyanotoxins, hepatoxic cyclic peptides collectively known as microcystins are the most commonly studied cyanotoxins, which cause a wide range of symptoms in humans [101]. Other studies have also shown that cyanotoxins, which include hepatotoxic microcystins and neurotoxic compounds such as anatoxin-a and beta-*N*-methylamino-alanine, can make their way to human and animal food chain from contaminated reservoirs [102–105]. A recent review has summarized the extent of the literature investigating the fate in soils, and agricultural crops [103]. It seems quite clear that toxins can accumulate in plants, including food crops and under some conditions can also inhibit plant growth [102,103,106–108]. Though there are many gaps, and only a handful of studies have investigated this route for exposure. There is evidence for human health effects through consumption of plants contaminated with cyanotoxins by irrigation using surface water sources impacted by cyanotoxins.

Mycotoxins are naturally occurring fungal toxins (chemicals), which can cause a variety of adverse health effects to both humans and livestock. A few mycotoxins are known or suspected carcinogens. Fungi do pose potential hazards to human health. However, there were relatively few studies of mycotoxins in water sources until recently. Fungal contamination has been observed in drinking water [109] and recently it was reported that untreated surface water can be breeding place for these fungi, generating mycotoxins [110]. Kolpin et al. led a broad scale study on the occurrence of mycotoxins across streams in the United States (US) [111]. Their study concluded that the ecotoxicological effects from long-term, low-level exposures to mycotoxins are poorly understood and would require further investigation. Mycotoxin uptake in rice has been studied [112] and these chemicals have been reported to be present in various food grains [113,114]. The prevalence of mycotoxins in surface water makes it an important consideration regarding modern water quality of irrigation as mycotoxin health hazards are widely reported.

3.2. Biological Contaminants: Bacteria, Virus and Antibiotic Resistance

3.2.1. Pathogens

Often because of its nutrient content and accessibility, untreated wastewater from municipal and domestic sources containing excessive levels of pathogens is often directly used for irrigation in developing countries [6]. Untreated wastewater generally carries a high pathogen load compared to other irrigation water sources. Risks from pathogen (bacteria, viruses, or protozoan or larger organisms) contamination to irrigation water quality will continue to be a topic of primary concern [115–117], and it is impossible to adequately address this topic in a few paragraphs. Pathogenic microorganisms in irrigation water likely pose the greatest acute risk to human health and will continue to be a concern especially in freshly-eaten produce. Pathogens are biological organisms that may influence modern-day irrigation water quality. Pathogen contamination is generally related to surface water sources, but groundwater may also be under threat if it is recharged with wastewater sources [116]. The complexity of reproducibly measuring microbiological contamination of irrigation water has made monitoring difficult. Several different types of pathogens have been detected in diverse irrigation water sources including bacteria (e.g., *Salmonella* and *Escherichia coli*), protozoa (e.g., *Cryptosporidium* and *Giardia*), as well as viruses (e.g., noroviruses)) [118,119]. Irrigation of food crops with surface water clearly has the highest potential for contaminating freshly eaten produce, and this topic has had the greatest research and regulatory effort in recent years.

There have been quite a few comprehensive reviews emphasizing irrigation water as a source of pathogenic microorganisms in fresh produce [120–123]. Between 1973 and 2012, the Centers for Disease Control and Prevention reported 606 leafy-vegetable associated pathogenic outbreaks (norovirus (55% of outbreaks), Shiga toxin-producing *Escherichia coli* (STEC) (18%), and *Salmonella* (11%)), with 20,003 associated illness and 19 deaths [124]. From 2013 to 2017, the number of outbreaks (mainly from norovirus (32%) STEC (23%), and *Salmonella* (32%)) associated with leafy greens and vegetables decreased to 21, with 699 illness and five deaths [125]. In 2018, 272 infections were reported from two outbreaks (*E. Coli*) associated with romaine lettuce resulting in five deaths [126,127], and another multi-state outbreak was linked to parasite *Cyclospora*, which reported 511 cases of infection [128]. However, a 2014 risk-based review conducted in California suggests that recycled water quality criteria, along with proper agricultural management practices do not lead to increased public health risk [129]. In the US, the Center for Produce Safety has published information on the factors that affect the microbiological safety of agricultural water [130]. The Foodborne Disease Outbreak Surveillance System (FDOSS) has an online tool, the National Outbreak Report System (NORS), which keeps track of outbreaks in the United States. Reports of pathogen contamination from inadequately treated wastewater have also been documented in developing countries [8]. The occurrence of pathogens in water used to irrigate food crops is considered a severe problem affecting human health both in both developing and even in developed countries [131]. Groundwater sources are generally considered less vulnerable to contamination by pathogenic microorganisms, while surface water and wastewater have a much higher potential for contamination. Farmers utilizing surface water for food crops, which are consumed raw should follow proper mitigation strategies to control contamination [132]. The method utilized for irrigation has a substantive role in pathogenic contamination of crops. For example, subsurface drip may have the lowest risk as the water is generally applied at the root zone, unlike other methods (e.g., sprinkler irrigation) where the edible portions of crops can come in contact with contaminated water [133]. New and more intensive monitoring approaches and potential disinfection and treatment techniques for surface water used to irrigate food crops are needed to improve food safety [8].

3.2.2. Antibiotic Resistance

The World Health Organization (WHO) has listed antibiotic resistance among today's biggest threats for global health, food safety, and development, as this threatens the ability to treat common infectious diseases [134]. Antibiotic resistome is defined as the sum of all genes directly or indirectly contributing to antibiotic resistance both in the clinics and the environment [135]. Aquatic ecosystems are regarded as a primary reservoir of antibiotic-resistant bacteria (ARB) [136]. The presence of ARB and their resistance determinants in surface water sources have been well documented and is generally linked to nearby wastewater treatment plant effluent [137–141]. Wastewater treatment plants enrich ARB and their resistance determinants as it favors exchange of antibiotic resistance genes (ARG) among bacteria and selection of resistant strains [142]. In a recent study, it was found that multidrug-resistant (MDR) bacteria were found to be more prevalent in surface waters than in treated wastewater [143].

Irrigation water is one of the major sources for contamination of fresh produce with antibiotic resistance bacteria [58,144,145]. Similar to pathogens, the incidence of ARB contamination is higher when using overhead sprinklers as water can directly come in contact with fresh produce. When fresh produce is consumed raw it can act as an ideal vector for exposure. The diversity of ARB present in fresh produce is significant and can have a severe impact on human health.

3.3. Inorganic Contaminants: Geogenic Source and Nanomaterials

3.3.1. Geogenic Contaminants in Irrigation Water

Selected naturally occurring geogenic contaminants, such as boron, arsenic and selenium, have been the subject of much previous work focused on irrigation water quality [146,147], especially in areas with extensive use of groundwater. With the exception of boron, trace element contaminants were not studied with respect to soil quality, crop productivity and phytotoxicity [9]. Boron is an essential trace element for plant growth, but elevated boron concentrations (>1 mg/L) in irrigation water can cause stunted growth and reduced productivity in sensitive crops such as wheat. The occurrence of geogenic contaminants is a growing contemporary issue because of the potential impact on food quality and human health. Concentrations of geogenic contaminants have likely been increasing over time in a variety of irrigation water sources, often due to increasing agricultural intensification [148,149]. Researchers have reported that groundwater in China has elevated levels of arsenic; this water is used for irrigating feed crops [150]. Similarly, studies in the United States have reported higher levels of uranium [15] and arsenic [151] in its aquifers. India, Bangladesh and Vietnam have widespread arsenic contamination in groundwater used for drinking and irrigation, especially in areas where the use of contaminated water has led to contaminated soils and crops [152–154]. Presently there are no federal guidelines regulating geogenic contaminant levels in irrigation water except in the case of direct wastewater reuse [155,156]. The increasing levels of contaminants in soils and irrigation water are a growing issue across the globe, and there is little work to date regarding strategies to mitigate accumulation in plants and food crops. Contaminant uptake by crops has been well studied and plant uptake has even been used as a remediation method, viz. phytoremediation. However, mitigation strategies for prevention of geogenic contaminant uptake by plants have received scant attention in the literature. Plant uptake and accumulation of specific trace elements may not affect plant growth, but accumulation and consumption may pose hazards to animal or humans.

Table 1 summarizes the estimated ranges of aqueous concentrations of many geogenic elements with respect to guidelines and recommendations of water use. Irrigation water guidelines compiled by the Food and Agriculture Organization (FAO) in 1976 are generally based on the toxic effects to crops and plant growth [157]. FAO has recommended using these values as guidelines for irrigation utilizing groundwater and surface water sources, but not for irrigation with wastewater containing measurable levels of trace elements [157]. Wastewater guidelines set by FAO (Table 1) are equivalent to irrigation water quality recommendations in 1976, though may contradict newer guidelines for special constituents of wastewater [157]. Early recommendations rarely consider uptake of trace elements by crops and the impact on food quality, which may affect human health. Recommended concentrations to ensure consistent food quality are absent, as even low concentrations of geogenic contaminants can impact food quality (Table 1) [158–172]. Continuous use of irrigation water with low concentrations of geogenic contaminants concentrations can result in soil enrichment and affect food crop quality [158–172].

Table 1. Common geogenic trace element contaminants in irrigation water with current regulatory and recommended levels (NA: Not Available, AL: Action level, CCC: Criterion continuous concentration, CMC: Criterion maximum concentration, SDWR: Secondary drinking water regulations). (* Few examples of plant uptake, ** in μg, *** greenhouse experiment concentration was equivalent to natural condition).

Trace Element	EPA Drinking Water Guideline (μg L^{-1}) [173]	Regulatory Limits for Wastewater (μg L^{-1}) [156,174,175]	Freshwater CCC, CMC Limits for Aquatic Life (μg L^{-1}) [176]	Recommended Maximum Concentrations of Trace Elements in Irrigation Waters (μg L^{-1}) [157]	Reported Ranges for Trace Elements in Glacier Aquifer System of USA (Includes Wells Used for Agriculture) (μg L^{-1}) [177]	Reported Ranges of Water Known to Impact Food and Soil Quality * (μg L^{-1})
Arsenic	10	100	340, 150	100	0.09–340	58.7 [158], 0.2–164 [159]
Boron	NA	NA	NA	1000	NA	3351–16,000 [165–167]
Cadmium	5	10	1.8, 0.72	10	0.018–1	<1–3200 [168]
Cobalt	NA	50	NA	50	0.007–95	0.21–0.81 [169]
Copper	1300 (AL)	200	NA	200	0.126–127	10–133 ** [170]
Chromium (III) or (VI)	100 (Cr)	100 (Cr)	570, 74 (III) 16, 11 (VI)	100 (Cr)	0.4–22 (Cr)	1–46 (VI) [171] ≤250 *** [172]
Iron	300 (SDWR)	5000	NA, 1000	5000	3–38,100	-
Lead	15 (AL)	5000	65, 2.5	5000	0.04–9.0	≤140 [160]
Lithium	NA	2500	NA	2500	0.040–126	-
Manganese	500 (SDWR)	200	NA	200	0.056–28,200	≤100 [160]
Nickel	-	200	470, 52	200	0.035–56	≤50 [160]
Selenium	50	20	NA	20	0.173–223	0.12–341 [161]
Silver	100 (SDWR)	NA	3.2, NA	NA	NA	-
Uranium	30	NA	NA	NA	0.009–162	1–1200 [162]
Zinc	5000 (SDWR)	2000	120, 120	2000	0.536–1000	~130 [168]

It is likely that irrigation with elevated levels of geogenic contaminants leads to contaminant enrichment in crops. For example, Bundschuh et al. [178] compiled data from different regions of South America known for high occurrences of arsenic in groundwater and surface water. Their study indicated that arsenic concentrations in edible plants and crops were associated with the elevated arsenic concentrations in soil and irrigation waters. The study showed that regions with high arsenic concentrations in surface water and groundwater relate directly to accumulation in plants, fish, livestock meat, milk, and milk products. Their study indicates that there is a need for more rigorous studies in evaluating pathways of arsenic exposure through the food chain in Latin America and other regions.

Factors such as arsenic speciation, type and composition of soil, and plant species also plays a significant role in crop uptake [179]. An interesting aspect of arsenic transformation is where arsenic present in soil may occur as oxidized arsenate (As(V)) but may become reduced to arsenite (As(III)) after uptake in crops [180]. Arsenite is regarded as 25–60 times more toxic to humans than arsenate [181]. Arsenic, by far, is the most studied geogenic contaminant in crops, especially rice, and is included in many review articles detailing its impact of food quality and human health [182–186]. Ongoing research is focused on managing arsenic uptake by crops [154,187,188]. Other geogenic contaminants in groundwater, such as uranium, are less well studied with respect to food contamination.

Elevated levels of selenium may be toxic, though selenium is an essential micronutrient for crops. Selenium is known to accumulate in crops grown on soils with high selenium content [161,189], and selenium enrichment has been reported in soils throughout the United States [190]. In their recent study, Wang et al. [191] found that volatile organic compounds released by plant growth-promoting rhizobacteria increases both selenium and iron uptake. Uranium is another geogenic, potentially toxic contaminant [192] and studies in nutrient culture show its uptake by crops [193,194]. Several studies confirmed that irrigation water contaminated with uranium has an impact on crop quality though less to soil contamination by uranium [195–197]. Lead, mercury [38], chromium [198] and cadmium [199] are also known to accumulate in crops grown on soils with high levels of these contaminants. While there is extensive information in the literature on uptake by plants, most studies have focused on the general type of contaminant, or its accumulation in crops, use in phytoremediation of soils and the pathway of uptake within the crop. Contaminated irrigation water is capable of enriching surface soil with these contaminants [200] and likely enhances availability for uptake by crops [197]. There are few federal guidelines for trace element limits in foods in the United States, other than in arsenic, lead, cadmium, and mercury in wastewater [201], and the evidence suggests there is an urgent need for irrigation water quality standards that include geogenic contaminants.

3.3.2. Engineered Nanomaterials

Earth is rich with natural nanomaterials and it is estimated that thousands of megatons move through the hydrosphere annually [14]. Natural nanoparticles in water can easily pass through conventional membrane filter pore sizes of 0.2-µm and may not be accounted for as nanoparticles [202], which adds to the complexity in understanding their impact on crop health. In the past decade, production and use of engineered nanoparticles have also risen significantly and continues to trend upwards. Nanoparticles are used in a wide variety of contemporary products, ranging from electronics and cosmetics to processed food. This proliferation in use of nanoparticles has paved their way to increasing occurrence in water sources [203,204]. Nanotechnology is also used widely for water treatment for both groundwater and surface water sources [205,206], but its repercussions are still not well understood [207]. In 2018, the occurrence of nanoparticle size plastics (nanoplastics) were critically reviewed with respect to human health and growing global occurrence in freshwater [208,209]. Analytical methods capable of detecting and quantifying nanoparticles in complex aqueous matrices are lacking, increasing the challenge in tracking fate and transport of these particles [207]. In 2005, Oberdörster et al. [210] reviewed the interaction of nanomaterials in regards to human health. Still, the full-scale toxicity of natural or engineered nanomaterials is not well understood in the context of complex biosystems [14]. The size of nanomaterials (100 nm or less in size) is the main factor [14] which makes studying nanoparticle contamination challenging. Therefore, the paradigm of risk assessment of nanoparticles needs to be reevaluated with respect to the unique challenges involved in monitoring environmental pathways and assessing impacts on human health.

Broadly, engineered nanoparticles can be classified into carbon and metal-based nanoparticles. Nanoparticles derived from carbon (e.g., carbon nanotubes) simultaneously act like particles and high molecular weight organic compounds. Metal oxides or metal nanoparticles must be evaluated for their metal-related chemotoxicity, and also toxicity arising due to their particulate form. These dual behaviors again add to the complexity of studying and assessing risk. Nanotoxicity can elicit significant effects to human health. Nanoparticles can be carcinogenic and may produce reactive species inside the body [211]. Ganguly et al. have recently reviewed the toxicity of nanomaterials, emphasizing major exposure pathways [211].

The nanoparticle life cycle is poorly understood. Therefore, the Environmental Protection Agency (EPA) has allowed limited manufacturing of new materials by administrative orders or new use rules under the Toxic Substances Control Act. These new rules have significantly expanded since their inception in 2010, which is on par with our increase in understanding. However, the use of and occurrence of nanomaterials in irrigation water is not monitored or regulated [206,212].

Nanotechnology usage is vital in modern civilization and will have a substantial impact on the world economy [213]. It is projected that the nanomaterials market will reach 55 billion US dollars by year 2022 [213]. Production of engineered nanoparticles is expected to rise in the coming years, making these contaminants more likely to occur in different water sources, both conventional (surface and/or groundwater) and in treated wastewater. Both carbon-based and metal-based nanoparticles are of concern and it has been reported these particles are persistent in water [214,215]. Nanomaterial transformation in fresh water systems is an active field of research [216]. Nanoparticle assessments have concluded that fine particulate matter occurs in a variety of water sources [217–221], so its impact on irrigation water, accumulation in soil, and potential for crop uptake is of paramount significance.

The growing availability of engineered nanomaterials/particles is relevant to understand how these nanosized particles may impact water and food quality in agriculture. The use of nanomaterials in agriculture is also increasing, and little data is available to understand how occurrence in irrigation water may influence crops [222,223]. There is recent work on uptake of both carbon and metal-based nanoparticles by plants. For example, multi-walled carbon nanotubes have been observed in broccoli, resulting in a positive impact on plant growth [224] but potentially creating health concerns. Carbon nanotubes are also known to act as a carrier for other contaminants (like organochlorine pesticide, etc.) in plants and enhance contaminant translocation [225]. These contrasting effects of carbon nanotubes have been well summarized in a recent review article by Vithanage et al. [226].

Metal oxide nanoparticles, which form a bulk of engineered nanomaterials, have been well studied under the purview of plant uptake [227]. Metal oxides like titanium dioxide, silver oxide, iron oxide, copper oxide and metal nanoparticles have been shown to accumulate in a variety of food crops and have even been detected in commercial produce [228–232]. A recent review by Ma et al. describes studies of nanoparticle uptake by crops and their occurrence in the final produce [40]. Similar to carbon-based nanoparticles, metal or metal oxide-based nanoparticles are also known to be beneficial to plant growth and have been marketed as nanofertilizers [233]. For example, iron oxide nanoparticles have shown to be a potential iron source for peanut crops [234]. Nanoparticles are also known to induce oxidative stress in crops [235]. A recent study by Liu et al. suggested that combinations of nanomaterials might have different impacts on the soil microcosm compared to a single nanomaterial [236]. In actual conditions, mixed nanomaterials will be more prevalent in soil and irrigation water. In addition to engineered nanomaterial, natural nanomaterials of silver are found in groundwater [237]. Life cycle assessment of engineered titanium dioxide nanoparticles showed that their impact is not just limited to crop uptake, but exhibit marine aquatic ecotoxicity and human toxicity [238,239]. Nanoscale is an important factor in ensuring uptake of fertilizer [240,241] but the long-term effects of nanoagrochemicals have yet to be studied [242]. While nanoagrochemicals may be beneficial to crops [243], they may have undesired effects on the environment and on human health [242].

There are many questions and few answers for understanding the effects of nanosized contaminants in irrigated agriculture. Future studies should be focused on understanding retention times and fate in water, plants and soils, including degradation and transformation rates, and biological effects of different forms. Naturally-formed nanoparticles can occur in irrigation water and in food crops and it is clear this route for exposure needs to be better understood and monitored. Are nanoparticles easily broken down in the environment or are they stable, do they form aggregates, and what is the accumulation in both soil and water sources? If they accumulate, how do they impact soil health and modern irrigation techniques? Nanoparticle occurrence and behavior in irrigation water sources, soils and plants is clearly an emerging area of research.

4. Changing Quality of Water Sources Used for Irrigation

This discussion on the changing composition and quality of different water sources for irrigation—be it conventional sources like surface- or groundwater, or nonconventional source like reclaimed water—signifies an urgent need for increased efforts to monitor the quality of agricultural water [244]. Recently, there have been various efforts to address changing water quality [9] by incorporating measurement of new contaminant types (e.g., arsenic, fluoride) [10,245,246] in water quality guidelines. Still, it is challenging to address the complexity of irrigation water quality covering different forms of geogenic and emerging contaminants. In the coming decades, the concentrations of specific contaminants will likely increase and continue to affect the quality of water sources for irrigation. Impending climate change may make the situation more extreme as drought and water scari]city may concentrate contaminants in water. The increasing prevalence of cyanotoxins may increase due to more intensive agriculture, fertilization and water use. Although there are strict guidelines for contaminant concentrations in drinking water, few guidelines exist for the use of water for any irrigation, including irrigation of food crops. The substantial expense of monitoring, including sampling and laboratory analysis, for "new" recently identified contaminants is a reason to implement a more practical approach for monitoring irrigation water quality, incomparison to the complex and expensive framework currently used for drinking water in developed countries.

Establishing regulations and clear guidelines for irrigation water quality in different countries or individual states of the United States for conventional sources of water is necessary, but not sufficient, to ensure healthy produce for human consumption. Proper food quality can only be insured if water sources are regulated and regularly tested. Testing can also be used to monitor accumulation of contaminants in soil. Moreover, plant tissue should also be checked periodically for contaminant uptake to ensure appropriate produce quality. Presently in the United States, reclaimed wastewater is regulated for irrigation usage [247,248] but almost any other source may be used without restriction. Guidelines are provided for pathogen levels in reclaimed wastewater [174]. There are recommendations for limits of geogenic contaminants in irrigation water in traditional sources [157,249] but there is a complete lack of recommendations or information on new and emerging organic contaminants or nanomaterials. There is a whole range of second and third generation nanomaterials proposed for commercial uses (e.g., nanocomposites and multi-element materials). These new nanomaterials will occur in waste streams and we have little understanding of their fate or toxicity. Moreover, the basis for existing recommendations is questionable, as concentrations far lower than some recommended values have been shown to be biomagnified in crops (see Table 1).

The present review focuses on the changing quality of water used for irrigation, and clearly there are many gaps to be addressed in future research. High priority should be given to research focused on improving our understanding and address the increasing occurrence of geogenic contaminants, pathogens and other biological contaminants in irrigation water, especially as they relate to food crops. Increased wastewater reuse for agricultural purposes will likely increase the occurrence of biologically active organic contaminants such as pharmaceuticals and antibiotics, and the effect on food crops and human health is also a major research gap. The dynamic nature of the chemical composition of different irrigation water makes it very important that relevant and periodic data is available for both conventional and nonconventional water sources. A more comprehensive set of water quality guidelines needs to be created incorporating our present understanding of the occurrence and effects of emerging contaminants. There should be specific recommendations for monitoring and tests at to check irrigation water quality used in agriculture. These recommendations can be soil specific, crop specific and water specific. The availability of new guidelines would help ensure better food quality, as the next generation will not only need larger quantities of irrigation water to feed the growing population, but health concerns may rise as we resort to the use of low-quality water to irrigate food and feed crops.

Author Contributions: A.M. led writing of this article, including extensive literature investigation on irrigation water quality, orginal draft preparation and sources on geogenic and nanomaterial contaminants. D.D.S. provided conceptualization of the need for a review on this topic, editing and input on literature sources for emerging contaminants. C.R. provided additional input on wastewater contaminants, nanomaterials and additional editing.

Acknowledgments: Malakar thanks NET for salary support. The author thank feedback on drafts by Jason White and Sushil Kanel, as well as editorial comments from Erin Haackker and Lacey Bodnar.

References

1. Dieter, C.A.; Maupin, M.A.; Caldwell, R.R.; Harris, M.A.; Ivahnenko, T.I.; Lovelace, J.K.; Barber, N.L.; Linsey, K.S. *Estimated Use of Water in the United States in 2015*; U.S. Geological Survey: Reston, VA, USA, 2018.
2. *World Agriculture: Towards 2015/2030 an FAO Perspective*; Bruinsma, J. (Ed.) Earthscan Publications Ltd.: London, UK, 2003; ISBN 9251048355.
3. FAO. How to Feed the World in 2050. *Insights Expert Meet. FAO* **2009**, *2050*, 1–35. [CrossRef]
4. Sauvé, S.; Desrosiers, M. A review of what is an emerging contaminant. *Chem. Cent. J.* **2014**, *8*, 15. [CrossRef] [PubMed]
5. National Research Council. *Identifying Future Drinking Water Contaminants*; National Academies Press: Washington, DC, USA, 1999; ISBN 978-0-309-06432-3.
6. WWAP (United Nations World Water Assessment Programme). *The United Nations World Water Development Report 2017. Wastewater: The Untapped Resource*; United Nations Educational, Scientific and Cultural Organization: Paris, France, 22 March 2017.
7. Hass, A.; Mingelgrin, U.; Fine, P. Heavy metals in soils irrigated with wastewater. In *Treated Wastewater in Agriculture: Use and Impacts on the Soil Environment and Crops*; Wiley-Blackwell: Hoboken, NJ, USA, 2010; ISBN 9781405148627.
8. Allende, A.; Monaghan, J. Irrigation water quality for leafy crops: A perspective of risks and potential solutions. *Int. J. Environ. Res. Public Health* **2015**, *12*, 7457–7477. [CrossRef] [PubMed]
9. Singh, S.; Ghosh, N.C.; Gurjar, S.; Krishan, G.; Kumar, S.; Berwal, P. Index-based assessment of suitability of water quality for irrigation purpose under Indian conditions. *Environ. Monit. Assess.* **2018**, *190*, 29. [CrossRef] [PubMed]
10. Islam, M.A.; Romić, D.; Akber, M.A.; Romić, M. Trace metals accumulation in soil irrigated with polluted water and assessment of human health risk from vegetable consumption in Bangladesh. *Environ. Geochem. Health* **2018**, *40*, 59–85. [CrossRef] [PubMed]
11. Seiler, R.L.; Skorupa, J.P.; Naftz, D.L.; Nolan, B.T. *Irrigation-Induced Contamination of Water, Sediment, and Biota in the Western United States—Synthesis of Data from the National Irrigation Water Quality Program*; U.S. Geological Survey Professional Paper1655; U.S. Geological Survey: Reston, VA, USA, 2003.
12. Servin, A.D.; De la Torre-Roche, R.; Castillo-Michel, H.; Pagano, L.; Hawthorne, J.; Musante, C.; Pignatello, J.; Uchimiya, M.; White, J.C. Exposure of agricultural crops to nanoparticle CeO_2 in biochar-amended soil. *Plant Physiol. Biochem.* **2017**, *110*, 147–157. [CrossRef] [PubMed]

13. Zhao, Q.; Ma, C.; White, J.C.; Dhankher, O.P.; Zhang, X.; Zhang, S.; Xing, B. Quantitative evaluation of multi-wall carbon nanotube uptake by terrestrial plants. *Carbon* **2017**, *114*, 661–670. [CrossRef]
14. Hochella, M.F.; Mogk, D.W.; Ranville, J.; Allen, I.C.; Luther, G.W.; Marr, L.C.; McGrail, B.P.; Murayama, M.; Qafoku, N.P.; Rosso, K.M.; et al. Natural, incidental, and engineered nanomaterials and their impacts on the Earth system. *Science* **2019**, *363*, eaau8299. [CrossRef] [PubMed]
15. Nolan, J.; Weber, K.A. Natural Uranium Contamination in Major U.S. Aquifers Linked to Nitrate. *Environ. Sci. Technol. Lett.* **2015**, *2*, 215–220. [CrossRef]
16. Mateo-Sagasta, J.; Marjani, S.; Turral, H.; Burke, J. *Water Pollution from Agriculture: A Global Review Executive Summary*; The Food and Agriculture Organization of the United Nations: Rome, Italy; The International Water Management Institute: Colombo, Sri Lanka, 2017.
17. Dexssen, A.; Dole, R.B. *Ground Water in LaSalle and McMullen Counties*; Texas. U.S. Geol. Surv. Water-Supply Paper 375-G; U.S. GPO: Washington, DC, USA, 1916.
18. Schwennesen, A.T.; Forbes, R.H. *Ground Water in San Simon Valley, Arizona and New Mexico*; Water-Supply Paper 425-A; U.S. GPO: Washington, DC, USA, 1917.
19. Clark, W.O. *Ground Water for Irrigation in the Morgan Hill Area, California*; Water-Supply Paper 400-E; U.S. GPO: Washington, DC, USA, 1917.
20. Scofield, C.S.; Headley, F.B. Quality of irrigation water in relation to land reclamation. *J. Agric. Res.* **1921**, *21*, 265–278. [CrossRef]
21. Sturdy, D.; Calton, W.E.; Milne, G. A chemical survey of the waters of Mount Meru, Tanganyika Territory, especially with regard to their qualities for irrigation. *J. East Afr. Uganda Nat. Hist. Soc.* **1932**, *45–46*, 1–38.
22. Taylor, E.M.; Puri, A.N.; Asghar, A.G. Soil deterioration in the canal-irrigated areas of the Punjab. I. Equilibrium between calcium and sodium ions in base-exchange reactions. *Res. Publ.* **1934**, *4*, 7.
23. Mados, L. The qualifications of irrigation waters. *Mezogazdasagi Kut* **1940**, *12*, 121–131.
24. Eaton, F.M.; McCallum, R.D.; Mayhugh, M.S. *Quality of Irrigation Waters of the Hollister area of California with Special Reference to Boron Content and Its Effect on Apricots and Prunes*; Technical Bulletin; United States Department Agriculture: Washington, DC, USA, 1941; Volume 746, p. 59.
25. Wilcox, L.V. The quality of water for irrigation use. *U.S. Dept. Agr. Tech. Bull.* **1948**, *962*, 40.
26. Pacheco, J.d.l.R.; Lopez-Rubio, F.B. Analysis of waters for agricultural uses. *Inf. Quim. Anal.* **1949**, *3*, 90–96.
27. Thorne, D.W.; Thorne, J.P. Changes in composition of irrigated soils as related to the quality of irrigation waters. *Soil Sci. Soc. Am. Proc.* **1949**, *18*, 92–97. [CrossRef]
28. Lewis, G.C.; Juve, R.L. Some effects of irrigation-water quality on soil characteristics. *Soil Sci.* **1956**, *81*, 125–137. [CrossRef]
29. Pearson, H.E.; Huberty, M.R. Response of citrus to irrigation with waters of different chemical characteristics. *Proc. Am. Soc. Hortic. Sci.* **1959**, *73*, 248–256.
30. Babcock, K.L.; Carlson, R.M.; Schulz, R.K.; Overstreet, R. A study of the Effect of irrigation water composition on soil properties. *Hilgardia* **1959**, *29*, 155–170. [CrossRef]
31. Longenecker, D.E.; Lyerly, P.J. Chemical characteristics of soils of west Texas as affected by irrigation water quality. *Soil Sci.* **1959**, *87*, 207–216. [CrossRef]
32. Bernstein, L. Quantitative assessment of irrigation water quality. In *Water Quality Criteria*; Bramer, H., Ed.; ASTM International: West Conshohocken, PA, USA, 1967; pp. 51–65.
33. Park, D.M.; White, S.A.; McCarty, L.B.; Menchyk, N.A. *Interpreting Irrigation Water Quality Reports*; CU-14-700; Clemson University Cooperative Extension: Clemson, SC, USA, 2014.
34. Frenkel, H. Reassessment of Water Quality Criteria for Irrigation. *Ecol. Stud. Anal. Synth.* **1984**, *51*, 142–172.
35. Bauder, T.; Waskom, R.; Davis, J.; Sutherland, P. Irrigation water quality criteria. *Crop Ser. Irrig. Fact Sheet* **2007**, *506*, 10–13.
36. Feltz, H.; Engberg, R.; Sylvester, M. Investigations of water quality, bottom sediment, and biota associated with irrigation drainage in the western United States. In Proceedings of the International Symposium on the Hydrologic Basis for Water Resources Management, Beijing, China, 23–26 October 1990. Publ. no. 197.
37. Seiler, R.L. Synthesis of data from studies by the national irrigation water-quality program. *J. Am. Water Resour. Assoc.* **1996**, *32*, 1233–1245. [CrossRef]
38. Tangahu, B.V.; Sheikh Abdullah, S.R.; Basri, H.; Idris, M.; Anuar, N.; Mukhlisin, M. A review on heavy metals (As, Pb, and Hg) uptake by plants through phytoremediation. *Int. J. Chem. Eng.* **2011**, *2011*, 31. [CrossRef]

39. Intawongse, M.; Dean, J.R. Uptake of heavy metals by vegetable plants grown on contaminated soil and their bioavailability in the human gastrointestinal tract. *Food Addit. Contam.* **2006**, *23*, 36–48. [CrossRef] [PubMed]
40. Ma, C.; White, J.C.; Zhao, J.; Zhao, Q.; Xing, B. Uptake of engineered nanoparticles by food crops: Characterization, mechanisms, and implications. *Annu. Rev. Food Sci. Technol.* **2018**, *9*, 129–153. [CrossRef]
41. Calderón-Preciado, D.; Matamoros, V.; Bayona, J.M. Occurrence and potential crop uptake of emerging contaminants and related compounds in an agricultural irrigation network. *Sci. Total Environ.* **2011**, *412–413*, 14–19. [CrossRef]
42. Sedlak, D.L.; Gray, J.L.; Pinkston, K.E. Peer reviewed: Understanding microcontaminants in recycled water. *Environ. Sci. Technol.* **2000**, *34*, 508A–515A. [CrossRef]
43. IOM (Institute of Medicine) and NRC (National Research Council). *A Framework for Assessing Effects of the Food System*; The National Academic Press: Washington, DC, USA, 2015; ISBN 978-0-309-30780-2.
44. World Health Organization. *Pharmaceuticals in Drinking Water*; WHO: Geneva, Switerland, 2012; ISBN 9789241502085.
45. Balakrishna, K.; Rath, A.; Praveenkumarreddy, Y.; Guruge, K.S.; Subedi, B. A review of the occurrence of pharmaceuticals and personal care products in Indian water bodies. *Ecotoxicol. Environ. Saf.* **2017**, *137*, 113–120. [CrossRef]
46. Yang, Y.; Ok, Y.S.; Kim, K.H.; Kwon, E.E.; Tsang, Y.F. Occurrences and removal of pharmaceuticals and personal care products (PPCPs) in drinking water and water/sewage treatment plants: A review. *Sci. Total Environ.* **2017**, *596–597*, 303–320. [CrossRef]
47. Fijalkowski, K.; Rorat, A.; Grobelak, A.; Kacprzak, M.J. The presence of contaminations in sewage sludge—The current situation. *J. Environ. Manag.* **2017**, *203*, 1126–1136. [CrossRef] [PubMed]
48. Subedi, B.; Balakrishna, K.; Joshua, D.I.; Kannan, K. Mass loading and removal of pharmaceuticals and personal care products including psychoactives, antihypertensives, and antibiotics in two sewage treatment plants in southern India. *Chemosphere* **2017**, *167*, 429–437. [CrossRef] [PubMed]
49. Subedi, B.; Lee, S.; Moon, H.B.; Kannan, K. Emission of artificial sweeteners, select pharmaceuticals, and personal care products through sewage sludge from wastewater treatment plants in Korea. *Environ. Int.* **2014**, *68*, 33–40. [CrossRef] [PubMed]
50. Madikizela, L.M.; Ncube, S.; Chimuka, L. Uptake of pharmaceuticals by plants grown under hydroponic conditions and natural occurring plant species: A review. *Sci. Total Environ.* **2018**, *636*, 477–486. [CrossRef] [PubMed]
51. Wu, X.; Dodgen, L.K.; Conkle, J.L.; Gan, J. Plant uptake of pharmaceutical and personal care products from recycled water and biosolids: A review. *Sci. Total Environ.* **2015**, *536*, 655–666. [CrossRef]
52. Tasho, R.P.; Cho, J.Y. Veterinary antibiotics in animal waste, its distribution in soil and uptake by plants: A review. *Sci. Total Environ.* **2016**, *563–564*, 366–376. [CrossRef] [PubMed]
53. Wu, X.; Conkle, J.L.; Ernst, F.; Gan, J. Treated wastewater irrigation: Uptake of pharmaceutical and personal care products by common vegetables under field conditions. *Environ. Sci. Technol.* **2014**, *48*, 11286–11293. [CrossRef]
54. Santiago, S.; Roll, D.M.; Ray, C.; Williams, C.; Moravcik, P.; Knopf, A. Effects of soil moisture depletion on vegetable crop uptake of pharmaceuticals and personal care products (PPCPs). *Environ. Sci. Pollut. Res.* **2016**, *23*, 20257–20268. [CrossRef]
55. Riemenschneider, C.; Al-Raggad, M.; Moeder, M.; Seiwert, B.; Salameh, E.; Reemtsma, T. Pharmaceuticals, their metabolites, and other polar pollutants in field-grown vegetables irrigated with treated municipal wastewater. *J. Agric. Food Chem.* **2016**, *64*, 5784–5792. [CrossRef]
56. Colon, B.; Toor, G.S. A review of uptake and translocation of pharmaceuticals and personal care products by food crops irrigated with treated wastewater. *Adv. Agron.* **2016**, *140*, 75–100.
57. Calderón-Preciado, D.; Jiménez-Cartagena, C.; Matamoros, V.; Bayona, J.M. Screening of 47 organic microcontaminants in agricultural irrigation waters and their soil loading. *Water Res.* **2011**, *45*, 221–231. [CrossRef] [PubMed]
58. Christou, A.; Agüera, A.; Bayona, J.M.; Cytryn, E.; Fotopoulos, V.; Lambropoulou, D.; Manaia, C.M.; Michael, C.; Revitt, M.; Schröder, P.; et al. The potential implications of reclaimed wastewater reuse for irrigation on the agricultural environment: The knowns and unknowns of the fate of antibiotics and antibiotic resistant bacteria and resistance genes—A review. *Water Res.* **2017**, *123*, 448–467. [CrossRef] [PubMed]

59. Williams-Nguyen, J.; Sallach, J.B.; Bartelt-Hunt, S.; Boxall, A.B.; Durso, L.M.; McLain, J.E.; Singer, R.S.; Snow, D.D.; Zilles, J.L. Antibiotics and antibiotic resistance in agroecosystems: State of the science. *J. Environ. Qual.* **2016**, *45*, 394–406. [CrossRef] [PubMed]
60. Durso, L.M.; Cook, K.L. Impacts of antibiotic use in agriculture: What are the benefits and risks? *Curr. Opin. Microbiol.* **2014**, *19*, 37–44. [CrossRef] [PubMed]
61. Kümmerer, K. Antibiotics in the aquatic environment—A review—Part I. *Chemosphere* **2009**, *75*, 417–434. [CrossRef] [PubMed]
62. Kümmerer, K. Antibiotics in the aquatic environment—A review—Part II. *Chemosphere* **2009**, *75*, 435–441. [CrossRef]
63. Behera, S.K.; Kim, H.W.; Oh, J.E.; Park, H.S. Occurrence and removal of antibiotics, hormones and several other pharmaceuticals in wastewater treatment plants of the largest industrial city of Korea. *Sci. Total Environ.* **2011**, *409*, 4351–4360. [CrossRef] [PubMed]
64. Zhou, L.J.; Ying, G.G.; Liu, S.; Zhao, J.L.; Yang, B.; Chen, Z.F.; Lai, H.J. Occurrence and fate of eleven classes of antibiotics in two typical wastewater treatment plants in South China. *Sci. Total Environ.* **2013**, *452–453*, 365–376. [CrossRef]
65. Aga, D.S.; Lenczewski, M.; Snow, D.; Muurinen, J.; Sallach, J.B.; Wallace, J.S. Challenges in the measurement of antibiotics and in evaluating their impacts in agroecosystems: A critical review. *J. Environ. Qual.* **2016**, *45*, 407–419. [CrossRef]
66. Burkholder, J.A.; Libra, B.; Weyer, P.; Heathcote, S.; Kolpin, D.; Thorne, P.S.; Wichman, M. Impacts of waste from concentrated animal feeding operations on water quality. *Environ. Health Perspect.* **2007**, *115*, 308–312. [CrossRef]
67. Chang, X.; Meyer, M.T.; Liu, X.; Zhao, Q.; Chen, H.; Chen, J.A.; Qiu, Z.; Yang, L.; Cao, J.; Shu, W. Determination of antibiotics in sewage from hospitals, nursery and slaughter house, wastewater treatment plant and source water in Chongqing region of Three Gorge Reservoir in China. *Environ. Pollut.* **2010**, *158*, 1444–1450. [CrossRef] [PubMed]
68. Duong, H.A.; Pham, N.H.; Nguyen, H.T.; Hoang, T.T.; Pham, H.V.; Pham, V.C.; Berg, M.; Giger, W.; Alder, A.C. Occurrence, fate and antibiotic resistance of fluoroquinolone antibacterials in hospital wastewaters in Hanoi, Vietnam. *Chemosphere* **2008**, *72*, 968–973. [CrossRef] [PubMed]
69. Hirsch, R.; Ternes, T.; Haberer, K.; Kratz, K.L. Occurrence of antibiotics in the aquatic environment. *Sci. Total Environ.* **1999**, *225*, 109–118. [CrossRef]
70. Kolpin, D.W.; Furlong, E.T.; Meyer, M.T.; Thurman, E.M.; Zaugg, S.D.; Barber, L.B.; Buxton, H.T. Pharmaceuticals, hormones, and other organic wastewater contaminants in U.S. streams, 1999–2000: A national reconnaissance. *Environ. Sci. Technol.* **2002**, *36*, 1202–1211. [CrossRef] [PubMed]
71. Yan, C.; Yang, Y.; Zhou, J.; Liu, M.; Nie, M.; Shi, H.; Gu, L. Antibiotics in the surface water of the Yangtze Estuary: Occurrence, distribution and risk assessment. *Environ. Pollut.* **2013**, *175*, 22–29. [CrossRef] [PubMed]
72. Deng, W.; Li, N.; Zheng, H.; Lin, H. Occurrence and risk assessment of antibiotics in river water in Hong Kong. *Ecotoxicol. Environ. Saf.* **2016**, *125*, 121–127. [CrossRef] [PubMed]
73. Zuccato, E.; Castiglioni, S.; Bagnati, R.; Melis, M.; Fanelli, R. Source, occurrence and fate of antibiotics in the Italian aquatic environment. *J. Hazard. Mater.* **2010**, *179*, 1042–1048. [CrossRef]
74. Ahmed, M.B.M.; Rajapaksha, A.U.; Lim, J.E.; Vu, N.T.; Kim, I.S.; Kang, H.M.; Lee, S.S.; Ok, Y.S. Distribution and accumulative pattern of tetracyclines and sulfonamides in edible vegetables of cucumber, tomato, and lettuce. *J. Agric. Food Chem.* **2015**, *63*, 398–405. [CrossRef]
75. Chitescu, C.L.; Nicolau, A.I.; Stolker, A.A.M. Uptake of oxytetracycline, sulfamethoxazole and ketoconazole from fertilised soils by plants. *Food Addit. Contam. Part A* **2013**, *30*, 1138–1146. [CrossRef]
76. Sallach, J.B.; Bartelt-Hunt, S.L.; Snow, D.D.; Li, X.; Hodges, L. Uptake of antibiotics and their toxicity to lettuce following routine irrigation with contaminated water in different soil types. *Environ. Eng. Sci.* **2018**, *35*. [CrossRef]
77. Azanu, D.; Mortey, C.; Darko, G.; Weisser, J.J.; Styrishave, B.; Abaidoo, R.C. Uptake of antibiotics from irrigation water by plants. *Chemosphere* **2016**, *157*, 107–114. [CrossRef] [PubMed]
78. Phillips, I.; Casewell, M.; Cox, T.; De Groot, B.; Friis, C.; Jones, R.; Nightingale, C.; Preston, R.; Waddell, J. Does the use of antibiotics in food animals pose a risk to human health? A critical review of published data. *J. Antimicrob. Chemother.* **2004**, *53*, 28–52. [CrossRef] [PubMed]

79. Kuppusamy, S.; Kakarla, D.; Venkateswarlu, K.; Megharaj, M.; Yoon, Y.E.; Lee, Y.B. Veterinary antibiotics (VAs) contamination as a global agro-ecological issue: A critical view. *Agric. Ecosyst. Environ.* **2018**, *257*, 47–59. [CrossRef]
80. Bedford, M. Removal of antibiotic growth promoters from poultry diets: Implications and strategies to minimise subsequent problems. *Worlds Poult. Sci. J.* **2000**, *56*, 347–365. [CrossRef]
81. Schuijt, T.J.; van der Poll, T.; de Vos, W.M.; Wiersinga, W.J. The intestinal microbiota and host immune interactions in the critically ill. *Trends Microbiol.* **2013**, *21*, 221–229. [CrossRef] [PubMed]
82. Davies, J.; Davies, D. Origins and evolution of antibiotic resistance. *Microbiol. Mol. Biol. Rev.* **2010**, *74*, 417–433. [CrossRef]
83. Zhang, F.S.; Xie, Y.F.; Li, X.W.; Wang, D.Y.; Yang, L.S.; Nie, Z.Q. Accumulation of steroid hormones in soil and its adjacent aquatic environment from a typical intensive vegetable cultivation of North China. *Sci. Total Environ.* **2015**, *538*, 423–430. [CrossRef] [PubMed]
84. Donk, S.v.; Biswas, S.; Kranz, W.; Snow, D.; Bartelt-Hunt, S.; Mader, T.; Shapiro, C.; Shelton, D.; Tarkalson, D.; Zhang, T.; et al. Transport of steroid hormones in the vadose zoneafter land application of beef cattle manure. *Biol. Syst. Eng. Pap. Publ.* **2013**, *56*, 1327–1338.
85. Bartelt-Hunt, S.; Snow, D.D.; Damon-Powell, T.; Miesbach, D. Occurrence of steroid hormones and antibiotics in shallow groundwater impacted by livestock waste control facilities. *J. Contam. Hydrol.* **2011**, *123*, 94–103. [CrossRef]
86. Torres, N.H.; Aguiar, M.M.; Ferreira, L.F.R.; Américo, J.H.P.; Machado, Â.M.; Cavalcanti, E.B.; Tornisielo, V.L. Detection of hormones in surface and drinking water in Brazil by LC-ESI-MS/MS and ecotoxicological assessment with Daphnia magna. *Environ. Monit. Assess.* **2015**, *187*, 379. [CrossRef]
87. Pauwels, B.; Noppe, H.; De Brabander, H.; Verstraete, W. Comparison of steroid hormone concentrations in domestic and hospital wastewater treatment plants. *J. Environ. Eng.* **2008**, *134*, 933–936. [CrossRef]
88. Servos, M.R.; Bennie, D.T.; Burnison, B.K.; Jurkovic, A.; McInnis, R.; Neheli, T.; Schnell, A.; Seto, P.; Smyth, S.A.; Ternes, T.A. Distribution of estrogens, 17β-estradiol and estrone, in Canadian municipal wastewater treatment plants. *Sci. Total Environ.* **2005**, *336*, 155–170. [CrossRef] [PubMed]
89. Andersen, H.; Siegrist, H.; Halling-Sørensen, B.; Ternes, T.A. Fate of estrogens in a municipal sewage treatment plant. *Environ. Sci. Technol.* **2003**, *37*, 4021–4026. [CrossRef] [PubMed]
90. Baronti, C.; Curini, R.; D'Ascenzo, G.; Di Corcia, A.; Gentili, A.; Samperi, R. Monitoring natural and synthetic estrogens at activated sludge sewage treatment plants and in a receiving river water. *Environ. Sci. Technol.* **2000**, *34*, 5059–5066. [CrossRef]
91. Sellin, M.K.; Snow, D.D.; Akerly, D.L.; Kolok, A.S. Estrogenic compounds downstream from three small cities in Eastern Nebraska: Occurrence and biological effect. *J. Am. Water Resour. Assoc.* **2009**, *45*, 14–21. [CrossRef]
92. Yarahmadi, H.; Duy, S.V.; Hachad, M.; Dorner, S.; Sauvé, S.; Prévost, M. Seasonal variations of steroid hormones released by wastewater treatment plants to river water and sediments: Distribution between particulate and dissolved phases. *Sci. Total Environ.* **2018**, *635*, 144–155. [CrossRef] [PubMed]
93. Zheng, W.; Wiles, K.N.; Holm, N.; Deppe, N.A.; Shipley, C.R. Uptake, Translocation, and Accumulation of Pharmaceutical and Hormone Contaminants in Vegetables. In *ACS Symposium Series*; American Chemical Society: Washington, DC, USA, 2014; Volume 1171, pp. 167–181.
94. Adeel, M.; Song, X.; Wang, Y.; Francis, D.; Yang, Y. Environmental impact of estrogens on human, animal and plant life: A critical review. *Environ. Int.* **2017**, *99*, 107–119. [CrossRef]
95. Rose, S.C.; Carter, A.D. Agrochemical leaching and water contamination. In *Conservation Agriculture*; Springer: Dordrecht, The Netherlands, 2003; pp. 417–424.
96. Jimoh, O.D.; Ayodeji, M.A.; Mohammed, B. Effects of agrochemicals on surface waters and groundwaters in the Tunga-Kawo (Nigeria) irrigation scheme. *Hydrol. Sci. J.* **2003**, *48*, 1013–1023. [CrossRef]
97. Yu, Y.; Hu, S.; Yang, Y.; Zhao, X.; Xue, J.; Zhang, J.; Gao, S.; Yang, A. Successive monitoring surveys of selected banned and restricted pesticide residues in vegetables from the northwest region of China from 2011 to 2013. *BMC Public Health* **2017**, *18*, 91. [CrossRef]
98. Retention, uptake, and translocation of agrochemicals in plants. In *ACS Symposium Series*; Myung, K.; Satchivi, N.M.; Kingston, C.K. (Eds.) American Chemical Society: Washington, DC, USA, 2014; Volume 1171, ISBN 0-8412-2972-4.
99. Juraske, R.; Castells, F.; Vijay, A.; Muñoz, P.; Antón, A. Uptake and persistence of pesticides in plants: Measurements and model estimates for imidacloprid after foliar and soil application. *J. Hazard. Mater.* **2009**, *165*, 683–689. [CrossRef]

100. Elmore, S.A.; Boorman, G.A. Environmental toxicologic pathology and human health. *Haschek Rousseaux's Handb. Toxicol. Pathol.* **2013**, 1029–1049. [CrossRef]
101. US Protection Agency. *Cyanobacteria and Cyanotoxins: Information for Drinking Water Systems*; EPA-810F11001; USEPA: Washington, DC, USA, 2014.
102. Crush, J.R.; Briggs, L.R.; Sprosen, J.M.; Nichols, S.N. Effect of irrigation with lake water containing microcystins on microcystin content and growth of ryegrass, clover, rape, and lettuce. *Environ. Toxicol.* **2008**, *23*, 246–252. [CrossRef] [PubMed]
103. Corbel, S.; Mougin, C.; Bouaïcha, N. Cyanobacterial toxins: Modes of actions, fate in aquatic and soil ecosystems, phytotoxicity and bioaccumulation in agricultural crops. *Chemosphere* **2014**, *96*, 1–15. [CrossRef] [PubMed]
104. Loftin, K.A.; Graham, J.L.; Hilborn, E.D.; Lehmann, S.C.; Meyer, M.T.; Dietze, J.E.; Griffith, C.B. Cyanotoxins in inland lakes of the United States: Occurrence and potential recreational health risks in the EPA National Lakes Assessment 2007. *Harmful Algae* **2016**, *56*, 77–90. [CrossRef] [PubMed]
105. Al-Sammak, M.A.; Hoagland, K.D.; Cassada, D.; Snow, D.D. Co-occurrence of the cyanotoxins BMAA, DABA and anatoxin-a in Nebraska reservoirs, fish, and aquatic plants. *Toxins* **2014**, *6*, 488–508. [CrossRef] [PubMed]
106. Miller, A.; Russell, C. Food crops irrigated with cyanobacteria-contaminated water: An emerging public health issue in Canada. *Environ. Heal. Rev.* **2017**, *60*, 58–63. [CrossRef]
107. Saqrane, S.; Oudra, B. CyanoHAB occurrence and water irrigation cyanotoxin contamination: Ecological impacts and potential health risks. *Toxins* **2009**, *1*, 113–122. [CrossRef] [PubMed]
108. Abeysiriwardena, N.M.; Gascoigne, S.J.L.; Anandappa, A. Algal bloom expansion increases cyanotoxin risk in food. *Yale J. Biol. Med.* **2018**, *91*, 129–142. [PubMed]
109. Al-Gabr, H.M.; Zheng, T.; Yu, X. Fungi contamination of drinking water. *Rev. Environ. Contam. Toxicol.* **2014**, *228*, 121–139. [PubMed]
110. Oliveira, B.R.; Mata, A.T.; Ferreira, J.P.; Barreto Crespo, M.T.; Pereira, V.J.; Bronze, M.R. Production of mycotoxins by filamentous fungi in untreated surface water. *Environ. Sci. Pollut. Res.* **2018**, *25*, 17519–17528. [CrossRef]
111. Kolpin, D.W.; Hoerger, C.C.; Meyer, M.T.; Wettstein, F.E.; Hubbard, L.E.; Bucheli, T.D. Phytoestrogens and mycotoxins in Iowa streams: An examination of underinvestigated compounds in agricultural basins. *J. Environ. Qual.* **2010**, *39*, 2089–2099. [CrossRef] [PubMed]
112. Rao, G.J.; Govindaraju, G.; Sivasithamparam, N.; Shanmugasundaram, E.R.B. Uptake, translocation and persistence of mycotoxins in rice seedlings. *Plant Soil* **1982**, *66*, 121–123. [CrossRef]
113. Mohammad-Hasani, F.; Mirlohi, M.; Mosharraf, L.; Hasanzade, A. Occurrence of aflatoxins in wheat flour specified for sangak bread and its reduction through fermentation and baking. *Qual. Assur. Saf. Crop. Foods* **2016**, *8*, 1–8. [CrossRef]
114. Tola, M.; Kebede, B. Occurrence, importance and control of mycotoxins: A review. *Cogent Food Agric.* **2016**. [CrossRef]
115. Tanaka, H.; Asano, T.; Schroeder, E.D.; Tchobanoglous, G. Estimating the safety of wastewater reclamation and reuse using enteric virus monitoring data. *Water Environ. Res.* **1998**, *70*, 39–51. [CrossRef]
116. Asano, T.; Cotruvo, J.A. Groundwater recharge with reclaimed municipal wastewater: Health and regulatory considerations. *Water Res.* **2004**, *38*, 1941–1951. [CrossRef] [PubMed]
117. Lothrop, N.; Bright, K.R.; Sexton, J.; Pearce-Walker, J.; Reynolds, K.A.; Verhougstraete, M.P. Optimal strategies for monitoring irrigation water quality. *Agric. Water Manag.* **2018**, *199*, 86–92. [CrossRef]
118. Jongman, M.; Chidamba, L.; Korsten, L. Bacterial biomes and potential human pathogens in irrigation water and leafy greens from different production systems described using pyrosequencing. *J. Appl. Microbiol.* **2017**, *123*, 1043–1053. [CrossRef]
119. Truchado, P.; Hernandez, N.; Gil, M.I.; Ivanek, R.; Allende, A. Correlation between *E. coli* levels and the presence of foodborne pathogens in surface irrigation water: Establishment of a sampling program. *Water Res.* **2018**, *128*, 226–233. [CrossRef]
120. Steele, M.; Odumeru, J. Irrigation water as source of foodborne pathogens on fruit and vegetables. *J. Food Prot.* **2004**, *67*, 2839–2849. [CrossRef]
121. Pachepsky, Y.; Shelton, D.R.; McLain, J.E.T.; Patel, J.; Mandrell, R.E. Irrigation waters as a source of pathogenic microorganisms in produce: A review. *Adv. Agron.* **2011**, *113*, 75–141. [CrossRef]

122. Park, S.; Szonyi, B.; Gautam, R.; Nightingale, K.; Anciso, J.; Ivanek, R. Risk factors for microbial contamination in fruits and vegetables at the preharvest level: A systematic review. *J. Food Prot.* **2012**, *75*, 2055–2081. [CrossRef] [PubMed]
123. Jongman, M.; Korsten, L. Irrigation water quality and microbial safety of leafy greens in different vegetable production systems: A review. *Food Rev. Int.* **2018**, *34*, 308–328. [CrossRef]
124. Herman, K.M.; Hall, A.J.; Gould, L.H. Outbreaks attributed to fresh leafy vegetables, United States, 1973–2012. *Epidemiol. Infect.* **2015**, *143*, 3011–3021. [CrossRef] [PubMed]
125. National Outbreak Reporting System (NORS) Dashboard | CDC. Available online: https://wwwn.cdc.gov/norsdashboard/ (accessed on 25 March 2019).
126. Multistate Outbreak of *E. coli* O157:H7 Infections Linked to Romaine Lettuce (Final Update) | Investigation Notice: Multistate Outbreak of *E. coli* O157:H7 Infections April 2018 | *E. coli* | CDC. Available online: https://www.cdc.gov/ecoli/2018/o157h7-04-18/index.html (accessed on 25 March 2019).
127. Outbreak of *E. coli* Infections Linked to Romaine Lettuce | *E. coli* Infections Linked to Romaine Lettuce | November 2018 | *E. coli* | CDC. Available online: https://www.cdc.gov/ecoli/2018/o157h7-11-18/index.html (accessed on 25 March 2019).
128. FDA Investigation of a Multistate Outbreak of Cyclospora Illnesses Linked to Fresh Express Salad Mix Served at McDonald's Ends | FDA. Available online: https://www.fda.gov/food/outbreaks-foodborne-illness/fda-investigation-multistate-outbreak-cyclospora-illnesses-linked-fresh-express-salad-mix-served#Cyclospora (accessed on 3 July 2019).
129. Olivieri, A.W.; Seto, E.; Cooper, R.C.; Cahn, M.D.; Colford, J.; Crook, J.; Debroux, J.-F.; Mandrell, R.; Suslow, T.; Tchobanoglous, G.; et al. Risk-based review of California's water-recycling criteria for agricultural irrigation. *J. Environ. Eng.* **2014**, *140*, 04014015. [CrossRef]
130. Leaman, S.; Gorny, J.; Wetherington, D.; Belkris, H. *Agricultural Water: Five Year Research Review*; Center for Produce Safety: Davis, CA, USA, 2014.
131. U.S. EPA. *Regulations Governing Agricultural Use of Municipal Wastewater and Sludge*; National Academy Press: Washington, DC, USA, 1996; ISBN 0309054796.
132. Jones, L.A.; Worobo, R.W.; Smart, C.D. UV light inactivation of human and plant pathogens in unfiltered surface irrigation water. *Appl. Environ. Microbiol.* **2014**, *80*, 849–854. [CrossRef] [PubMed]
133. Uyttendaele, M.; Jaykus, L.A.; Amoah, P.; Chiodini, A.; Cunliffe, D.; Jacxsens, L.; Holvoet, K.; Korsten, L.; Lau, M.; McClure, P.; et al. Microbial hazards in irrigation water: Standards, norms, and testing to manage use of water in fresh produce primary production. *Compr. Rev. Food Sci. Food Saf.* **2015**, *14*, 336–356. [CrossRef]
134. World Health Organization World Health Organization (WHO): Antibiotic Resistance—Fact Sheet. Available online: https://www.who.int/en/news-room/fact-sheets/detail/antibiotic-resistance (accessed on 26 June 2019).
135. Perry, J.A.; Wright, G.D. The antibiotic resistance "mobilome": Searching for the link between environment and clinic. *Front. Microbiol.* **2013**, *4*, 138. [CrossRef]
136. Marti, E.; Variatza, E.; Balcazar, J.L. The role of aquatic ecosystems as reservoirs of antibiotic resistance. *Trends Microbiol.* **2014**, *22*, 36–41. [CrossRef]
137. Bergeron, S.; Brown, R.; Homer, J.; Rehage, S.; Boopathy, R. Presence of antibiotic resistance genes in different salinity gradients of freshwater to saltwater marshes in southeast Louisiana, USA. *Int. Biodeterior. Biodegrad.* **2016**, *113*, 80–87. [CrossRef]
138. Pepper, I.; Brooks, J.P.; Gerba, C.P. Antibiotic resistant bacteria in municipal wastes: Is there reason for concern? *Environ. Sci. Technol.* **2018**, *52*, 3949–3959. [CrossRef] [PubMed]
139. Zhang, X.X.; Zhang, T.; Fang, H.H.P. Antibiotic resistance genes in water environment. *Appl. Microbiol. Biotechnol.* **2009**, *82*, 397–414. [CrossRef] [PubMed]
140. Fahrenfeld, N.; Ma, Y.; O'Brien, M.; Pruden, A. Reclaimed water as a reservoir of antibiotic resistance genes: Distribution system and irrigation implications. *Front. Microbiol.* **2013**, *4*, 130. [CrossRef] [PubMed]
141. Aslan, A.; Cole, Z.; Bhattacharya, A.; Oyibo, O.; Aslan, A.; Cole, Z.; Bhattacharya, A.; Oyibo, O. Presence of antibiotic-resistant *Escherichia coli* in wastewater treatment plant effluents utilized as water reuse for irrigation. *Water* **2018**, *10*, 805. [CrossRef]
142. Gekenidis, M.-T.; Qi, W.; Hummerjohann, J.; Zbinden, R.; Walsh, F.; Drissner, D. Antibiotic-resistant indicator bacteria in irrigation water: High prevalence of extended-spectrum beta-lactamase (ESBL)-producing Escherichia coli. *PLoS ONE* **2018**, *13*, e0207857. [CrossRef] [PubMed]

143. Farkas, A.; Bocoş, B.; Butiuc-Keul, A. Antibiotic resistance and intI1 carriage in waterborne enterobacteriaceae. *Water Air Soil Pollut.* **2016**, *227*, 251. [CrossRef]
144. Olaimat, A.N.; Holley, R.A. Factors influencing the microbial safety of fresh produce: A review. *Food Microbiol.* **2012**, *32*, 1–19. [CrossRef]
145. Vital, P.G.; Zara, E.S.; Paraoan, C.E.M.; Dimasupil, M.A.Z.; Abello, J.J.M.; Santos, I.T.G.; Rivera, W.L. Antibiotic resistance and extended-spectrum beta-lactamase production of escherichia coli isolated from irrigationwaters in selected urban farms in Metro Manila, Philippines. *Water* **2018**, *10*, 548. [CrossRef]
146. Fipps, G. *Irrigation Water Quality Standards and Salinity Management Strategies*; Texas Agriculture Extension Service 7-96 Rev edition; Texas A&M University System: College Station, TX, USA, 1996.
147. Stoner, J.D. *Water-Quality Indices for Specific Water Uses*; Circulur 770; United States Department of the Interior, Geological Survey: Arlington, VA, USA, 1978. [CrossRef]
148. Ayotte, J.D.; Gronberg, J.A.M.; Apodaca, L.E. *Trace Elements and Radon in Groundwater across the United States, 1992–2003*; Scientific Investigations Report 2011–5059; U.S. Geological Survey: Reston, VA, USA, 2011; Volume i–xi, pp. 1–115.
149. Welch, A.H.; Westjohn, D.B.; Helsel, D.R.; Wanty, R.B. Arsenic in ground water of the United States: Occurrence and geochemistry. *Ground Water* **2000**, *38*, 589–604. [CrossRef]
150. Rodriǵuez-Lado, L.; Sun, G.; Berg, M.; Zhang, Q.; Xue, H.; Zheng, Q.; Johnson, C.A. Groundwater arsenic contamination throughout China. *Science* **2013**, *341*, 866–868. [CrossRef]
151. Selck, B.J.; Carling, G.T.; Kirby, S.M.; Hansen, N.C.; Bickmore, B.R.; Tingey, D.G.; Rey, K.; Wallace, J.; Jordan, J.L. Investigating anthropogenic and geogenic sources of groundwater contamination in a semi-arid alluvial basin, Goshen Valley, UT, USA. *Water Air Soil Pollut.* **2018**, *229*, 186. [CrossRef]
152. Signes-Pastor, A.J.; Mitra, K.; Sarkhel, S.; Hobbes, M.; Burló, F.; De Groot, W.T.; Carbonell-Barrachina, A.A. Arsenic speciation in food and estimation of the dietary intake of inorganic arsenic in a rural village of West Bengal, India. *J. Agric. Food Chem.* **2008**, *56*, 9469–9474. [CrossRef] [PubMed]
153. Erban, L.E.; Gorelick, S.M.; Fendorf, S. Arsenic in the multi-aquifer system of the Mekong Delta, Vietnam: Analysis of large-scale spatial trends and controlling factors. *Environ. Sci. Technol.* **2014**, *48*, 6081–6088. [CrossRef] [PubMed]
154. Huhmann, B.; Harvey, C.F.; Uddin, A.; Choudhury, I.; Ahmed, K.M.; Duxbury, J.M.; Ellis, T.; van Geen, A. Inversion of high-arsenic soil for improved rice yield in Bangladesh. *Environ. Sci. Technol.* **2019**, *53*, 3410–3418. [CrossRef] [PubMed]
155. Pick, T. *Assessing Water Quality for Human Consumption, Agriculture, and Aquatic Life Uses*; United States Department of Agriculture: Washington, DC, USA, 2011.
156. U.S. EPA. *Guidelines for Water Reuse 2012*; US Agency for International Development: Washington, DC, USA, 2012; p. 643.
157. Ayers, R.S.; Westcot, D.W. *Water Quality for Agriculture*; Food and Agricultural Organization, United Nations: Rome, Italy, 1976.
158. Kandakji, T.; Udeigwe, T.K.; Dixon, R.; Li, L. Groundwater-induced alterations in elemental concentration and interactions in semi-arid soils of the Southern High Plains, USA. *Environ. Monit. Assess.* **2015**, *187*, 665. [CrossRef] [PubMed]
159. Scanlon, B.R.; Nicot, J.P.; Reedy, R.C.; Kurtzman, D.; Mukherjee, A.; Nordstrom, D.K. Elevated naturally occurring arsenic in a semiarid oxidizing system, Southern High Plains aquifer, Texas, USA. *Appl. Geochem.* **2009**, *24*, 2061–2071. [CrossRef]
160. Malan, M.; Müller, F.; Cyster, L.; Raitt, L.; Aalbers, J. Heavy metals in the irrigation water, soils and vegetables in the Philippi horticultural area in the Western Cape Province of South Africa. *Environ. Monit. Assess.* **2015**, *187*, 1–8. [CrossRef] [PubMed]
161. Gupta, M.; Gupta, S. An overview of selenium uptake, metabolism, and toxicity in Plants. *Front. Plant Sci.* **2017**, *7*, 1–14. [CrossRef]
162. Hakonson-Hayes, A.C.; Fresquez, P.R.; Whicker, F.W. Assessing potential risks from exposure to natural uranium in well water. *J. Environ. Radioact.* **2002**, *59*, 29–40. [CrossRef]
163. Islam, S.M.A.; Fukushi, K.; Yamamoto, K. Contamination of agricultural soil by arsenic containing irrigation water in Bangladesh: Overview of status and a proposal for novel biological remediation. *WIT Trans. Biomed. Heal.* **2006**, *6*, 295–316. [CrossRef]

164. Jeambrun, M.; Pourcelot, L.; Mercat, C.; Boulet, B.; Pelt, E.; Chabaux, F.; Cagnat, X.; Gauthier-Lafaye, F. Potential sources affecting the activity concentrations of 238U, 235U, 232Th and some decay products in lettuce and wheat samples. *J. Environ. Monit.* **2012**, *14*, 2902–2912. [CrossRef] [PubMed]
165. Bañuelos, G.S.; Ajwa, H.A.; Caceres, L.; Dyer, D. Germination responses and boron accumulation in germplasm from Chile and the United States grown with boron-enriched water. *Ecotoxicol. Environ. Saf.* **1999**, *43*, 62–67. [CrossRef] [PubMed]
166. Rhoades, J.D.; Bingham, F.T.; Letey, J.; Hoffman, G.J.; Dedrick, A.R.; Pinter, P.J.; Replogle, J.A. Use of saline drainage water for irrigation: Imperial Valley study. *Agric. Water Manag.* **1989**, *16*, 25–36. [CrossRef]
167. Hopkins, B.G.; Horneck, D.A.; Stevens, R.G.; Ellsworth, J.W.; Sullivan, D.M. *Managing Irrigation Water Quality for Crop Production in the Pacific Northwest*; PNW597-E; USDA: Washington, DC, USA, 2007. Available online: https://catalog.extension.oregonstate.edu/sites/catalog/files/project/pdf/pnw597.pdf (accessed on 19 March 2019).
168. Hem, J.D. Chemistry and occurrence of cadmium and zinc in surface water and groundwater cadmium is reported of compounds in rice. *Water Resour. Res.* **1972**, *8*, 661–679. [CrossRef]
169. Alexakis, D. Assessment of water quality in the Messolonghi-Etoliko and Neochorio region (West Greece) using hydrochemical and statistical analysis methods. *Environ. Monit. Assess.* **2011**, *182*, 397–413. [CrossRef] [PubMed]
170. Irmak, S. Copper correlation of irrigation water, soils and plants in the Cukurova Region of Turkey. *Int. J. Soil Sci.* **2009**, *4*, 46–56. [CrossRef]
171. Manning, A.H.; Mills, C.T.; Morrison, J.M.; Ball, L.B. Insights into controls on hexavalent chromium in groundwater provided by environmental tracers, Sacramento Valley, California, USA. *Appl. Geochem.* **2015**, *62*, 186–199. [CrossRef]
172. Stasinos, S.; Zabetakis, I. The uptake of nickel and chromium from irrigation water by potatoes, carrots and onions. *Ecotoxicol. Environ. Saf.* **2013**, *91*, 122–128. [CrossRef]
173. USEPA. *2018 Edition of the Drinking Water Standards and Health Advisories*; USEPA: Washington, DC, USA, 2018.
174. Gurel, M.; Iskender, G.; Ovez, S.; Arslan-Alaton, I.; Tanik, A.; Orhon, D. A global overview of treated wastewater guidelines and standards for agricultural reuse. *Fresenius Environ. Bull.* **2007**, *16*, 590–595.
175. Pescod, M.B. *Wastewater Treatment and Use in Agriculture*; Food and Agricultural Organization, United Nations: Rome, Italy, 1992; ISBN 9251031355.
176. US EPA. National Recommended Water Quality Criteria—Aquatic Life Criteria Table. Available online: https://www.epa.gov/wqc/national-recommended-water-quality-criteria-aquatic-life-criteria-table (accessed on 27 December 2018).
177. Groschen, G.E.; Arnold, T.L.; Morrow, W.S.; Warner, K.L. *Occurrence and Distribution of Iron, Manganese, and Selected Trace elements in Ground Water in the Glacial Aquifer System of the Northern United States*; U.S. Geological Survey Scientific Investigations Report 2009–5006; USGS: Reston, VA, USA, 2008.
178. Bundschuh, J.; Nath, B.; Bhattacharya, P.; Liu, C.W.; Armienta, M.A.; Moreno López, M.V.; Lopez, D.L.; Jean, J.S.; Cornejo, L.; Lauer Macedo, L.F.; et al. Arsenic in the human food chain: The Latin American perspective. *Sci. Total Environ.* **2012**, *429*, 92–106. [CrossRef]
179. Meharg, A.A.; Rahman, M. Arsenic contamination of Bangladesh paddy field soils: Implications for rice contribution to arsenic consumption. *Environ. Sci. Technol.* **2003**, *37*, 229–234. [CrossRef] [PubMed]
180. Finnegan, P.M.; Chen, W. Arsenic toxicity: The effects on plant metabolism. *Front. Physiol.* **2012**, *3*, 182. [CrossRef] [PubMed]
181. Kim, J.Y.; Davis, A.P.; Kim, K.W. Stabilization of available arsenic in highly contaminated mine tailings using iron. *Environ. Sci. Technol.* **2003**, *37*, 189–195. [CrossRef] [PubMed]
182. Bakhat, H.F.; Zia, Z.; Fahad, S.; Abbas, S.; Hammad, H.M.; Shahzad, A.N.; Abbas, F.; Alharby, H.; Shahid, M. Arsenic uptake, accumulation and toxicity in rice plants: Possible remedies for its detoxification: A review. *Environ. Sci. Pollut. Res.* **2017**, *24*, 9142–9158. [CrossRef] [PubMed]
183. Arslan, B.; Djamgoz, M.B.A.; Akün, E. ARSENIC: A review on exposure pathways, accumulation, mobility and transmission into the human food chain. *Rev. Environ. Contam. Toxicol.* **2017**, *243*, 27–51. [PubMed]
184. Zhao, F.J.; Ma, J.F.; Meharg, A.A.; McGrath, S.P. Arsenic uptake and metabolism in plants. *New Phytol.* **2009**, *181*, 777–794. [CrossRef] [PubMed]
185. Zhao, F.-J.; McGrath, S.P.; Meharg, A.A. Arsenic as a food chain contaminant: Mechanisms of plant uptake and metabolism and mitigation strategies. *Annu. Rev. Plant Biol.* **2010**, *61*, 535–559. [CrossRef] [PubMed]

186. Chakraborty, S.; Alam, M.O.; Bhattacharya, T.; Singh, Y.N. Arsenic accumulation in food crops: A potential threat in Bengal Delta Plain. *Water Qual. Expo. Heal.* **2014**, *6*, 233–246. [CrossRef]
187. Brammer, H. Mitigation of arsenic contamination in irrigated paddy soils in South and South-East Asia. *Environ. Int.* **2009**, *35*, 856–863. [CrossRef]
188. Polizzotto, M.L.; Birgand, F.; Badruzzaman, A.B.M.; Ali, M.A. Amending irrigation channels with jute-mesh structures to decrease arsenic loading to rice fields in Bangladesh. *Ecol. Eng.* **2015**, *74*, 101–106. [CrossRef]
189. Winkel, L.H.E.; Johnson, C.A.; Lenz, M.; Grundl, T.; Leupin, O.X.; Amini, M.; Charlet, L. Environmental selenium research: From microscopic processes to global understanding. *Environ. Sci. Technol.* **2012**, *46*, 571–579. [CrossRef] [PubMed]
190. USGS. *Geochemical and Mineralogical Data for Soils of the Conterminous United States*; USGS: Reston, VA, USA, 2014.
191. Wang, J.; Zhou, C.; Xiao, X.; Xie, Y.; Zhu, L.; Ma, Z. Enhanced iron and selenium uptake in plants by volatile emissions of *Bacillus amyloliquefaciens* (BF06). *Appl. Sci.* **2017**, *7*, 85. [CrossRef]
192. Mitchell, N.; Pérez-Sánchez, D.; Thorne, M.C. A review of the behaviour of U-238 series radionuclides in soils and plants. *J. Radiol. Prot.* **2013**, *33*, R17–R48. [CrossRef] [PubMed]
193. Soudek, P.; Petrova, T.; Benesova, D.; Dvorakova, M.; Vanek, T. Uranium uptake by hydroponically cultivated crop plants. *J. Environ. Radioact.* **2011**, *102*, 598–604. [CrossRef] [PubMed]
194. Boghi, A.; Roose, T.; Kirk, G.J.D. A model of uranium uptake by plant roots allowing for root-induced changes in the soil. *Environ. Sci. Technol.* **2018**, *52*, 3536–3545. [CrossRef] [PubMed]
195. Hayes, A.C.; Fresquez, P.R.; Whicker, W.F. *Uranium Uptake Study, Nambe, New Mexico: Source Document*; Los Alamos National Laboratory: Los Alamos, NM, USA, 2000.
196. Neves, O.; Abreu, M.M. Are uranium-contaminated soil and irrigation water a risk for human vegetables consumers? A study case with *Solanum tuberosum* L., *Phaseolus vulgaris* L. and *Lactuca sativa* L. *Ecotoxicology* **2009**, *18*, 1130–1136. [CrossRef] [PubMed]
197. Neves, M.O.; Abreu, M.M.; Figueiredo, V. Uranium in vegetable foodstuffs: Should residents near the Cunha Baixa uranium mine site (Central Northern Portugal) be concerned? *Environ. Geochem. Health* **2012**, *34*, 181–189. [CrossRef]
198. Gomes, M.A.d.C.; Hauser-Davis, R.A.; Suzuki, M.S.; Vitória, A.P. Plant chromium uptake and transport, physiological effects and recent advances in molecular investigations. *Ecotoxicol. Environ. Saf.* **2017**, *140*, 55–64. [CrossRef]
199. Song, Y.; Jin, L.; Wang, X. Cadmium absorption and transportation pathways in plants. *Int. J. Phytoremediation* **2017**, *19*, 133–141. [CrossRef]
200. Amrhein, C.; Mosher, P.A.; Brown, A.D. The effects of redox on Mo, U, B, V, and As solubility in evaporation pond soils. *Soil Sci.* **1993**, *155*, 249–255. [CrossRef]
201. Metals-U.S. Food & Drug Administration. Available online: https://www.fda.gov/Food/FoodborneIllnessContaminants/Metals/default.htm (accessed on 18 February 2019).
202. Lapworth, D.J.; Stolpe, B.; Williams, P.J.; Gooddy, D.C.; Lead, J.R. Characterization of suboxic groundwater colloids using a multi-method approach. *Environ. Sci. Technol.* **2013**, *47*, 2554–2561. [CrossRef] [PubMed]
203. Praetorius, A.; Scheringer, M.; Hungerbühler, K. Development of environmental fate models for engineered nanoparticles—A case study of TiO_2 nanoparticles in the Rhine River. *Environ. Sci. Technol.* **2012**, *46*, 6705–6713. [CrossRef] [PubMed]
204. González-Gálvez, D.; Janer, G.; Vilar, G.; Vílchez, A.; Vázquez-Campos, S. *The Life Cycle of Engineered Nanoparticles*; Springer: Cham, Switzerland, 2017; pp. 41–69.
205. Thomé, A.; Reddy, K.R.; Reginatto, C.; Cecchin, I. Review of nanotechnology for soil and groundwater remediation: Brazilian perspectives. *Water Air Soil Pollut.* **2015**, *226*, 121. [CrossRef]
206. Gehrke, I.; Geiser, A.; Somborn-Schulz, A. Innovations in nanotechnology for water treatment. *Nanotechnol. Sci. Appl.* **2015**, *8*, 1–17. [CrossRef] [PubMed]
207. Troester, M.; Brauch, H.-J.; Hofmann, T. Vulnerability of drinking water supplies to engineered nanoparticles. *Water Res.* **2016**, *96*, 255–279. [CrossRef] [PubMed]
208. Alimi, O.S.; Farner Budarz, J.; Hernandez, L.M.; Tufenkji, N. Microplastics and nanoplastics in aquatic environments: Aggregation, deposition, and enhanced contaminant transport. *Environ. Sci. Technol.* **2018**, *52*, 1704–1724. [CrossRef] [PubMed]
209. Lehner, R.; Weder, C.; Petri-Fink, A.; Rothen-Rutishauser, B. Emergence of nanoplastic in the environment and possible impact on human health. *Environ. Sci. Technol.* **2019**, *53*, 1748–1765. [CrossRef] [PubMed]

210. Oberdörster, G.; Oberdörster, E.; Oberdörster, J. Nanotoxicology: An emerging discipline evolving from studies of ultrafine particles. *Environ. Health Perspect.* **2005**, *113*, 823–839. [CrossRef]
211. Ganguly, P.; Breen, A.; Pillai, S.C. Toxicity of nanomaterials: Exposure, pathways, assessment, and recent advances. *ACS Biomater. Sci. Eng.* **2018**, *4*, 2237–2275. [CrossRef]
212. Environmental Protection Agency. *Technical Fact Sheet—Nanomaterials*; USEPA: Washington, DC, USA, 2017. Available online: https://www.epa.gov/sites/production/files/2014-03/documents/ffrrofactsheet_emergingcontaminant_nanomaterials_jan2014_final.pdf (accessed on 2 February 2019).
213. Inshakova, E.; Inshakov, O. World market for nanomaterials: Structure and trends. *MATEC Web Conf.* **2017**. [CrossRef]
214. Hyung, H.; Kim, J.-H. Dispersion of C60 in natural water and removal by conventional drinking water treatment processes. *Water Res.* **2009**, *43*, 2463–2470. [CrossRef] [PubMed]
215. Zhang, Y.; Chen, Y.; Westerhoff, P.; Hristovski, K.; Crittenden, J.C. Stability of commercial metal oxide nanoparticles in water. *Water Res.* **2008**, *42*, 2204–2212. [CrossRef] [PubMed]
216. Hedberg, J.; Blomberg, E.; Odnevall Wallinder, I. In the search for nanospecific effects of dissolution of metallic nanoparticles at freshwater-like conditions: A critical review. *Environ. Sci. Technol.* **2019**, *53*, 4030–4044. [CrossRef] [PubMed]
217. Gottschalk, F.; Sonderer, T.; Scholz, R.W.; Nowack, B. Modeled environmental concentrations of engineered nanomaterials (TiO_2, ZnO, Ag, CNT, Fullerenes) for different regions. *Environ. Sci. Technol.* **2009**, *43*, 9216–9222. [CrossRef] [PubMed]
218. Blaser, S.A.; Scheringer, M.; MacLeod, M.; Hungerbühler, K. Estimation of cumulative aquatic exposure and risk due to silver: Contribution of nano-functionalized plastics and textiles. *Sci. Total Environ.* **2008**, *390*, 396–409. [CrossRef] [PubMed]
219. Brar, S.K.; Verma, M.; Tyagi, R.D.; Surampalli, R.Y. Engineered nanoparticles in wastewater and wastewater sludge—Evidence and impacts. *Waste Manag.* **2010**, *30*, 504–520. [CrossRef] [PubMed]
220. Baalousha, M.; Yang, Y.; Vance, M.E.; Colman, B.P.; McNeal, S.; Xu, J.; Blaszczak, J.; Steele, M.; Bernhardt, E.; Hochella, M.F. Outdoor urban nanomaterials: The emergence of a new, integrated, and critical field of study. *Sci. Total Environ.* **2016**, *557–558*, 740–753. [CrossRef] [PubMed]
221. Min Park, C.; Hoon Chu, K.; Her, N.; Jang, M.; Baalousha, M.; Heo, J.; Yoon, Y. Occurrence and removal of engineered nanoparticles in drinking water treatment and wastewater treatment processes. *Sep. Purif. Rev.* **2017**, *46*, 255–272. [CrossRef]
222. Kaphle, A.; Navya, P.N.; Umapathi, A.; Daima, H.K. Nanomaterials for agriculture, food and environment: Applications, toxicity and regulation. *Environ. Chem. Lett.* **2018**, *16*, 43–58. [CrossRef]
223. Pourzahedi, L.; Pandorf, M.; Ravikumar, D.; Zimmerman, J.B.; Seager, T.P.; Theis, T.L.; Westerhoff, P.; Gilbertson, L.M.; Lowry, G.V. Life cycle considerations of nano-enabled agrochemicals: Are today's tools up to the task? *Environ. Sci. Nano* **2018**, *5*, 1057–1069. [CrossRef]
224. Martínez-Ballesta, M.C.; Zapata, L.; Chalbi, N.; Carvajal, M. Multiwalled carbon nanotubes enter broccoli cells enhancing growth and water uptake of plants exposed to salinity. *J. Nanobiotechnol.* **2016**, *14*, 42. [CrossRef] [PubMed]
225. Chen, G.; Qiu, J.; Liu, Y.; Jiang, R.; Cai, S.; Liu, Y.; Zhu, F.; Zeng, F.; Luan, T.; Ouyang, G. Carbon nanotubes act as contaminant carriers and translocate within plants. *Sci. Rep.* **2015**, *5*, 15682. [CrossRef] [PubMed]
226. Vithanage, M.; Seneviratne, M.; Ahmad, M.; Sarkar, B.; Ok, Y.S. Contrasting effects of engineered carbon nanotubes on plants: A review. *Environ. Geochem. Health* **2017**, *39*, 1421–1439. [CrossRef] [PubMed]
227. Lv, J.; Christie, P.; Zhang, S. Uptake, translocation, and transformation of metal-based nanoparticles in plants: Recent advances and methodological challenges. *Environ. Sci. Nano* **2019**, *6*, 41–59. [CrossRef]
228. Larue, C.; Khodja, H.; Herlin-Boime, N.; Brisset, F.; Flank, A.M.; Fayard, B.; Chaillou, S.; Carrière, M. Investigation of titanium dioxide nanoparticles toxicity and uptake by plants. *J. Phys. Conf. Ser.* **2011**, *304*, 012057. [CrossRef]
229. Siddiqi, K.S.; Husen, A. Plant response to engineered metal oxide nanoparticles. *Nanoscale Res. Lett.* **2017**, *12*, 92. [CrossRef] [PubMed]
230. Rastogi, A.; Zivcak, M.; Sytar, O.; Kalaji, H.M.; He, X.; Mbarki, S.; Brestic, M. Impact of metal and metal oxide nanoparticles on plant: A critical review. *Front. Chem.* **2017**, *5*, 78. [CrossRef]
231. Zhu, H.; Han, J.; Xiao, J.Q.; Jin, Y. Uptake, translocation, and accumulation of manufactured iron oxide nanoparticles by pumpkin plants. *J. Environ. Monit.* **2008**, *10*, 713–717. [CrossRef]

232. Tripathi, D.K.; Tripathi, A.; Shweta; Singh, S.; Singh, Y.; Vishwakarma, K.; Yadav, G.; Sharma, S.; Singh, V.K.; Mishra, R.K.; et al. Uptake, Accumulation and toxicity of silver nanoparticle in autotrophic plants, and heterotrophic microbes: A concentric review. *Front. Microbiol.* **2017**, *8*, 7. [CrossRef]
233. Dimkpa, C.O.; Bindraban, P.S. Nanofertilizers: New products for the industry? *J. Agric. Food Chem.* **2018**, *66*, 6462–6473. [CrossRef]
234. Rui, M.; Ma, C.; Hao, Y.; Guo, J.; Rui, Y.; Tang, X.; Zhao, Q.; Fan, X.; Zhang, Z.; Hou, T.; et al. Iron oxide nanoparticles as a potential iron fertilizer for peanut (*Arachis hypogaea*). *Front. Plant Sci.* **2016**, *7*, 815. [CrossRef] [PubMed]
235. Motyka, O.; Štrbová, K.; Olšovská, E.; Seidlerová, J. Influence of Nano-ZnO Exposure to Plants on L-Ascorbic Acid Levels: Indication of Nanoparticle-Induced Oxidative Stress. *J. Nanosci. Nanotechnol.* **2019**, *19*, 3019–3023. [CrossRef] [PubMed]
236. Liu, J.; Williams, P.C.; Goodson, B.M.; Geisler-Lee, J.; Fakharifar, M.; Gemeinhardt, M.E. TiO_2 nanoparticles in irrigation water mitigate impacts of aged Ag nanoparticles on soil microorganisms, Arabidopsis thaliana plants, and Eisenia fetida earthworms. *Environ. Res.* **2019**, *172*, 202–215. [CrossRef] [PubMed]
237. Hu, G.; Cao, J. Occurrence and significance of natural ore-related Ag nanoparticles in groundwater systems. *Chem. Geol.* **2019**, *515*, 9–21. [CrossRef]
238. Céspedes, C.; Yeo, M.-K. Life cycle assessment of a celery paddy macrocosm exposed to manufactured Nano-TiO_2. *Toxicol. Environ. Health Sci.* **2018**, *10*, 288–296. [CrossRef]
239. Wu, F.; Zhou, Z.; Hicks, A.L. Life cycle impact of titanium dioxide nanoparticle synthesis through physical, chemical, and biological routes. *Environ. Sci. Technol.* **2019**, *53*, 4078–4087. [CrossRef] [PubMed]
240. Solanki, P.; Bhargava, A.; Chhipa, H.; Jain, N.; Panwar, J. Nano-fertilizers and their smart delivery system. In *Nanotechnologies in Food and Agriculture*; Springer International Publishing: Cham, Switzerland, 2015; pp. 81–101.
241. Rai, M.; Ribeiro, C.; Mattoso, L.; Duran, N. Nanotechnologies in food and agriculture. *Nanotechnol. Food Agric.* **2015**, 1–347. [CrossRef]
242. Kah, M.; Kookana, R.S.; Gogos, A.; Bucheli, T.D. A critical evaluation of nanopesticides and nanofertilizers against their conventional analogues. *Nat. Nanotechnol.* **2018**, *13*, 677–684. [CrossRef]
243. White, J.C.; Gardea-Torresdey, J. Achieving food security through the very small. *Nat. Nanotechnol.* **2018**, *13*, 627–629. [CrossRef]
244. Jackson, R.B.; Carpenter, S.R.; Dahm, C.N.; McKnight, D.M.; Naiman, R.J.; Postel, S.L.; Running, S.W. Water in a changing world. *Ecol. Appl.* **2001**, *11*, 1027–1045. [CrossRef]
245. RamyaPriya, R.; Elango, L. Evaluation of geogenic and anthropogenic impacts on spatio-temporal variation in quality of surface water and groundwater along Cauvery River, India. *Environ. Earth Sci.* **2018**, *77*, 1–17. [CrossRef]
246. Etheridge, A.B.; MacCoy, D.E.; Weakland, R.J. *Water-Quality and Biological Conditions in Selected Tributaries of the Lower Boise River, Southwestern Idaho, Water Years 2009—12 Scientific Investigations Report 2014—5132*; U.S. Geological Survey: Reston, VA, USA, 2014; pp. 1–70.
247. *Use of Reclaimed Water and Sludge in Food Crop Production*; National Academies Press: Washington, DC, USA, 1996; ISBN 978-0-309-05479-9.
248. Crook, J.; Surampalli, R.Y. Water reclamation and reuse criteria in the U.S. *Water Sci. Technol.* **1996**, *33*, 451–462. [CrossRef]
249. Bortolini, L.; Maucieri, C.; Borin, M. A tool for the evaluation of irrigation water quality in the arid and semi-arid regions. *Agronomy* **2018**, *8*, 23. [CrossRef]

7

Groundwater Table Effects on the Yield, Growth and Water use of Canola (*Brassica napus* L.) Plant

Hakan Kadioglu [1], Harlene Hatterman-Valenti [2], Xinhua Jia [3], Xuefeng Chu [4], Hakan Aslan [5] and Halis Simsek [2,*]

1 School of Natural Resource Sciences, North Dakota State University, Fargo, ND 58105, USA
2 Agricultural & Biosystems Engineering, North Dakota State University, Fargo, ND 58105, USA
3 Plant Science, North Dakota State University, Fargo, ND 58105, USA
4 Civil & Environmental Engineering, North Dakota State University, Fargo, ND 58105, USA
5 Farm Structure and Irrigation, Ondokuz Mayis University, 55270 Samsun, Turkey
* Correspondence: halis.simsek@ndsu.edu

Abstract: Lysimeter experiments were conducted under greenhouse conditions to investigate canola (*Brassica napus* L.) plant water use, growth, and yield parameters for three different water table depths of 30, 60, and 90 cm. Additionally, control experiments were conducted, and only irrigation was applied to these lysimeters without water table limitations. The canola plant's tolerance level to shallow groundwater was determined. Results showed that groundwater contributions to canola plant for the treatments at 30, 60, and 90 cm water table depths were 97%, 71%, and 68%, respectively, while the average grain yields of canola were 4.5, 5.3, and 6.3 gr, respectively. These results demonstrate that a 90 cm water table depth is the optimum depth for canola plants to produce a high yield with the least amount of water utilization.

Keywords: lysimeter; canola; water table; water use efficiency; root distribution; evapotranspiration

---

1. Introduction

As the global population grows, the demand for fresh water in many regions has increased dramatically. These population increases have caused more water stress for agriculture, the production of energy, industrial uses, and human consumption. Even though many countries currently have not faced a lack of water, water can no longer be considered an infinite source. Numerous regions have water use restrictions, so additional strategies to decrease the impact of water crises across the globe are needed [1–3].

One strategy for agricultural water management would encourage farmers to use shallow groundwater. Approximately 80% of available water resources in the world are being used in agricultural applications, and, therefore, the gap between adequate water availability and water needs is increasing [2]. Hence, the management of groundwater utilization in agriculture may be an acceptable alternative strategy to reduce freshwater demand. Therefore, surface water and shallow groundwater resources have become important for water demands.

Water use efficiency (WUE) is defined as a grain crop yield or total crop biomass per unit of water use [4]. Improved and well-managed WUE in agricultural water management systems is an important strategy to increase the productivity and reliability of crop yields. The consumption of groundwater is an extremely significant part of WUE. However, describing WUE for irrigation is complicated [5].

Good quality groundwater is a supplemental irrigation water source that can supply crops' water demands. When managed correctly, shallow groundwater can reduce both drainage and irrigation requirements. Some crops, such as canola (*Brassica napus* L.), soybean (*Glycine max*), and safflower (*Carthamus tinctorius*), are able to use moderate saline groundwater and could help to increase the utilization of groundwater and decrease the utilization of surface irrigation water [1,6,7]. In addition, there are obvious relationships between water table management (WTM), crop productivity, and environmental pollution. The environmental and economic benefits of WTM could decrease environmental pollution and increase crop productivity and irrigation intervals. However, WTM must be utilized correctly to supply sufficient soil moisture content to the crops [8].

The consumption of shallow groundwater as a crop water supply depends on several factors, such as groundwater table depths, groundwater availability and quality, crop species, distribution of the plant root system, weather conditions, and soil types [7,9]. The quantity and quality of groundwater are also affected by the irrigation method and management practices, as an excessive amount of irrigation water will increase groundwater utilization. It is impossible to control all these factors under field conditions because groundwater contributions are highly variable and difficult to estimate. Therefore, lysimeters are often used to evaluate a single parameter at a time [10].

Mejia et al. [8] utilized lysimeters to determine the effect of two different water table depths (50 and 75 cm) on corn and soybean grain yields. A free drainage system was installed 100 cm below the soil surface for both treatments. In the first year, corn yield was determined to be 13.8% higher with the free drainage treatment compared to the treatment without drainage at the 50 cm water table depth. However, only a 2.8% corn yield increase was observed at the 75 cm water table depth. In the second year, corn yield increases with the free drainage treatment compared to no drainage were measured as 6.6% at the 50 cm water table depth and 6.9% at the 75 cm water table depth. Similar results were observed for soybean. The authors concluded that the 75 cm water table depth with a free drainage system for corn and soybean was the most efficient water table depth.

Luo and Sophocleous [10] used lysimeters to evaluate the influence of the groundwater evaporation's contribution to winter wheat crop water use. Different water table depths, climates, and irrigation conditions were used to determine the amount of crop water use from the desired groundwater table levels. The relationship between wheat crop water use and water table depth varied. Winter wheat was supplied with 75% of crop water-use from a 100 cm groundwater depth without an irrigation application, while 3% of crop water use was supplied from the 300 cm groundwater level with three irrigation applications. The results showed that the water table contribution was affected not only by the water table depth, but also by the soil profile, rainfall, irrigation, and climatic variations.

Plant water uptake from shallow groundwater is affected by water table depth, plant salt tolerance, and plant root characteristics, the soil's hydraulic properties, the salinity level of the groundwater, and the presence of irrigation and drainage systems. Plant salt tolerance is the leading factor affecting water extraction from shallow groundwater. Each plant has a different tolerance to salinity, and plant tolerance differs in each growth stage. All the plants tend to be more susceptible to salinity in their early stages [11,12].

Fidantemiz et al. [13] used lysimeters under a controlled environment condition to determine the effect of different groundwater table levels (30, 50, 70, and 90 cm) on soybean growth. The highest grain yield and WUE results were obtained from 90 cm water table depth with 17.2 g/lys and 0.31 g/lys./c, respectively. In terms of WUE, grain yield and root distribution, both 70 and 90 cm water table depths were optimum for soybean yield in the experiments conducted without surface irrigation.

In this current study, canola plants are grown in the lysimeters. Canola can be grown with inadequate irrigation and weather conditions and, therefore, is highly adapted to cold weather conditions with insufficient water availability. High temperatures may cause abiotic stress on canola plant and influences its growth. Canola's sensitivity to high temperatures is higher in the flowering period than the podding period. During the blooming season of the canola plant, heat stress may shorten the flowering period. Two common types of canola, winter (*B. rapa*) and spring (*B. napus*) canola, can be grown in North Dakota. Although winter canola can be produced in ND and northwestern Minnesota, ND farmers mainly prefer to plant spring canola since spring canola can survive under the harsh winter condition, and its yield growth is higher than that of winter canola [14–16].

The main scope of this study was to determine an optimum shallow groundwater depth to achieve a high yield for canola plants. The lysimeter experiment was conducted to: (1) determine the optimum groundwater depth for canola growth and yield parameters for water table depths of 30, 60, and 90 cm without irrigation, (2) to quantify the amount of water consumption for water table depths of 30, 60, and 90 cm during canola growth, and (3) to determine the canola plant root distribution at water table depths of 30, 60, and 90 cm.

2. Materials and Methods

2.1. *Lysimeter Design and Preparation*

A greenhouse located in the North Dakota State University campus, Fargo, ND was used for the lysimeter study. Four treatments at 30 cm (T_{30}), 60 cm (T_{60}), and 90 cm (T_{90}) water table depths with no irrigation application and a control treatment ($T_{control}$, no water table) with a surface irrigation application were used. These three different water table depths were selected because they represent the elevated water table conditions in the fields where canola is normally grown. Each treatment had eight replications, so a total of 32 lysimeters were used. For the control treatment, 50% of the total available moisture (TAM) was considered as readily available moisture (RAM) in the soil profile. RAM is defined as the portion of the available water (field capacity minus permanent wilting point) before growth and yield are affected. RAM varies with crop and the evapotranspiration (ET) rates. According to Huffman et al. [17], 50% of the TAM for canola and a maximal ET rate of 6 mm/day was recommended. Tap water was used for both the groundwater and irrigation water sources. All the lysimeters in the greenhouse were distributed using a randomized complete block design method with eight replications.

Amber colored class bottles were used as Mariotte bottles to prevent algal growth and connected to the 24 lysimeters used for the water table depth treatments. The volume of the Mariotte bottles were 4 L, and four adjustable shelves were used to adjust the desired water table depth. The variation of the water volume in the Mariotte bottles was measured to determine the water consumption of canola. The Mariotte bottles were connected to the lysimeters from the bottom and continuously fed the lysimeters with a constant flow rate (Figure 1). The water reduction on the Mariotte bottles was monitored, and the difference was considered as the canola water consumption that supplied from the groundwater. Graduated cylinders were used for replenishment in the Mariotte bottles to obtain reliable measured water use.

2.2. *Soil Packing and Sensor Installation*

The loam soil was used to pack all the lysimeters. Bulk soil samples were obtained from an agricultural field in Fergus Falls, MN. The soil texture was classified as a loam soil based on the USDA/FAO texture classification system. The soil was then air-dried and sieved through a 2 mm screen and packed into the lysimeters. At the beginning of the study, the soil compaction problem in the lysimeters was observed, and 300 g of sand was added to 1.0 kg of the soil to deal with this problem. According to the laboratory analysis, the packed soil field capacity, readily available water, permanent wilting point, and bulk density were 0.32 cm^3/cm^3, 0.27 cm^3/cm^3, 0.21 cm^3/cm^3, and 1.14 Mg/m^3, respectively. All these parameters were measured using the combined HYPROP (Data Evaluation Software) and WP4 method [18]. Gravel (8 cm) was packed at the bottom of the lysimeters, sand (8 cm) was then packed, and finally the processed loam soil (100 cm) was used to fill the lysimeters (Figure 1). All lysimeters were packed identically. Each lysimeter's diameter, wall thickness, and height were 152.4 mm, 5 mm, and 1260 mm, respectively. The lysimeters were made of Schedule-40 PVC material. The bottoms of the lysimeters were closed with a cap and glued to prevent leaking.

In the control treatment lysimeters ($T_{control}$), three soil water potential sensors (TEROS-21, METER Group, Inc., Pullman, WA, USA) were used to determine (i) the irrigation timing and (ii) the water needed for irrigation. Water potential sensors were installed at depths of 15, 45, and 75 cm in the lysimeters. For the remaining treatments (T_{30}, T_{60}, and T_{90}), six soil water potential sensors were used and placed at the appropriate depths. One soil water potential sensor was placed at a depth of 15 cm from the top of the soil surface in the T_{30} lysimeter. Two soil water potential sensors were placed at depths of 15 and 45 cm in the T_{60} lysimeter, and three soil water potential sensors were placed at the depths of 15, 45, and 75 cm in the T_{90} lysimeter [13]. To ensure hydraulic contact between sensors and moisture in the soil, all the sensors were placed horizontally in the lysimeters. All 9 water potential sensors were plugged into two Em50G (Decagon Inc.) dataloggers, and the data recording time interval was selected as 10 min.

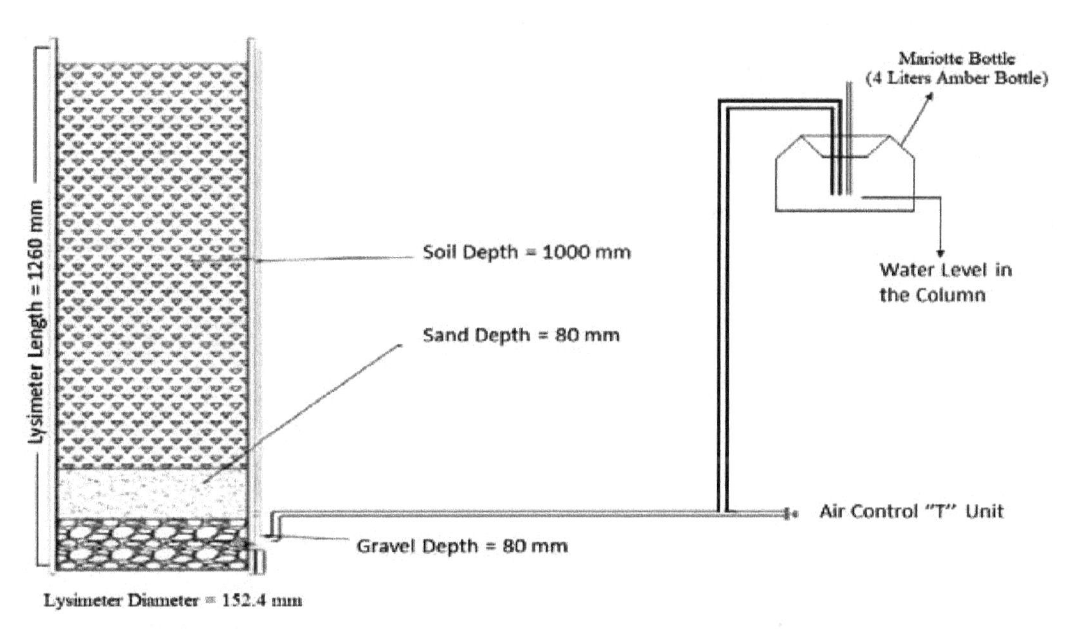

Figure 1. Schematic diagram of a lysimeter and Mariotte bottle system.

Two ETgage model E atmometers (C&M Meteorological Supply, Colorado Springs, CO, USA) were used to measure reference crop evapotranspiration (ET_0) in the greenhouse and recorded daily using HOBO Pendant Event Data Loggers (Onset Computer, Bourne, MA, USA) between 4 November 2018 (planting) and 4 February 2019 (harvesting). In addition, air temperature, barometric pressure, relative humidity, and vapor pressure were measured using an Atmos 14 sensor (Decagon Devices, Inc., Pullman, WA, USA). The device was connected to an Em50G datalogger to transfer the data to a computer.

2.3. *Planting and Harvesting of Canola*

Canola seeds (NDOLA-01) were planted on 4 November 2018 and harvested from 4 February to 10 February 2019, according to the plant harvest stages (Table 1). Ten seeds were sowed from 1 to 3 cm soil depths and thinned so that the three healthiest seedlings remained in each lysimeter. All the planted seeds were germinated in eight days. Although iron deficiency was observed at the beginning of the experiment, beneficial nematodes, supplements, and chemicals were not applied during the experiment. Similarly, no fertilizer was used in this study.

Table 1. Canola harvesting dates.

Treatments	Replications							
	1	2	3	4	5	6	7	8
$T_{control}$	5 February	5 February	6 February	6 February	5 February	5 February	5 February	5 February
T_{30}	9 February	10 February	10 February	9 February	10 February	9 February	9 February	10 February
T_{60}	7 February	6 February	6 February	7 February	6 February	7 February	7 February	7 February
T_{90}	5 February	4 February	5 February	4 February	5 February	4 February	4 February	4 February

To provide identical water curve conditions at the beginning of the experiment, all the lysimeters were filled with water to the soil surface in the lysimeters. Then, the valves at the bottom of the lysimeters were opened, and water in the lysimeters was drained. Approximately 30 h later, the valves were closed to maintain adequate moisture in the lysimeters for germination and the Mariotte bottles were connected and adjusted for the desired water table level for each lysimeter. For the control experiments, surface irrigation was applied based on the data obtained from the sensors, regardless of the germination stage in the lysimeters. Thus, starting from the first irrigation application, the irrigation timing and the amount of water needed for irrigation were determined by considering only the sensors' outcomes. Therefore, the germination stages in the columns were not monitored. As explained earlier, our goal was to maintain the packed soil field capacity at 0.32 cm^3/cm^3 and readily available water content at 0.27 cm^3/cm^3.

2.4. Calculation of Crop Water Use from Groundwater and Irrigation Water

After plant harvest, four randomly selected lysimeters from each treatment were cut vertically to determine the canola plant root's dry mass. In order to analyze the entire root distribution in each treatment, lysimeters were cut from the top through the bottom using electric saw. During the soil extraction process, three plant root depth intervals (0–30, 30–60, and 60–90 cm) were selected based on three water table depths. The soil in the lysimeters was washed, and the roots were separated gently from the soil. The roots were air-dried for 24 h before weighing to determine the root distribution and dry mass at each depth interval. Evapotranspiration in each lysimeter was calculated using Equation (1):

$$(\Delta S) = (I + Cr) - (Dp + ET) \quad (1)$$

where Cr is the water inflow due to capillarity, I is the irrigation, Dp is the deep percolation, ET is the evapotranspiration, and ΔS is the change in soil water content. Precipitation, runoff, and deep percolation were not applicable in this study since the experiments was performed in a controlled greenhouse. Irrigation was only applied to the control experiments. After evaluation of the controlled environment's conditions, the soil water balance equation was used to determine ET for each treatment (Equation (2)):

$$ET = Cr + S_1 - S_2 \quad (2)$$

where S_1 is the initial soil water storage (soil moisture) and S_2 is the final soil water storage in the lysimeters. Water reduction in the Mariotte bottles was measured every 15 days to determine the capillary water inflow in the lysimeters. The amount of water used by the canola was calculated using the soil water balance equation (Equation (1)) [19].

To determine the initial moisture conditions of the lysimeters at the beginning of the experiment, soil water potential sensors were used. After cutting the sixteen lysimeters, soil water content was measured, and the final moisture conditions of the sixteen lysimeters were determined. The soil water release curve was used to consider 50% of the total available moisture as the RAM in the soil profile of control treatment. The irrigation water depths for the lysimeters were calculated by using Equation (3) [20]:

$$d = \sum_{i=1}^{n} \frac{F_{ci} - M_{bi}}{100} \times A_{si} \times D_i \quad (3)$$

where d is the equivalent depth of water in cm, F_{ci} is the field capacity of the soil layer in percent by weight, M_{bi} is the current water content of the soil layer in percent by weight, A_{si} is the apparent specific gravity (bulk density), D_i is the depth of each soil layer, and n is the total number of soil layers.

To determine the soil water retention curve, the water in each lysimeter was drained out through a valve at the bottom of the lysimeter until 50% readily available soil moisture content was obtained in the lysimeter. For the control treatment, supplemental water was applied at the surface of the lysimeters to maintain the soil field capacity at 0.32 cm^3/cm^3.

WUE was calculated for both grain yield (harvested seed weight) and total biomass (harvested total dry matter). Since sixteen lysimeters were cut, the grain yield and total biomass values of sixteen lysimeters were used for grain yield and biomass WUE calculations. The same statistical difference in the grain yield and total biomass WUE results of thirty-two lysimeters was extrapolated by using the data of sixteen lysimeters in response to different WTDs.

2.5. Statistical Analysis

A randomized complete block design method was used in this study. The effect of different groundwater levels on canola growth and yield parameters (crop water use, plant height, seed weight, pod weight, total biomass, root-shoot ratio, and root distribution) were analyzed by using a one-way analysis of variance (ANOVA) with a $P \leq 0.05$ level of significance. The Statistical Package for the Social Sciences version 25 (SPSS) and Duncan homogeneous test comparisons with the $P = 0.05$ probability level were used to conduct mean separation tests, when appropriate.

3. Results and Discussions

3.1. Evapotranspiration and Climate Conditions in the Greenhouse

To determine the relationship between evapotranspiration and temperature in the greenhouse and interpret the temperature and ET_0 changes during the different canola growing stages (germination, growing, and harvesting), daily average ET_0 rates and temperature data were collected between 4 November 2018 (planting) and 4 February 2019 (first harvesting). According to the result obtained from ETgages, the lowest and highest temperature in the greenhouse were determined as 15.5 °C and 29.5 °C, respectively (Figure 2). The lowest temperatures were observed during the first 10 days after planting because of extreme cold ambient temperatures. The temperatures in the greenhouse were 25 ± 5 °C from 14 November 2018 to 4 February 2019.

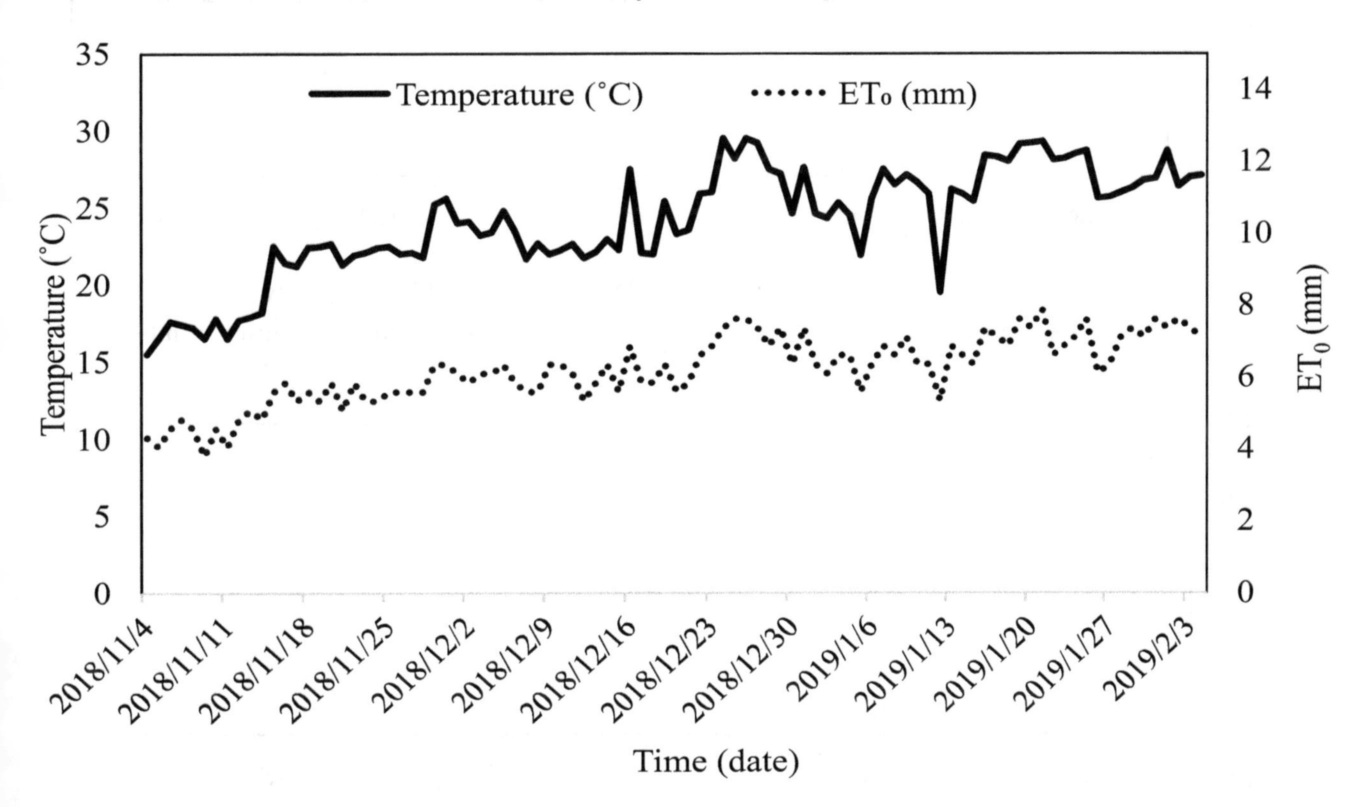

Figure 2. Measured daily air temperature (°C) and ET_0 values in the greenhouse. (ET_0 is reference crop evapotranspiration).

The lowest daily ET_0 was measured as 3.80 mm during the germination stage of canola (Figure 2). After 10 days of planting (when the canola was germinating and emerging), ET_0 rates fluctuated, with the highest ET_0 measured at 7.80 mm. Cumulative ET_0 was calculated as 577 mm during the entire experimental period (92 days). Fluctuations in air temperature influenced evapotranspiration. When the greenhouse temperature dropped from time to time, ET_0 also decreased accordingly.

3.2. Canola Irrigation Water Use

In the control treatment, the water content was kept between the field capacity (0.32 cm^3/cm^3) and the RAM (0.27 cm^3/cm^3) in order to prevent water stress and the application of an excessive amount of irrigation water. Three different plant root depths were considered for water requirement calculations. The canola root depth was projected 30 cm between 4 November and 4 December 2018 for the calculation of crop water requirements. Between 4 December 2018 and 4 January 2019, the control plants were irrigated up to a 60 cm root depth. After 4 January 2019, the crop water requirement was calculated for a 90 cm root depth. The volumetric water content for the specified root depths of the control plants was always maintained between field capacity and RAM with supplemental watering (Figure 3).

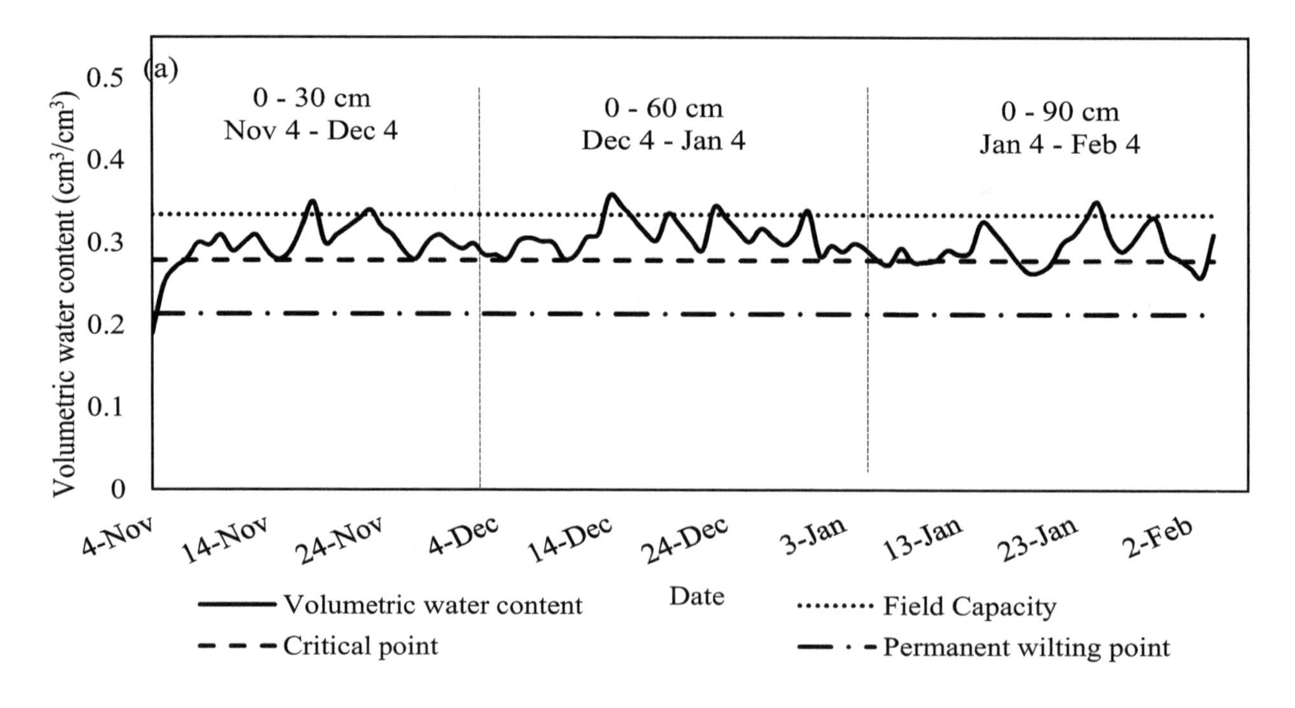

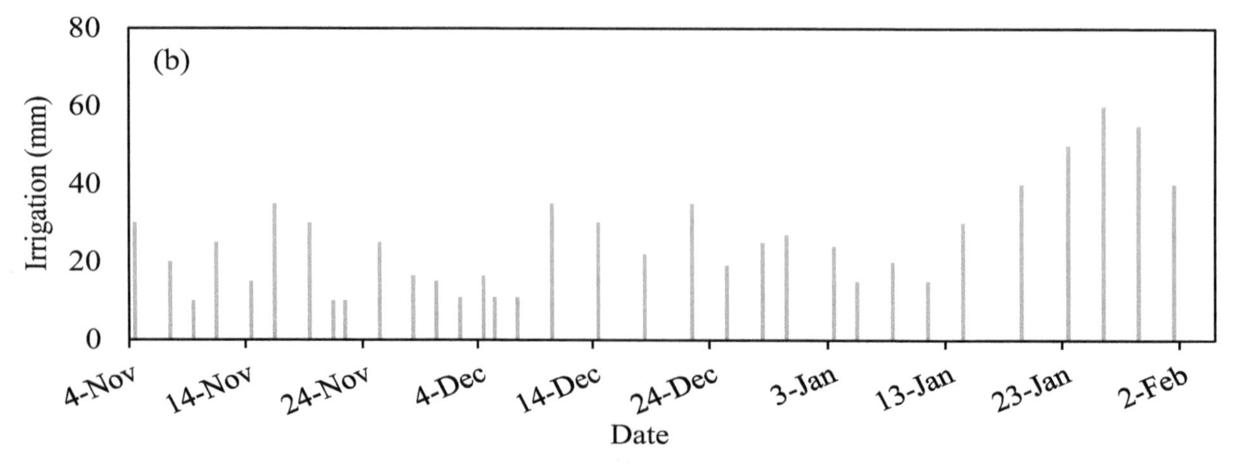

Figure 3. (**a**) Soil moisture content measurements for the control treatment at the desired depth of soil profile and (**b**) irrigation water applied to the lysimeters.

Control lysimeters received 752 mm of cumulative irrigation water. The calculated cumulative canola plant water use varied from 733 to 749 mm, with an average ET_c of 740 mm for the control treatments (Table 2).

Table 2. Total canola water-use for the control treatments.

Lysimeter Number	Initial Condition	Cumulative Irrigation Water	Final Condition	Cumulative ET_c	Mean ET_c
#	mm	mm	mm	mm	mm
R_3-$T_{control}$	162	752	181	733	740
R_4-$T_{control}$	162	752	174	740	
R_6-$T_{control}$	162	752	165	749	
R_7-$T_{control}$	162	752	176	738	

3.3. *Canola Groundwater Use*

The total canola plant water use and the groundwater contribution for the 12 lysimeters (3 lysimeters from each treatment) are presented in Table 3. Data collected from the water potential sensors were used to calculate soil water content. Initial soil water content was 350 mm in all the lysimeters. Similar to the control treatments, the canola evapotranspiration values in the water table treatments were measured. Each root depth had different ET_c values because evapotranspiration was influenced by WTD (Table 3). According to the ET_c value comparisons from different WTDs, the 30 cm soil profile had the highest ET_c, 717 mm. A significant difference in ET_c was observed between the 30 and 90 cm soil profiles. These results showed an inverse relationship between WTD and evapotranspiration. Additionally, the same inverse relationship was obtained between WTD and groundwater contribution. When the water table depth increased from 30 to 90 cm, the amount of water use from the groundwater also increased.

Table 3. Total canola water use from the groundwater for different depths.

Lysimeter Number	Depth	Initial Condition	Water Use from GW	Final Condition	ET_c	Mean ET_c	Mean ET_c
#	cm	mm	mm	mm	mm	mm	% $T_{control}$
R_3	30	350	632	241	741	717	97
R_4	30	350	615	268	697		
R_6	30	350	643	245	748		
R_7	30	350	607	275	682		
R_3	60	350	485	279	556	527	71
R_4	60	350	426	289	487		
R_6	60	350	433	271	512		
R_7	60	350	454	251	553		
R_3	90	350	402	235	517	501	68
R_4	90	350	355	225	480		
R_6	90	350	379	214	515		
R_7	90	350	341	198	493		

Note: R denotes replication. The initial condition is assumed to be identical for all lysimeters.

3.4. *Growth and Yield Parameters*

Plants in the T_{90} treatments were taller than the plants in the $T_{control}$, T_{30}, and T_{60} treatments, with a mean plant height of 134.6 cm for plants in the T_{90} treatment, while the shortest plants (113.3 cm) were in the T_{30} treatment. An inverse relationship was observed between the mean plant height and WTD, which was similar to the relationship between the WTD and the groundwater contribution.

When the water level was increased from 90 to 60 cm (measured from the soil surface) in the lysimeters, the canola plant height decreased from 134.6 to 113.3 cm.

Treatment differences were significant for total biomass, pod weight, and seed weight per plant (Table 4). The highest mean total biomass, pod weight, and seed weight were 22.1, 12.6, and 6.3 gr, respectively, for the T_{90} treatment. The lowest mean total biomass, pod weight, and seed weight results were 15.1, 9.5, and 4.8 gr, respectively, for the T_{30} treatment. These results indicate that the shallower water table depth decreased canola harvesting. Overall, the statistical results suggest a negative correlation between the canola plant growth and the yield parameters and WTD. A similar correlation between the WTD and crop harvesting results was reported by Mejia et al. [8]. According to the 2 year lysimeter experiment results, corn and soybean grain yield weights increased when the WTD decreased from 50 cm to 75 cm [8].

Plants in the control treatment consumed the most water (Tables 2 and 3), while plants in the T_{90} treatment had the highest yield values compared to other treatmnets (Table 4). As explained earlier, plants in$T_{control}$ used the optimum amount of irrigation water, 740 mm, while plants in T_{90} consumed 501 mm of water from the groundwater. However, better growth and yield results were obtained from plants in the T_{90} treatment.

Table 4. Statistical analysis of the canola growth and yield parameters.

Treatment #	Plant Height cm/plant	Total Biomass g/plant	Pod Weight g/plant	Seed Weight g/plant
$T_{control}$	118.0 ^{a}	19.4 ^{c}	11.7 ^{c}	5.8 ^{c}
T_{30}	113.2 ^{a}	15.1 ^{a}	9.5 ^{a}	4.7 ^{a}
T_{60}	113.8 ^{a}	17.8 ^{b}	10.8 ^{b}	5.3 ^{b}
T_{90}	134.6 ^{b}	22.0 ^{d}	12.5 ^{d}	6.3 ^{d}

Note: Lowercase letters; ^{a}, ^{b}, ^{c}, and ^{d} show statistical significance at $p = 0.05$. The values followed by the same letter in each column are not significantly different.

3.5. Water Use Efficiency (WUE)

The relationship between WUE for canola total biomass and grain yield was significant. The correlation of both WUE values was also determined (Figure 4). The effects of different WTD levels on both the grain yield and total biomass WUEs were also significant. The highest grain yield and biomass WUE values for T_{90} were 0.0126 and 0.0449 g lys^{-1} mm^{-1}, respectively. The lowest WUE values for both parameters occurred with the T_{30} treatment (Table 5).

After cutting the 16 lysimeters as mentioned earlier, each soil profile was divided into three different layers: 0–30, 30–60, and 60–90 cm (measured from the top) to determine the percentage of the root mass distribution in terms of WTD (Table 6). Overall, the highest root–mass ratio was found with plants in $T_{control}$ at the 0–30 cm soil layer (4.52 g and 54.6%). There was an inverse relationship between soil depth and root mass distribution for $T_{control}$ plants. When WTD changed from 90 to 30 cm, the root weight increased from 0.73 to 4.52 g. Since $T_{control}$ did not have WTD, a lower amount of roots was found in deeper soil layers. Significant differences were observed in 0–30 cm between T_{90} and other treatments, and a greater root mass was observed at 60–90 cm in treatment T_{90}. The highest average root weight was 7.97 g in the third layer for plants in the T_{90} treatment (Table 6). Similar to the inverse relationship between the soil depth and root mass for $T_{control}$, an inverse correlation was observed between WTD and the root mass for T_{90}.

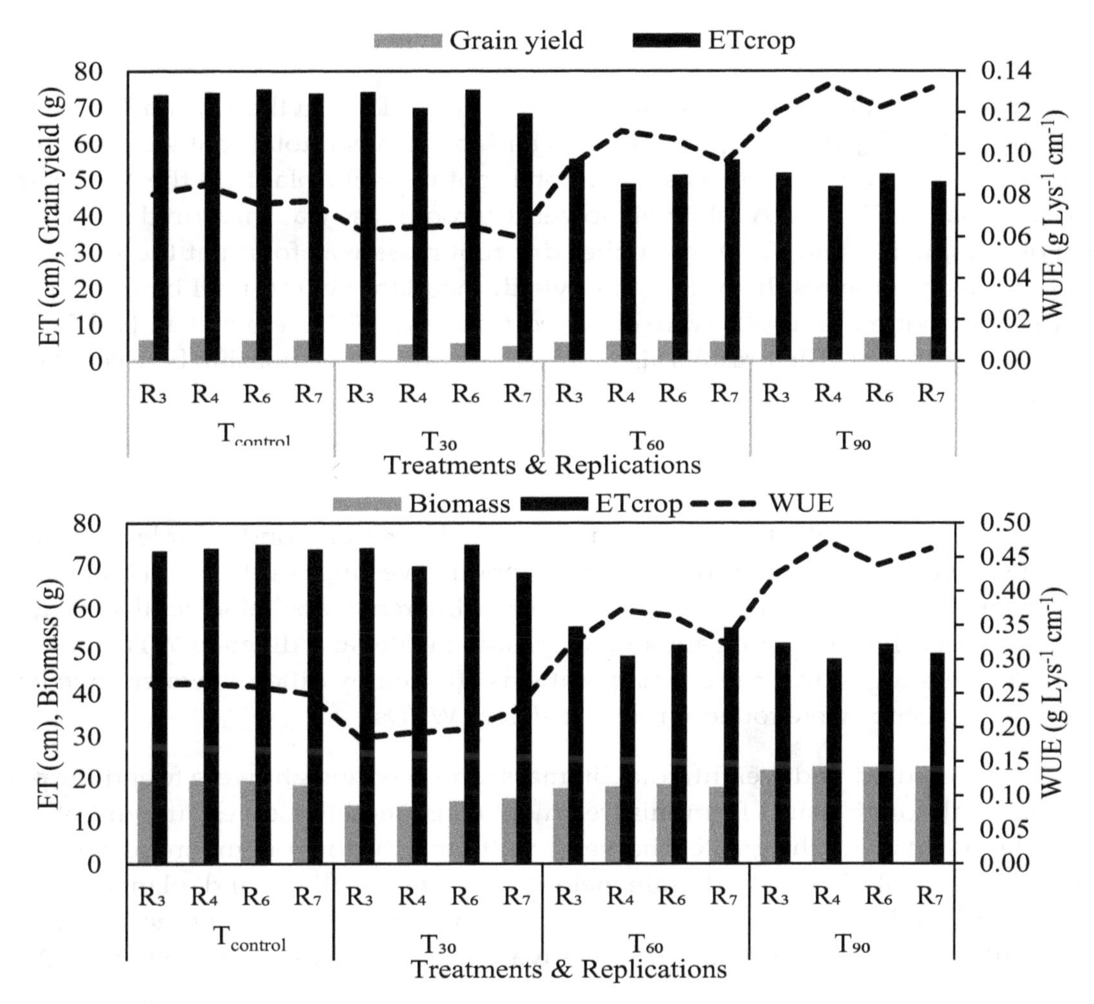

Figure 4. Canola water use efficiency with treatments: (**a**) grain yield water use efficiency (WUE) and (**b**) total biomass WUE.

Table 5. Statistical analysis results of the canola's mean grain yield, total biomass, water use, and water use efficiency.

Treatment	Mean Grain Yield (g/lys)	Mean Total Biomass (g/lys)	Mean Crop Water Use (mm)	Mean Grain Yield WUE (g/lys·mm)	Mean Total Biomass WUE (g/lys·mm)
$T_{control}$	5.9 ^{c}	19.2 ^{c}	740.0 ^{b}	0.0079 ^{b}	0.0259 ^{b}
T_{30}	4.5 ^{a}	14.3 ^{a}	717.0 ^{b}	0.0063 ^{a}	0.0199 ^{a}
T_{60}	5.3 ^{b}	18.0 ^{b}	527.0 ^{a}	0.0101 ^{c}	0.0344 ^{c}
T_{90}	6.3 ^{d}	22.5 ^{d}	501.2 ^{a}	0.0126 ^{d}	0.0449 ^{d}

Note: Lowercase letters; ^{a}, ^{b}, ^{c}, and ^{d} show statistical significance at $p = 0.05$. The values followed by the same letter in each column are not significantly different.

Table 6. Average root mass and proportions of roots.

Layers	Depth	Average Root Mass and Percentage							
		$T_{control}$		T_{30}		T_{60}		T_{90}	
	cm	g	%	g	%	g	%	g	%
1th	0–30	4.52 ^{b}	54.6	4.4 ^{b}	47.8	4.60 ^{b}	42.9	3.13 ^{a}	19.1
2nd	30–60	3.02 ^{a}	36.7	2.95 ^{a}	31.9	4.02 ^{b}	37.6	5.22 ^{c}	32.1
3rd	60–90	0.73 ^{a}	8.7	1.87 ^{b}	20.3	2.08 ^{b}	19.5	7.97 ^{c}	48.8
Total		8.27 ^{a}	100	9.22 ^{a}	100	10.7 ^{b}	100	16.32 ^{c}	100

Note: Lowercase letters; ^{a}, ^{b}, ^{c}, and ^{d} show statistical significance at $p = 0.05$. The values followed by the same letter in each column are not significantly different.

3.6. Root Mass Distribution

There is no significant total root mass difference between plants in the T_{30} and $T_{control}$ treatments. However, plants in the T_{30} and $T_{control}$ treatments had lower mean total root weights compared to plants in the T_{60} and T_{90} treatments. The mean total root mass for plants in the T_{90} treatment was always two-fold that for $T_{control}$. A relatively lower dry root mass was measured at the second and third layers of $T_{control}$, T_{30}, and T_{60} but a higher dry root mass was found at the second and third layers of T_{90}. Similar to the results of the grain yield, the plant height, total biomass, pod weight, and WUE, the best root mass results were obtained from T_{90}. Fidantemiz et al. [13] found similar results for the T90 treatment. Their results also showed the inverse relationship between the WTD and root mass distribution.

4. Conclusions

In this study, the canola plant height, water use from different groundwater levels, total biomass WUE, grain yield WUE, root mass, root-shoot ratio, and harvesting results (total biomass, pod and seed weight) were determined and compared with three different water table depths in a greenhouse. In addition, the effects of the optimum amount of irrigation water and different WTDs on canola were exanimated. Results suggest that the canola plant was affected by different water table levels since inverse linear relationships were found with the different WTDs.

The highest measured pod weight, total biomass, and seed weight were found for plants with the T_{90} treatment, although plants from this treatment consumed the lowest amount of water from the groundwater. Plants with the greatest harvest results and the lowest amount of water utilization also had the greatest total biomass and grain yield WUEs. On the other hand, plants with the lowest harvest results and the highest crop water use occurred when plants were at the 30 cm water table depth. As a result, a high WTD level (30 cm) negatively impacted the canola growth.

Significant statistical differences were found between the root distribution and soil layers. In addition, stronger and heavier roots were found near the water table level. In contrast, the total root weight was affected by WTD, and significant statistical differences were observed among the treatments. The total root weight of the 90 cm lysimeter was significantly higher than that of the other treatments. It was projected that canola in a drier lysimeter developed its root structure very well, since canola plants have a tendency to reach the water. Overall, the results from this study can be used to guide water management through drainage water management, in order to achieve the best yield potential.

Author Contributions: Conceptualization, H.S., X.C. and X.J.; methodology, H.S., X.J. and H.A.; investigation, H.S.; resources, H.H.-V., H.A.; writing—original draft preparation, H.K.; writing—review and editing, H.H.-V., X.J., X.C., H.S.; supervision, H.S.; funding acquisition, H.S.

Acknowledgments: We thank our colleagues Hans Kendal and Mukhlesur Rahman from Plant Science Department and Aaron L.M. Daigh from Soil Science Department at North Dakota State University for their valuable suggestions throughout the study. We are also immensely grateful to Izzet Kadioglu from Plant Protection Department at Gaziosmanpasa University, Tokat, Turkey for his support and encouragement during our experiments. Any opinions, findings, conclusions, or recommendations expressed in this material are those of the author(s) and do not necessarily reflect the views of the funding institutions.

References

1. Hamdy, A.; Ragab, R.; Scarascia-Mugnozza, E. Coping with water scarcity: Water saving and increasing water productivity. *Irrig. Drain.* **2003**, *52*, 3–20. [CrossRef]

2. Condon, A.G.; Richards, R.; Rebetzke, G.; Farquhar, G. Breeding for high water-use efficiency. *J. Exp. Bot.* **2004**, *55*, 2447–2460. [CrossRef] [PubMed]
3. Ripoll, J.; Urban, L.; Staudt, M.; Lopez-Lauri, F.; Bidel, L.P.; Bertin, N. Water shortage and quality of fleshy fruits—Making the most of the unavoidable. *J. Exp. Bot.* **2014**, *65*, 4097–4117. [CrossRef] [PubMed]
4. Sinclair, T.R.; Tanner, C.; Bennett, J. Water-use efficiency in crop production. *Bioscience* **1984**, *34*, 36–40. [CrossRef]
5. Howell, T.A. Enhancing water use efficiency in irrigated agriculture. *Agron. J.* **2001**, *93*, 281–289. [CrossRef]
6. Yang, J.; Wan, S.; Deng, W.; Zhang, G. Water fluxes at a fluctuating water table and groundwater contributions to wheat water use in the lower Yellow River flood plain, China. *Hydrol. Process. Int. J.* **2007**, *21*, 717–724. [CrossRef]
7. Ghamarnia, H.; Golamian, M.; Sepehri, S.; Arji, I.; Rezvani, V. Groundwater contribution by safflower (*Carthamus tinctorius* L.) under high salinity, different water table levels, with and without irrigation. *J. Irrig. Drain. Eng.* **2011**, *138*, 156–165. [CrossRef]
8. Mejia, M.; Madramootoo, C.; Broughton, R. Influence of water table management on corn and soybean yields. *Agric. Water Manag.* **2000**, *46*, 73–89. [CrossRef]
9. Huo, Z.; Feng, S.; Huang, G.; Zheng, Y.; Wang, Y.; Guo, P. Effect of groundwater level depth and irrigation amount on water fluxes at the groundwater table and water use of wheat. *Irrig. Drain.* **2012**, *61*, 348–356. [CrossRef]
10. Luo, Y.; Sophocleous, M. Seasonal groundwater contribution to crop-water use assessed with lysimeter observations and model simulations. *J. Hydrol.* **2010**, *389*, 325–335. [CrossRef]
11. Kruse, E.; Champion, D.; Cuevas, D.; Yoder, R.; Young, D. Crop water use from shallow, saline water tables. *Trans. ASAE* **1993**, *36*, 697–707. [CrossRef]
12. Talebnejad, R.; Sepaskhah, A. Effect of different saline groundwater depths and irrigation water salinities on yield and water use of quinoa in lysimeter. *Agric. Water Manag.* **2015**, *148*, 177–188. [CrossRef]
13. Fidantemiz, Y.F.; Jia, X.; Daigh, A.L.; Hatterman-Valenti, H.; Steele, D.D.; Niaghi, A.R.; Simsek, H. Effect of Water Table Depth on Soybean Water Use, Growth, and Yield Parameters. *Water* **2019**, *11*, 931. [CrossRef]
14. Johnston, A.M.; Tanaka, D.L.; Miller, P.R.; Brandt, S.A.; Nielsen, D.C.; Lafond, G.P.; Riveland, N.R. Oilseed crops for semiarid cropping systems in the northern Great Plains. *Agron. J.* **2002**, *94*, 231–240. [CrossRef]
15. Kandel, H. *Soybean Production Field Guide for North Dakota and Northwestern Minnesota*; North Dakota State University: Fargo, ND, USA, 2010.
16. Kutcher, H.; Warland, J.; Brandt, S. Temperature and precipitation effects on canola yields in Saskatchewan, Canada. *Agric. For. Meteorol.* **2010**, *150*, 161–165. [CrossRef]
17. Huffman, R.L.; Fangmeier, D.D.; Elliot, W.J.; Workman, S.R. Conservation and the Environment. In *Soil and Water Conservation Engineering*, 7th ed.; Cengage Learning: Boston, MA, USA, 2012; pp. 1–7.
18. Roy, D.; Jia, X.; Steele, D.D.; Lin, D. Development and comparison of soil water release curves for three soils in the red river valley. *Soil Sci. Soc. Am. J.* **2018**, *82*, 568–577. [CrossRef]
19. Hillel, D. *Environmental Soil Physics: Fundamentals, Applications, and Environmental Considerations*; Elsevier: Amsterdam, The Netherlands, 1998.
20. Majumdar, D.K. *Irrigation Water Management: Principles and Practice*; Prentice-Hall India Pvt. Ltd.: Delhi, India, 2004; p. 500.

8

Simulation of Crop Growth and Water-Saving Irrigation Scenarios for Lettuce

Pinnara Ket [1,2,*], Sarah Garré [3], Chantha Oeurng [1], Lyda Hok [4] and Aurore Degré [2]

1. Faculty of Hydrology and Water Resources Engineering, Institute of Technology of Cambodia, Russian Federation Bd, P.O. Box 86, Phnom Penh 12156, Cambodia; oeurng_chantha@yahoo.com
2. BIOSE, Gembloux Agro-Bio Tech, Liège University, Passage des Déportés 2, Gembloux 5030, Belgium; aurore.degre@uliege.be
3. TERRA, Gembloux Agro-Bio Tech, Liège University, Passage des Déportés 2, Gembloux 5030, Belgium; sarah.garre@uliege.be
4. Department of Soil Science, Faculty of Agronomy, Royal University of Agriculture, P.O. Box 2696, Phnom Penh 12401, Cambodia; hoklyda@rua.edu.kh

* Correspondence: ket.pinnara@gmail.com

Abstract: Setting up water-saving irrigation strategies is a major challenge farmers face, in order to adapt to climate change and to improve water-use efficiency in crop productions. Currently, the production of vegetables, such as lettuce, poses a greater challenge in managing effective water irrigation, due to their sensitivity to water shortage. Crop growth models, such as AquaCrop, play an important role in exploring and providing effective irrigation strategies under various environmental conditions. The objectives of this study were (i) to parameterise the AquaCrop model for lettuce (*Lactuca sativa* var. *crispa L.*) using data from farmers' fields in Cambodia, and (ii) to assess the impact of two distinct full and deficit irrigation scenarios in silico, using AquaCrop, under two contrasting soil types in the Cambodian climate. Field observations of biomass and canopy cover during the growing season of 2017 were used to adjust the crop growth parameters of the model. The results confirmed the ability of AquaCrop to correctly simulate lettuce growth. The irrigation scenario analysis suggested that deficit irrigation is a "silver bullet" water saving strategy that can save 20–60% of water compared to full irrigation scenarios in the conditions of this study.

Keywords: crop growth; lettuce; AquaCrop; water saving; water productivity; deficit irrigation

---

1. Introduction

Humanity's environmental footprint is unsustainable within the Earth's limited natural resources and assimilative capacity [1]. Climate change and growth in the global population are increasing pressure on these scarce environmental resources, notably water [2–4]. Particularly, increasing relative evapotranspiration from flow regulation and irrigation over the past century raises the global human water consumption and footprint [5]. Improving food production with less water and benchmarking efficiency of resource use is therefore a great challenge of our time, and urgently needed to ensure food security [1,6,7].

Cambodia is considered to be the country most vulnerable to climate change in Southeast Asia [8]. In recent decades, extreme events, such as floods and droughts, have negatively affected the livelihoods of farmers, especially in terms of the loss of crop production [9]. Cambodian farmers are generally conscious of these changes and challenges [9]. Guidelines for agricultural adaptation to improve crop productivity and the sustainability of the farming system and to minimise vulnerability to climate change, are therefore crucial [8,10]. Currently, the production of vegetables, like lettuce, poses more challenges in term of managing irrigation water efficiently, due to the crop's sensitivity to water shortage [11–13]. Lettuce, the most widely consumed leaf vegetable, is also one of the most widely cultivated vegetables in the world [14]. It is also an important to local vegetable production in Cambodia [15,16]. Improving strategies for vegetable farming productivity, including lettuce, for Cambodian farmers, is being increasingly considered [17].

Many irrigation strategies have been investigated for improving irrigation water productivity (IWP) during recent decades, with IWP defined as the ratio of agricultural output to the amount of irrigation water use [18]. Full irrigation via water application with the crop evapotranspiration requirements (ETc) method is an effective irrigation practice for crop production [19–22]. In traditional irrigation scheduling, a technique to meet full irrigation, as well, the soil moisture in the root zone is allowed to fluctuate between an upper limit approximating "field capacity" and the lower limit of the readily accessible water (RAW), referred to as "the threshold", somewhat above where a crop begins to experience water stress [23,24]. These methods have been applied to improve crop water productivity in various regions of the world, including Asian regions [25–30]. Nevertheless, deficit irrigation, as an adaptation strategy for regions with limited water resources or prone to drought, has been proven to be worth considering [31,32].

Deficit irrigation is an irrigation practice whereby a crop is irrigated with an amount of water below the full requirement for optimal plant growth, thereby saving water and minimising the economic impact on the harvest [18,19]. By limiting water applications to drought-sensitive growth stages such as, the vegetative stages and the late ripening period, the aims of this approach is to maximise water productivity and to stabilise, rather than maximise yields [33]. Water deficit can be defined at five levels: severe deficit (with soil moisture (SM) less than 50% of field capacity (FC)), moderate deficit (SM < 50–60% of FC), mild deficit (SM < 60–70% of FC), no deficit or full irrigation (SM > 70% of FC), and overirrigation (application above water requirements) [34]. Crops under deficit irrigation will experience some level of water stress, and often have lower yields than fully irrigated plants [35]. Deficit irrigation can allow irrigation water savings of up to 20–40% at yield reductions below 10% [36], and has been widely investigated in dry regions [36]. Deficit irrigation can be based on applying irrigation water under crop evapotranspiration. Patanè et al. [37] found that deficit irrigation at 50% of ETc for tomato plants resulted in no biomass (B) loss and high irrigation water-use efficiency. Experimental results obtained by Abd El-Wahed et al. [38] suggested that deficit irrigation at 85% of ETc is favourable to save 15% of water provided, with no reduction in the bean crop. The study results of Samperio et al. [39] offered deficit irrigation at 20% and 60% of ETc during stage II and postharvest, respectively, to "Angeleno" Japanese plum as a water-saving strategy, without negatively affecting crop yield. Results from Yang et al. [40] confirmed that the yield loss for cotton was less than 10% under deficit irrigation of 70% of ETc and 85% of ETc. Meanwhile, crop sensitivity to water deficit can be affected by many factors, including climatic conditions, crop species and cultivars, and agronomic management practices, amongst others [34]. Payero et al. [41] suggested that deficit irrigation is not a good strategy for improving the crop water productivity of maize in a semi-arid climate. A study on deficit irrigation treatment on lettuce showed that water stress caused by deficit irrigation at 20% and 40% of ETc significantly reduced leaf number, leaf area index, and dry matter accumulation [42]. Final fresh weight was reduced by 20% to 30% when compared with full irrigation. Kuslu et al. [43] concluded that for lettuce grown in semi-arid regions, full irrigation should be used under no water shortage, and deficit irrigation by 60% of ETc could be used for 40% water saving with a 35.8% yield loss where irrigation water supplies are limited.

Elaborating irrigation strategies merely on the basis of field research is difficult and time consuming [44]. Crop models are effective decision-support tools to investigate irrigation scenarios and to develop improved irrigation strategies [7,45,46]. They can provide a rapid and reasonable accurate prediction of the response of agriculture over a range of environmental conditions [47]. The model AquaCrop, developed by the Food and Agricultural Organisation of the United Nations

(FAO), is a water-driven crop model that simulates daily crop growth (e.g., canopy cover and biomass production) and final crop yield, with a balance between accuracy, simplicity, and robustness in incorporating various agronomy practices [48,49]. It is considered as a valuable tool for improving irrigation water productivity in crop production planning [6,50]. AquaCrop has been calibrated and parameterised to various crops under various environmental and irrigation conditions, including barley [51], soybean [52], sunflower [53], cotton [54,55], corn [56], sugar beet [57], wheat [58,59], potato [60,61], cabbage [62], and rice [63]. However, this has not yet been done in the case of lettuce. Most of these studies proved that the model is capable of accurately simulating crop growth and yield. However, some case studies still report some flaws in simulation of crop evolution and yield, especially under severe deficit irrigation and heat stress conditions. Adeboye et al. [64] found that biomass of soybean simulated by AquaCrop was overestimated under deficit irrigation conditions. Zeleke et al. [65] found that AquaCrop simulated the canopy cover and biomass growth of canola well, but the model was less satisfactory under severe water stress conditions in a semi-arid region. Similarly, a reduction in model reliability in biomass and canopy cover prediction for maize under the severe stress conditions of deficit irrigation in a tropical environment was indicated in a study of Greaves et al. [66]. AquaCrop performed well in biomass simulation of potato in the experiment under deficit irrigation at 120, 100, 80, and 60% of ETc [67]. However, the potato yield simulation was overestimated due to the heat stress, with the authors suggesting the incorporation of a temperature stress coefficient into AquaCrop when a crop is affected by high temperatures. Further research is therefore required to improve the performance of AquaCrop. Furthermore, its performance simulating lettuce growth in Cambodian conditions has not yet been tested. The main objective of this study is to improve the water productivity of lettuce under limited irrigations in the Cambodian climate. More specific objectives are (i) to parameterise the crop model AquaCrop using data from farmer fields, since lettuce is not yet available in the AquaCrop catalogue; and (ii) to assess the impact of water-saving scenarios in full and deficit irrigation in silico using this calibrated model.

2. Materials and Methods

2.1. Experimental Sites

The field experiments were conducted with lettuce plants (*Lactuca sativa* var. *crispa L.*) which are widely used in the study area, during a period from August to September 2017 in two experimental sites located in the villages of Chea Rov (site S1) (104°38′54.442″ E 12°9′15.482″ N) and Ou Roung (site S2) (104°37′16.24″ E 12°11′52.518″ N) in the province of Kampong Chhnang, Cambodia (Figure 1).

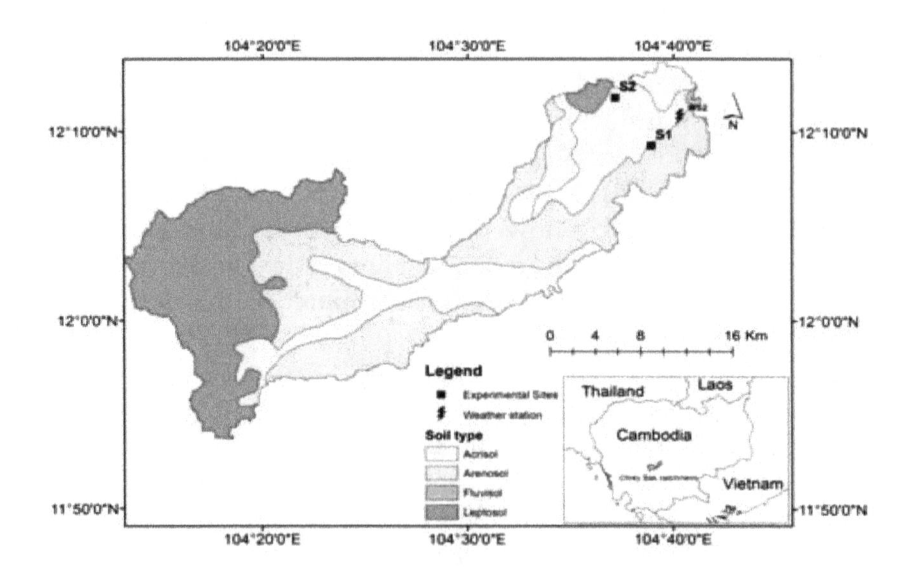

Figure 1. Experimental sites at Chearov (S1) and Ou Roung (S2), located in the Chrey Bak Catchment.

The total land area of the plots was 400 m^2. Lettuce seeds were sown in standard trays (with 123 holes). After 15 days, seedlings were transplanted into raised bed rows (0.30 m in height and with bed tops 0.50 cm wide) and covered with plastic mulch with a planting density of 12 plants m^{-2}. The compost was basally applied at the rate of 20 ton ha^{-1} before transplantation.

Irrigation was carried out using a drip system, with emitters of constructor maximum discharge of 3 L h^{-1} spaced 0.10 m apart. A plastic cover was used to protect the crops from heavy rainfall. Nevertheless, due to the intense rain which flowed between the crop rows, water ponding at 20 cm below the top bed row level was observed between the lettuce rows at both sites during almost the entire growing period. This ponding kept the soil wet during the growing period, and had to be factored into the calibration of the lettuce growing curve. At site S2, irrigation was not applied after a week after planting, due to the benefit of water ponding. At site S1, even though there was also water ponding in the field, the irrigation was applied every other day. The irrigation was determined by checking soil moisture (SM) using the feel and appearance method of Klocke et al. [68]. The irrigation was done when the SM was depleted below field capacity in the root zone at 5 cm, as lettuce have a root depth between 5–10 cm.

2.2. Data Collection and Measurement

2.2.1. Climate Data

Weather data for the experimental sites were collected from a local meteorological station (104°40′21.767″ E; 12°10′45.965″ N) (Figure 1). Daily maximum and minimum temperature, relative humidity, wind speed, rainfall, and solar radiation were recorded automatically at a five minute time step. The daily reference evapotranspiration (ETo) for the growing season, used as input data in AquaCrop, was calculated using the ETo calculator based on the FAO's Penman–Monteith method [69] (Figure 2).

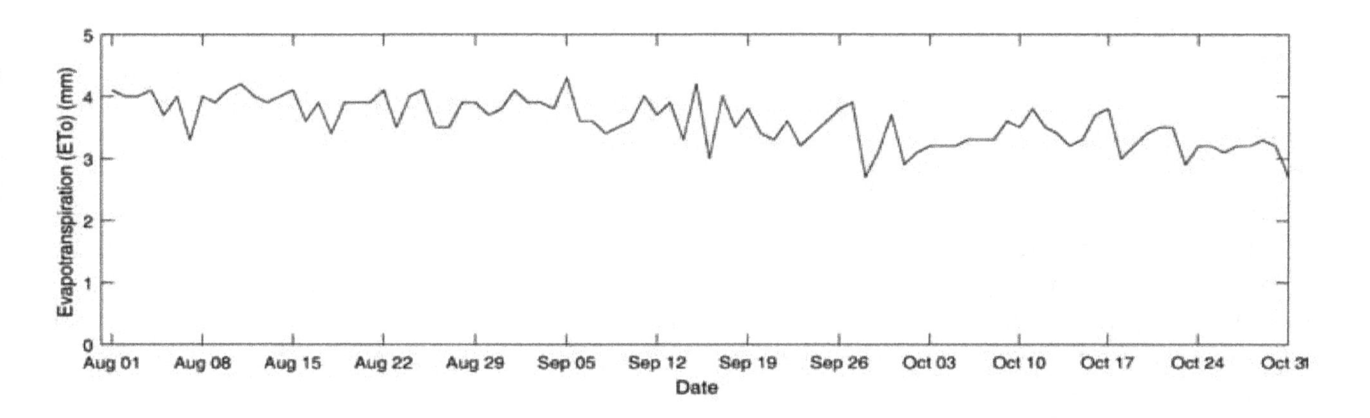

Figure 2. Daily potential evapotranspiration (ETo) during the growing season 2017.

2.2.2. Soil Data

The physical and chemical soil characteristics which were measured are listed in Tables 1 and 2. The soil texture was measured using the pipette method [70]. The bulk density was measured by the core method [71]. Field capacity, wilting point, and saturated hydraulic conductivity were derived from tension and soil moisture monitoring, using MPS2 and 10HS probes and using inverse modelling as presented in [72].

Table 1. Measured physical soil characteristics.

Parameters	Experimental Sites	
	Chea Rov (S1)	Ou Roung (S2)
Texture	Sand	Loam
Clay (%)	4.39	7.80
Silt (%)	9.56	41.15
Sand (%)	86.03	51.04
Bulk density (g cm^{-3})	1.5	1.5
Field capacity (m^3 m^{-3}) (sand: at −10 kPa, Loam: at −33 kPa)	0.11	0.14
Wilting point (m^3 m^{-3}) (at 150 kPa)	0.05	0.06
Soil saturation (m^3 m^{-3})	0.27	0.43
Available water content (AWC) (mm m^{-1})	62.48	81.43

Table 2. Measured chemical soil characteristics.

Site	Sampling Time	pH-H_2O	EC (uS cm^{-1})	OM (%)	N (%)	P (ppm)	K (meg 100 g^{-1})	CEC (cmol kg^{-1})
S1	Before transplanting	6.28	108	20.31	0.098	13.29	0.77	2.80
	At harvest	6.84	97.4	20.85	0.126	17.08	0.4	4.40
S2	Before transplanting	6.7	223	19.51	0.238	24.07	2.31	7.60
	At harvest	6.8	218	19.78	0.126	15.91	1.45	5.40

Note: EC is electrical conductivity; OM is organic matter content; N is total nitrogen; P is available phosphorous; K is exchangeable potassium; CEC is cation exchange capacity.

2.2.3. Crop Data

Canopy cover was measured at three-day intervals during the growing stage. Four pictures of 1 m^2 were taken randomly using a digital compact camera (Nikon Coolpix p600, Tokyo, Japan) at a fixed height of 1 m above ground level. The canopy cover was analysed using image processing with ImageJ® software (https://imagej.nih.gov). Aboveground dry biomass was determined by harvesting 10 heads at the surface level of each site, oven-drying plant samples at 70 °C for 48 h, and weighing them [73].

2.3. AquaCrop Model

The AquaCrop model is a crop water-driven productivity model developed by the FAO in 2009. A detailed description is presented in [49]. Water is the key limiting factor for crop production in this model [74]. Inputs for the AquaCrop model consist of weather data, crop, and soil characteristics (soil profile and groundwater), and field management practice or irrigation management practices [49].

Canopy cover is a crucial feature of AquaCrop [49]. Under unstressed condition, the exponential growth equation to simulate canopy development for the vegetative stage is

$$CC = CC_o e^{CGC \times t} \quad (1)$$

where CC is the canopy cover at time t and is expressed as fraction of ground covered, CC_o is initial canopy cover size (at t = 0) as a fraction (%), and CGC is the canopy growth coefficient in fraction per growing degree day (GDD), a constant for a crop under optimal conditions, but modulated by stresses.

In the condition of water stress, the CGC is multiplied by a water stress coefficient of expansive growth (Ks_{exp}) (Equation (2)).

$$CGC_{adj} = Ks_{exp}.CGC \quad (2)$$

where Ks_{exp} ranges from 1 to 0, canopy growth begins to slow down below the maximum rate when soil water depletion reaches the upper threshold, and stops completely when the depletion reaches the lower threshold.

Crop transpiration is proportional to the canopy cover and given by

$$Tr = Ks_{sto} Kc_{Tr} ET_o \tag{3}$$

Ks_{sto} is the stress coefficient for stomatal closure. Kc_{Tr} is the crop transpiration coefficient (determined by canopy cover and $Kc_{Tr,x}$), $Kc_{Tr,x}$ is the coefficient for maximum crop transpiration, and ETo is reference evapotranspiration (mm).

Biomass production is computed from crop transpiration and crop water productivity normalised for ETo and CO_2 (Equation (4)). The extreme effect of low temperature on crop phenology, biomass accumulation, and harvest index, is considered with adjustment factors [67,75].

$$B = K_{s_b}.f_{WP}.WP^*.\frac{Tr}{ETo} \tag{4}$$

where B is biomass, Tr is crop transpiration (mm day^{-1}), ETo is reference evapotranspiration (mm day^{-1}), and K_{s_b} is the stress coefficient for low-temperature effects on biomass production. f_{WP} is the adjustment factor to account for differences, if any exist, in the chemical composition of the vegetative biomass and harvestable organs. WP^* is normalised crop water productivity, defined as the ratio of biomass produced to water transpired, normalised for the evaporative demand and CO_2 concentration of the atmosphere.

The AquaCrop stress indicators include water storage (not enough water), waterlogging (too much water), air temperature (too high or too low), and soil salinity stress (too high).

2.4. Model Parameterisation

The process of parameterisation is illustrated in Figure 3. The vegetative stage of lettuce refers to the growing period of lettuce growth after germination until harvest. A growing period during the vegetative stage of 59 days after transplanting was simulated in this study.

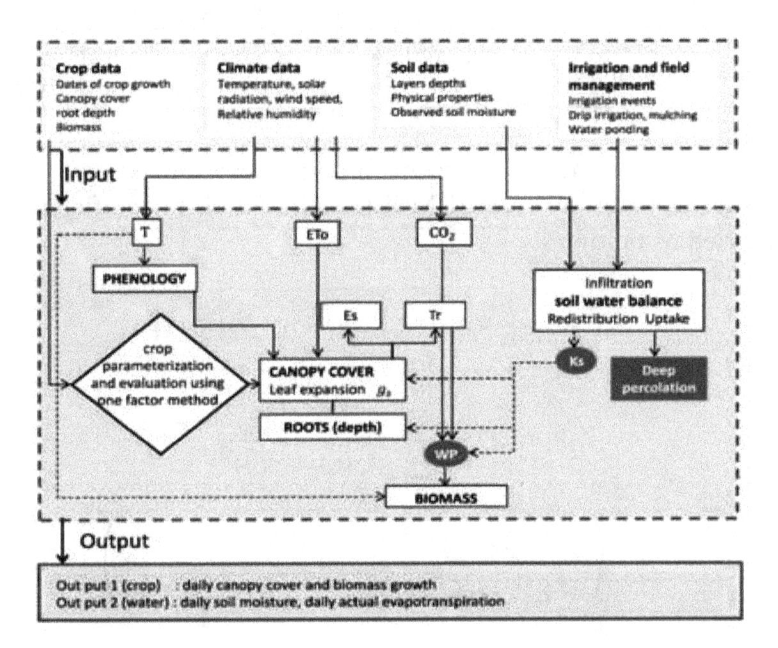

Figure 3. Flow chart of parameterisation of AquaCrop in this study (adjusted from [76]). T is temperature, ETo is potential evapotranspiration, gs is stomatal conductance, WP is water productivity coefficient, Ks is stress coefficient, Es is soil temperature, Tr is crop transpiration.

As lettuce is a crop which is not yet parameterised in AquaCrop, calibration of the model involved adjusting the model parameters to make them match the observed data [54,77].

The primary variables of lettuce growth, e.g., canopy cover and aboveground biomass were parameterised. For the calibration of the curves, the measured data in two experimental fields at the Chearov site (S1) (having sand soil) and Ourong site (S2) (with loam soil) were used, during the growing season in 2017. The AquaCrop model does not allow the use of observed data to build the canopy cover and biomass curves, but allows the data to be used to calibrate the canopy cover and biomass curves [78].

Canopy cover curves are a plot of the development of leaf expansion response to growing time per day, based on Equation (1). Biomass curves are a relationship plot of the growth of lettuce biomass response to growing time per day, based on Equation (4). The calibration of simulated canopy and biomass curves is based on one-at-a-time (OAT) methods (i.e., changing one parameter at a time while holding others constant) [79] and adjusting the parameters by trial and error, by comparing simulated and observed field data, and minimising the function of root mean square error.

We parameterised the canopy cover curve, which is important to the model for transpiration and evaporation [78]. The main parameters of Equation (1), e.g., CCo and CGC for canopy cover curve determination, were adjusted to match the observed canopy cover data. In addition, adjusting the maximum canopy cover (CCx), time to reach maximum canopy cover, and time to recover, is crucial in order to obtain correct simulations of canopy cover growth. Subsequently, the focus was on adjusting the biomass curve of Equation (4). WP* and $Kc_{Tr,x}$ (coefficient for maximum crop transpiration) are the main parameters for regulating biomass curves in AquaCrop [74]. As lettuce is a C3 crop type [80], the recommended values for WP* lie between 15 and 20 g m^{-2}. All calibrated crop parameters are shown in Table 3.

Table 3. Calibrated parameters of lettuce growth.

No	Calibration Step	Calibrated Parameters
1	Canopy cover calibration	Time to recover of transplant, Time to reach the maximum canopy cover, Initial canopy cover (CCo), Canopy growth coefficient (CGC), Maximum canopy cover growth coefficient (CCx)
2	Biomass calibration	Coefficient for maximum crop transpiration ($Kc_{Tr,x}$), Normalised biomass water productivity (WP*)

The model performance for canopy cover and biomass simulation was evaluated using statistic indicators, including root mean square error (RMSE), Nash–Sutcliffe coefficient (N), and coefficient of determination (R^2), defined as below.

$$\text{RMSE} = \sqrt{\frac{\sum_{i=1}^{n}(O_i - S_i)^2}{n}} \tag{5}$$

$$N = 1 - \frac{\sum_{i=1}^{n}(O_i - S_i)^2}{\sum_{i=1}^{n}(O_i - \overline{O})^2} \tag{6}$$

$$R^2 = \left(\frac{\sum_{i=1}^{n}(O_i - \overline{O})(O_i - \overline{S})}{\sqrt{\sum_{i=1}^{n}(O_i - \overline{O})^2 \sum_{i=1}^{n}(O_i - \overline{S})^2}}\right)^2 \tag{7}$$

where O and S are the observed and simulated values at time i, respectively, and n is the total amount of the data. When N and R^2 are close to 1, it is considered to be satisfactory [81]. RMSE should be close to 0.

AquaCrop requires the selection of inputs related to the irrigation method, such as sprinkler, drip, or surface. These methods determine the fraction of the soil surface made wet by irrigation [82] and the impact on irrigation efficiency [83].

Default AquaCrop settings for field management include mulching, and use an adjusted factor for the effect of mulches on soil evaporation. It varied between 0.5 for mulches derived from plant material, and 1.0 for plastic mulch [75].

The drip irrigation method with plastic mulch was applied as the input for field management in the model during the parameterisation, as this is the actual practice of the experiment in this study.

The soil water balance calculation, including soil moisture simulation in AquaCrop, is based on the storage capacity of the soil layers, described in Raes et al. [84], and previously in the BUDGET model [85].

During the experimental period, water ponding at 15 cm and 20 cm below the bed soil at site S1 and S2 respectively, which was observed during the experiment, was taken into account as a boundary condition during the parameterisation of the model. This water ponding resulted in wet soil during the growing period. The values of physical soil available data in the Section 2.2.2 were adopted to simulate soil moisture in this study.

It was noted that the plantation experiment was during the rainy season when irrigation was not needed. The crop parameters obtained after parameterisation are important for the investigation of the irrigation scenarios for water saving when irrigation is necessary, especially during the dry season.

2.5. Irrigation Scenarios

In the current study, AquaCrop was used to simulate the full and deficit irrigation scenarios described below (and in Table 4), in order to identify the optimal water use efficiency for lettuce.

Table 4. Irrigation Scenarios.

Scenario Code		Short Description
S1 (Sand)	**S2 (Loam)**	
Varied readily available water (RAW) threshold irrigation scenarios		
S0RAW	L0RAW	irrigate at 0% of RAW and refill to field capacity (FC)
S50RAW	L50RAW	irrigate at 50% of RAW and refill to FC
S80RAW	L80RAW	irrigate at 80% of RAW and refill to FC
S100RAW	L100RAW	irrigate at 100% of RAW and refill to FC
S120RAW	L120RAW	irrigate at 120% of RAW and refill to FC
S130RAW	L130RAW	irrigate at 130% of RAW and refill to FC
S150RAW	L150RAW	irrigate at 150% of RAW and refill to FC
S180RAW	L180RAW	irrigate at 180% of RAW and refill to FC
S200RAW	L200RAW	irrigate at 200% of RAW and refill to FC
Varied field capacity threshold irrigation scenarios		
S100FC	L100FC	full irrigation-daily irrigation at 100% of field capacity (FC)
S70FC	L70FC	deficit irrigation at 70% of FC
S60FC	L60FC	deficit irrigation at 60% of FC
S50FC	L50FC	deficit irrigation at 50% of FC
S40FC	L40FC	deficit irrigation at 40% of FC

2.5.1. Varied RAW Threshold Irrigation Scenarios

Figure 4 presents the calculation process of varied RAW threshold irrigation scenarios.

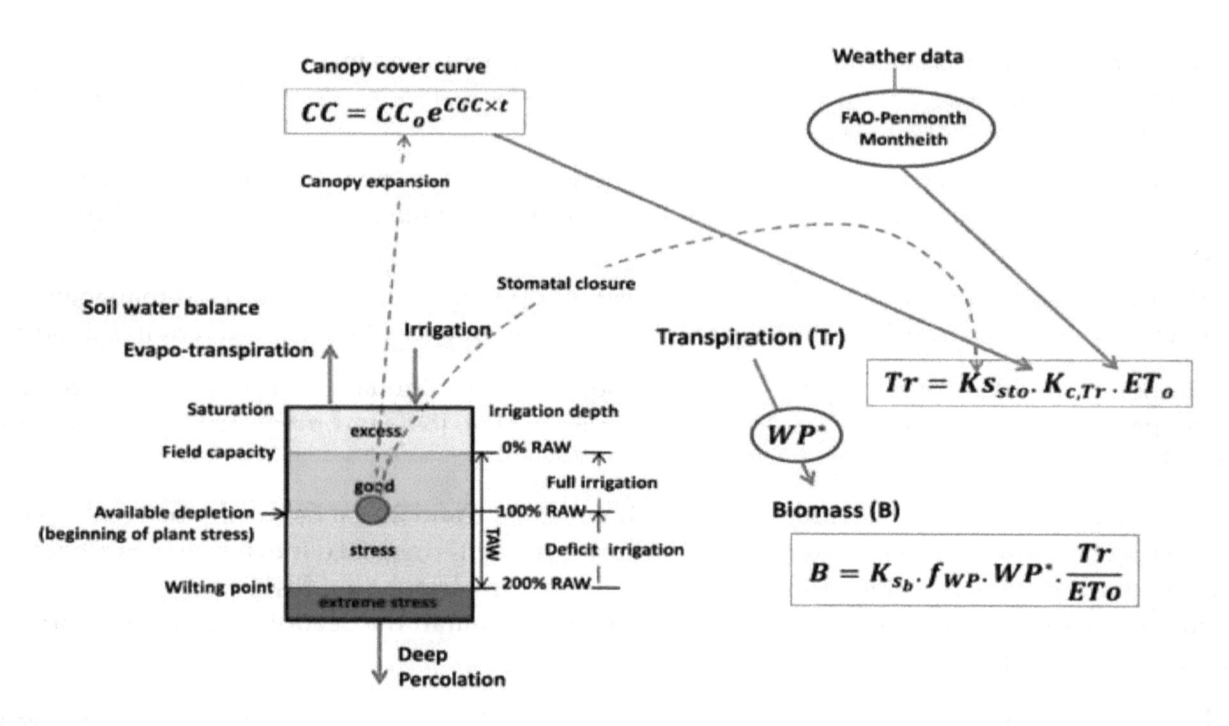

Figure 4. Schematic representation of the crop response to varied RAW threshold irrigation scenarios simulated by AquaCrop (adjusted from [77]). RAW is readily available water content, TAW is total available water content, CC is the simulated canopy cover, CC_o is initial canopy cover size, CGC is canopy growth coefficient in fraction per growing degree day (GDD), Ks_{sto} is the water stress for stomatal closure, Kc_{Tr} is the crop transpiration coefficient (determined by CC and $Kc_{Tr,x}$ at maximum canopy cover), ETo is the reference evapotranspiration, Ks_b is the stress coefficient for low-temperature effects on biomass production, f_{WP} is the adjustment factor to account for differences in chemical composition of the vegetative biomass and harvestable organs, WP* is the normalised water productivity.

These irrigation scenarios applied irrigation scheduling based on soil moisture depletion [86] by applying readily available water depletion in the default option in AquaCrop. The time and irrigation dose were calculated with the criteria below:

1. Soil water content depleted until a fixed lower threshold (RAW) and refill to field capacity (time criteria).
2. Irrigation dose can be determined by the following Equation (8) [87].

$$\mathrm{ID} = \mathrm{AD} \times \mathrm{RAW} \tag{8}$$

where ID is irrigation depth (mm), $\mathrm{RAW} = \mathrm{p\ TAW} = \mathrm{p}\ 1000(\mathrm{FC} - \mathrm{PWP})Z_r$, p is soil water depletion threshold, set to 0.3 for lettuce recommended by [69], and Z_r is root depth (m). TAW is the amount of water that a crop can extract from its root zone [88]. FC is field capacity, that is, the amount of water well-drained soil should hold against gravitational forces ($m^3\ m^{-3}$) [88]. PWP is permanent wilting point, referring to soil water content when a plant fails to recover its turgidity on watering ($m^3\ m^{-3}$) [88]. RAW is readily available soil water, referring to the fraction of TAW that a crop can extract from the root zone without suffering water stress [88]. AD is allowable depletion, defined as the percentage of RAW that can be depleted before irrigation water has to be applied.

Full irrigation scenarios with varied RAW thresholds were simulated by selecting allowable depletion levels at 0, 50, 80, 100% in AquaCrop, that avoid drought stress during the growth stage [41]. The irrigation schedule is generated by selecting a so-called "time" and "depth" criterion, with "back to field capacity" and "allowable depletion", respectively. In other words, the different full irrigation scenarios result in decreasing irrigation frequency.

Deficit irrigation scenarios with varied RAW thresholds were similar to the full irrigation scenario criteria, but applied allowable depletion levels at 120, 130, 150, 180, and 200%. These levels result in drought stress during the growing stage, since soil moisture can decrease to a level below RAW before an irrigation event is triggered [41].

2.5.2. Varied Field Capacity Threshold Irrigation Scenarios

Figure 5 illustrated concept of the varied field capacity threshold irrigation scenarios.

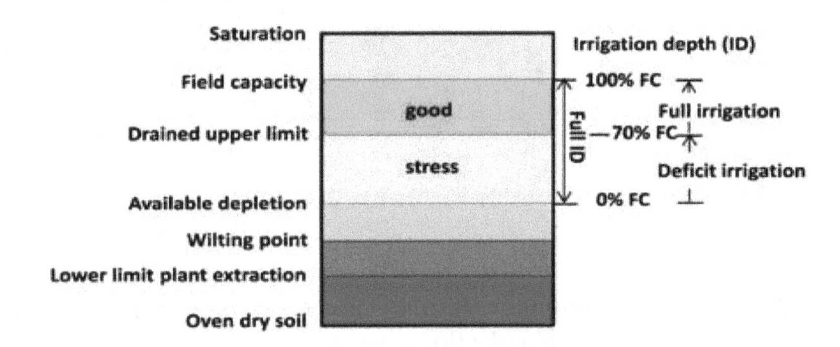

Figure 5. Schematic illustration of the soil water reservoir concepts of varied irrigation depth under field capacity irrigation scenarios (adjusted from [89]). FC is field capacity, full ID is full irrigation depth.

The full irrigation scenario, based on a fixed irrigation frequency maintained the soil moisture in the root zone at field capacity on a daily basis, since the literature claims this is the optimal status to maximise lettuce yield [90]. The irrigation schedule was generated with a fixed time interval (daily) (time criteria) and refill to field capacity (depth criteria).

Deficit irrigation scenarios with varied field capacity threshold reduce the irrigation dose below the dose at field capacity but keeping the same irrigation frequency, as in full irrigation scenario. Daily generated irrigation doses obtained in full irrigation scenario were reduced by 70, 60, 50, and 40%.

Irrigation water productivity (IWP) was used to evaluate the irrigation scenarios for efficient irrigation water use [31,91]. IWP is the ratio between the yield and the irrigation water use [31].

$$\text{IWP} = \frac{\text{Y}}{\text{I}} \tag{9}$$

where IWP is irrigation water productivity (kg m^{-3}), Y is simulated yield (kg ha^{-1}) and interest yield in this study is biomass, and I is irrigation water use (mm).

The adjusted crop parameters obtained from the parameterisation process were used in the scenario simulation under the same weather conditions, using no soil surface cover in model field management, and no ground water at bottom soil profile boundary condition.

3. Results

3.1. Plant Growth and Soil Moisture Status

Figure 6 shows both the lettuce growth measurement and simulation by AquaCrop. Biomass accumulated at a very low rate during the first two weeks of the growing season, and increased sharply in the final week. This trend accords with results obtained by Gallardo et al. [73].

The measured canopy cover and biomass yields were 34% and 0.11 ton ha^{-1}, respectively, at site S1 with sand soil, and 18.5% and 0.11 ton ha^{-1}, respectively, at site S2, which has loam soil. The measured results are comparable with Fazilah et al. [92], who found observed canopy cover of 33% and biomass yields of 0.22 ton ha^{-1} for lettuce under similar tropical conditions. Zhang et al. [93] found higher measured biomass for lettuce with a range of 0.33 to 0.63 ton ha^{-1} under lower temperatures of 20–25 °C. Thus, high day temperatures above 23 °C often limit lettuce production [94]. Optimum growth for lettuce occurs between 15–20 °C [12]. Unfavourable weather conditions, of high average temperature 33/25 °C (day/night) during the experiment, can be the reason of the low measured biomass yields for this study.

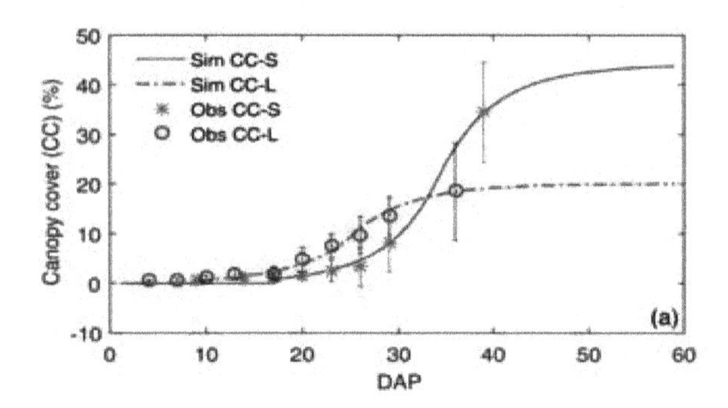

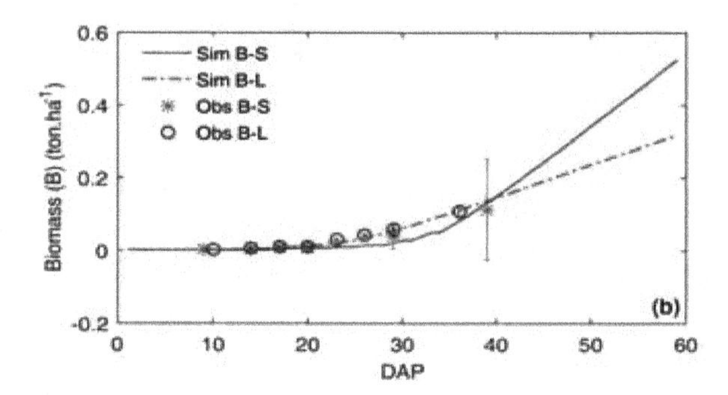

Figure 6. Observed (Obs) data and simulation (Sim) of lettuce growth of AquaCrop: (**a**) canopy cover at site S1 (CC-S) (sand soil) and site S2 (CC-L) (loam soil); (**b**) aboveground biomass at site S1 (B-S) and site S2 (B-S). The error bars were based on 10 biomass samples, except the last observed, which was based on 60 samples at harvest time. Sim CC-S is simulated canopy cover at site S1, Sim CC-L is simulated canopy cover at site S2, Obs CC-S is observed canopy cover at site S1, Obs CC-L is observed canopy cover at site S2, Sim B-S is simulated biomass at site S1, Sim B-L is simulated biomass at site S2, Obs B-S is observed biomass at site S1, Obs B-L is observed biomass at site S2.

3.2. Model Parameterisation and Evaluation

The primary crop variables calibrated for daily lettuce growth were canopy cover and biomass, with the daily soil moisture simulated by AquaCrop, by adapting available physical soil data.

Table 5 presents the adjusted model parameters for canopy cover and biomass curve simulation of lettuce growth. The time to recovery of transplant, the time to reach the maximum canopy cover, the initial canopy cover (CCo), the maximum canopy cover growth coefficient (CCx), the coefficient for maximum crop transpiration ($Kc_{Tr,x}$), and the normalised biomass water productivity (WP*) were mainly calibrated.

Table 5. AquaCrop variables parameterised.

Parameters	Symbol and Unit	Value				Sources
		S1		S2		
		Initial	Calibrated	Initial	Calibrated	
Crop Phenology						
Time to recovered transplant (C)	(GDD)	52	280	52	147	Default
Time to maximum canopy cover (C)	(GDD)	563	859	563	727	Default
Crop Growth						
Plant density (NC)	dp (plants m^{-2})	12	-	12	-	Measure
Initial canopy cover (NC)	CCo (%)	0.72	0.84	0.5	0.6	Default
Maximum effective rooting depth	Zr (m)	0.1	-	0.1	-	Measure
Maximum canopy cover (C)	CCx (%)	34	44	18	20	Measure
Canopy growth coefficient	CGC	22.7	18.5		16.8	Default
Base temperature (C)	Tbase (°C)	4	-	4	-	[95]
Upper temperature(C)	Tupper (°C)	28	-	28	-	[96]
Canopy size of transplanted seeding (C)	CC (cm^2 plant^{-1})	6	-	5	-	Measure
Coefficient for maximum crop transpiration (NC)	$Kc_{Tr,x}$	1.25	0.65	1.25	0.5	Default
Water productivity, (C)	WP* (g m^{-2})	15	16	15	16	Default

Note: C = conservative, NC = non-conservative.

WP* was adjusted at 16 gm^{-2} for both sites, within the recommended range. $Kc_{Tr,x}$ was adjusted at 0.65 and 0.5 for site S1 and S2, respectively. These adjusted $Kc_{Tr,x}$ are lower than crop coefficient for the mid-season ($K_{cb,mid}$ = 1) proposed by FAO-56. The difference between the values proposed by FAO-56 and the adjusted $Kc_{Tr,x}$ values is due to the fact that the FAO crop coefficients were obtained for specific agroclimatic conditions, which are different from the conditions of this study [78].

In addition, $Kc_{Tr,x}$ is a major requisite for estimating crop transpiration and biomass. The low adjusted value of this parameter resulted in low simulated biomass yields to fit to measured values.

High temperature stress observed during the experiment could be the reason for the low observed lettuce biomass production [12]. This observation leads to a recommendation for further development of a heat stress factor in relation to canopy cover and biomass simulations for lettuce.

The minimum root depth cannot be adjusted under 0.1 m, while the root development of lettuce was under this limit. Thus, root development in the model requires further modification [91].

The crop growth simulation of canopy cover and biomass fitted the observed data well (Figure 6). The statistical values for model evaluation in Table 6 were satisfactory, resulting in R^2 = 0.99, RMSE < 0.8%, N < 4.6 for canopy cover, and R^2 > 0.98, RMSE < 0.01 ton ha^{-1}, N < −0.07 for biomass. Thus, the model has ability to simulate well the growth of lettuce in both soil types at the two experimental sites.

Table 6. Statistical evaluation of model simulation.

Statistical Criteria	Sites	Canopy Cover (%)	Biomass (ton ha^{-1})
RMSE	S1	0.69	0.012
	S2	0.84	0.01
R^2	S1	0.99	0.98
	S2	0.99	0.99
N	S1	1.1	−0.015
	S2	4.6	−0.07

The measured and simulated soil moisture, at both soil depths of 5 and 15 cm in both sites, also matched well (Figure 7). The soil moisture simulation resulted in good accuracy with low RMSE of 0.18 and 0.14 m^3 m^{-3} at depths of 5 and 15 cm, respectively, at site S1, and 0.05 and 0.06 m^3 m^{-3} at depths of 5 and 15 cm, respectively, at site S2.

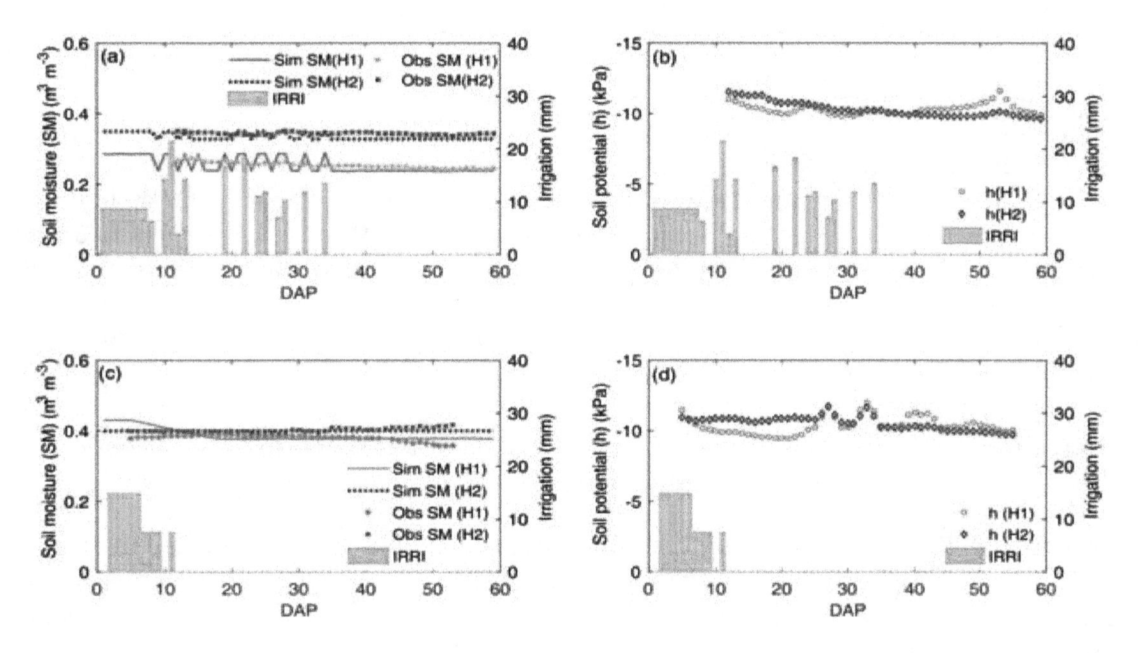

Figure 7. Simulated soil moisture and observed soil moisture data measured at depths of 5 cm (H1) and 15 cm (H2) using soil moisture sensor 10HS and soil potential MPS-2: (**a**) soil moisture at site S1; (**b**) soil potential at site S1; (**c**) soil moisture at site S2; (**d**) soil potential at site S2. DAP is day after planting, Sim SM is simulated soil moisture, Obs SM is observed soil moisture, IRRI is irrigation, h is soil potential.

3.3. Irrigation Scenarios

3.3.1. Irrigation and Soil Moisture Response

The cumulative irrigation in Figure 8, and the fluctuation of the soil moisture depletion in Figure 9, reflect the interaction between irrigation frequency and amount of water applied.

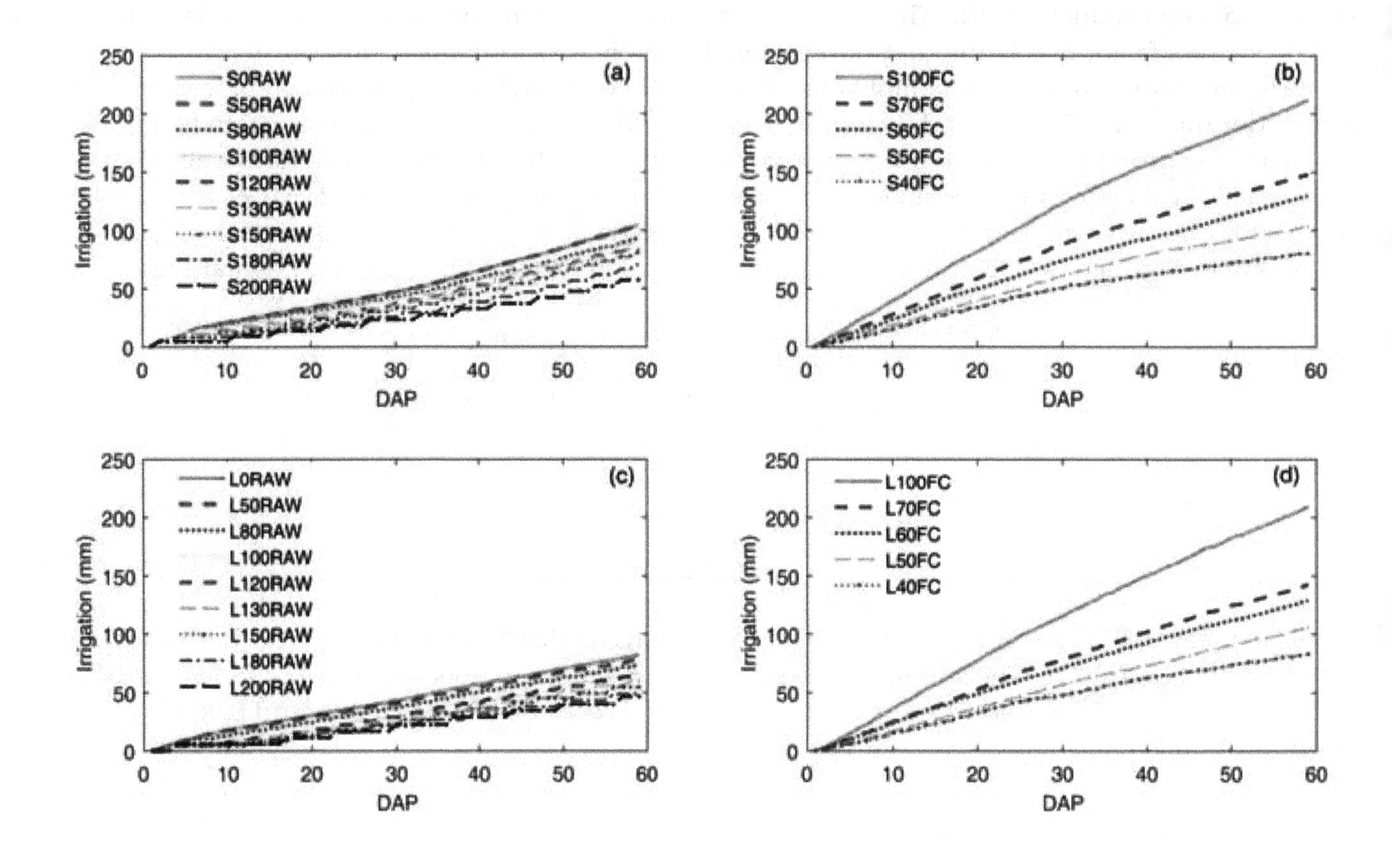

Figure 8. Irrigation accumulation response to different scenarios: (**a**) varied RAW threshold irrigation scenarios at site S1 (sand soil); (**b**) varied field capacity threshold irrigation scenarios at site S1; (**c**) varied RAW threshold irrigation scenarios at site S2 (loam soil); (**d**) varied field capacity threshold irrigation scenarios at site S2. RAW is readily available water content, S0RAW-S200RAW refers to irrigation scenarios with irrigation at 0–200% of RAW for sand soil. L0RAW-L200RAW refers to irrigation scenarios with irrigation at 0–200% of RAW for loam soil. S40FC-S100FC refers to deficit irrigation at 40–100% of field capacity for sand soil. L40FC-L100FC refers to deficit irrigation at 40–100% of field capacity for loam soil.

In both varied RAW and field capacity threshold irrigation scenarios, the irrigation frequency decreased together with decreasing the amount of water applied per irrigation event.

In varied RAW threshold irrigation scenarios, the simulation of irrigation resulted in irrigation depths which ranged from 57 to 104 mm in site S1 (sand soil) and 46–82 mm in site S2 (loam soil) (Figure 8a,c). In varied field capacity threshold irrigation scenarios, irrigation depths ranged from 81–201 mm in site S1 and 83–209 mm in site S2 (Figure 8b,d).

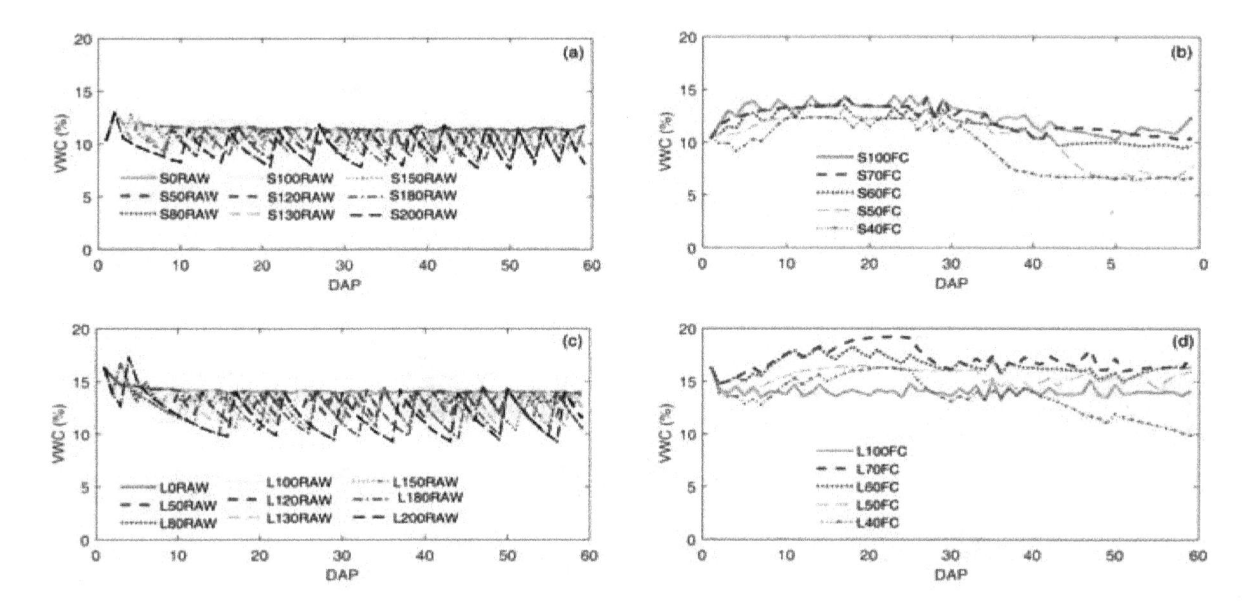

Figure 9. Daily soil moisture (VWC) response to different scenarios: (**a**) varied RAW threshold irrigation scenarios at site S1 (sand soil); (**b**) varied field capacity threshold irrigation scenarios at site S1; (**c**) varied RAW threshold irrigation scenarios at site S2 (loam soil); (**d**) varied field capacity threshold irrigation scenarios at site S2. RAW is readily available water content, SORAW-S200RAW refers to irrigation scenarios with irrigation at 0–200% of RAW for sand soil. LORAW-L200RAW refers to irrigation scenarios with irrigation at 0–200% of RAW for loam soil. S40FC-S100FC refers to deficit irrigation at 40–100% of field capacity for sand soil. L40FC-L100FC refers to deficit irrigation at 40–100% of field capacity for loam soil.

3.3.2. Crop Evapotranspiration and Biomass Growth Response

Figures 10 and 11 illustrate the cumulative crop evapotranspiration (ETc) and cumulative biomass of lettuce, respectively, under various irrigation scenarios simulated with AquaCrop calibrated for lettuce.

In varied RAW threshold irrigation scenarios, total simulated ETc ranged from 60 to 100 mm in site S1, and from 53 to 85 mm in site S2 (Figure 10a,c). The main reason for the higher ETc yield in site S1 is the higher adjusted transpiration characteristic of lettuce in sand soil as compared to loam soil. The simulated values of ETc fall within the range reported by Abdullah et al. [97] for lettuce, which varied from 43 mm to 285 mm in response to their different irrigation applications between 0 and 267 mm for open surface soil.

In varied field capacity threshold irrigation scenarios, simulated total crop evapotranspiration ranged from 77 to 205 mm in site S1, and from 83 to 211 mm in site S2 (Figure 10b,d). In both irrigation scenario classes, it was noted that while reducing irrigation events, crop evapotranspiration decreased simultaneously.

Figure 11 shows the response of biomass to the different irrigation scenarios. The varied RAW threshold irrigation scenarios (Figure 11a,c) resulted in biomass yield range from 0.88–1.77 ton ha^{-1} at site S1, and 0.44–0.91 ton ha^{-1} at site S2. By definition, biomass growth is closely related to crop evapotranspiration. Thus, the difference between biomass yields in the two experimental sites is due to the difference in the $Kc_{Tr,x}$ (coefficient for maximum crop transpiration) and CCx (maximum canopy cover) parameters between both sites.

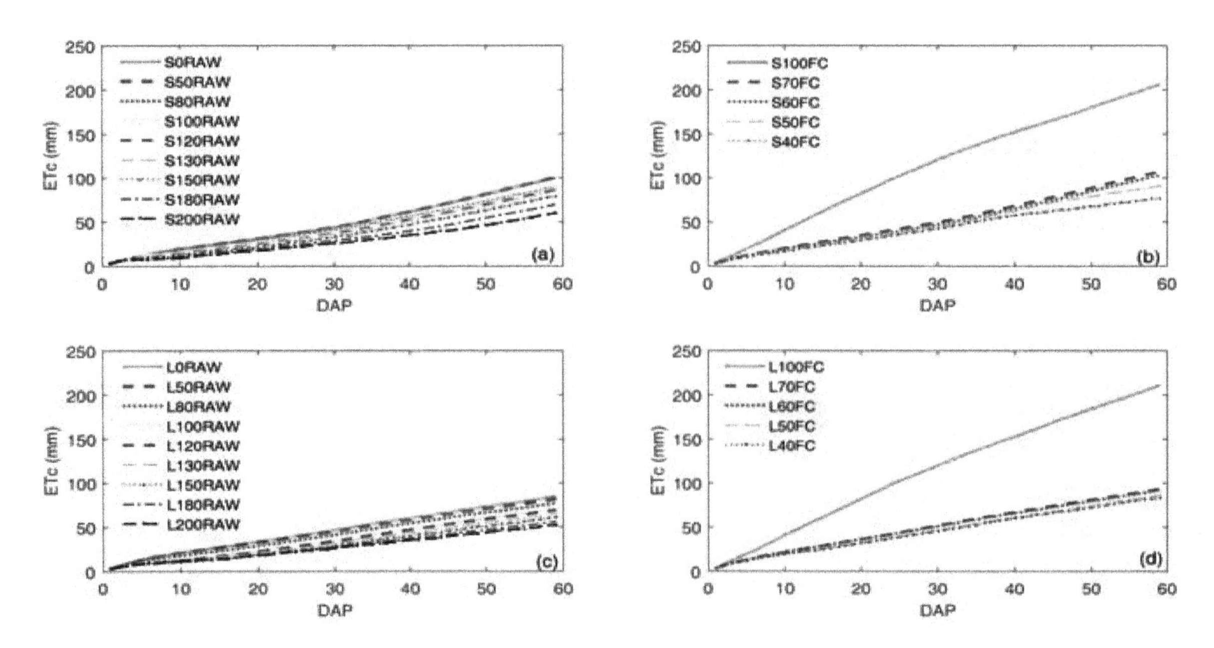

Figure 10. Crop evapotranspiration accumulation responses to different scenarios: (**a**) varied RAW threshold irrigation scenarios at site S1 (sand soil); (**b**) varied field capacity threshold irrigation scenarios at site S1; (**c**) varied RAW threshold irrigation scenarios at site S2 (loam soil); (**d**) varied field capacity threshold irrigation scenarios at site S2. RAW is readily available water content, S0RAW-S200RAW refers to irrigation scenarios with irrigation at 0–200% of RAW for sand soil. L0RAW-L200RAW refers to irrigation scenarios with irrigation at 0–200% of RAW for loam soil. S40FC-S100FC refers to deficit irrigation at 40–100% of field capacity for sand soil. L40FC-L100FC refers to deficit irrigation at 40–100% of field capacity for loam soil.

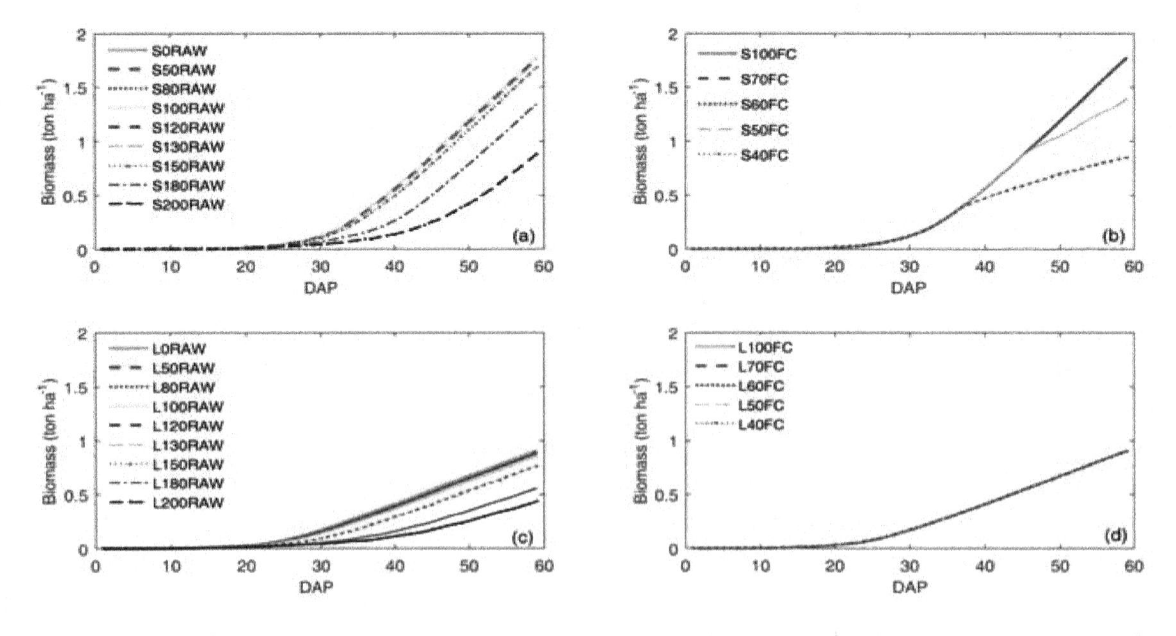

Figure 11. Biomass accumulation responses to different scenarios: (**a**) varied RAW threshold irrigation scenarios at site S1 (sand soil); (**b**) varied field capacity threshold irrigation scenarios at site S1; (**c**) varied RAW threshold irrigation scenarios at site S2 (loam soil); (**d**) varied field capacity threshold irrigation scenarios at site S2. RAW is readily available water content, S0RAW-S200RAW refers to irrigation scenarios with irrigation at 0–200% of RAW for sand soil. L0RAW-L200RAW refers to irrigation scenarios with irrigation at 0–200% of RAW for loam soil. S40FC-S100FC refers to deficit irrigation at 40–100% of field capacity for sand soil. L40FC-L100FC refers to deficit irrigation at 40–100% of field capacity for loam soil.

As expected, in varied RAW threshold irrigation scenarios, the simulations maintained biomass yield at 1.77 ton ha^{-1} at site S1 and 0.90 ton ha^{-1} at site S2 in the full irrigation scenarios with allowable depletion from 0–100% of RAW (e.g., S0RAW to S100RAW for site S1 and L0RAW to L100RAW for site S2), that is due to no-water stress condition. As the water stress started below the RAW line [41], with available depletion from 120–200% of RAW thresholds, the biomass yields decreased up to 50% in the S200RAW (200% of RAW threshold) scenario at site S1 and 52% in L200RAW scenario at site S2.

In varied field capacity threshold irrigation scenarios (Figure 11b,d), biomass yields ranged from 0.85 to 1.77 ton ha^{-1} at site S1, and 0.89 to 0.90 ton ha^{-1} at site S2. At site S1, reducing deficit irrigation at 50% of field capacity (S50FC scenario), the biomass yield started to decrease with 22% and deficit irrigation at 40% of field capacity (S40FC scenario), biomass yields decreased up to 51% compared to full irrigation scenario (S100FC). For site 2, deficit irrigation up to 40% of field capacity (L40FC) did not affect biomass yield.

3.3.3. Relationship between Water Productivity and Irrigation Scenarios

The responses of biomass yield and irrigation water productivity to irrigation depths in various scenarios are presented in Figure 12. Simulated water productivity of varied RAW threshold irrigation scenarios ranged from 1.5 to 2.1 kg m^{-3} for site S1 and 0.9 to 1.4 kg m^{-3} for site S2. In varied field capacity irrigation scenarios, simulated irrigation water productivity (IWP) ranged from 0.8 to 1.36 kg m^{-3} for site S1 and 0.43–1.08 kg m^{-3} for site S2. The simulated irrigation water productivity results are comparable with other studies found in the literature. For instance, Gallardo et al. [98] found a measured IWP for lettuce dry matter of 1.86 kg m^{-3}.

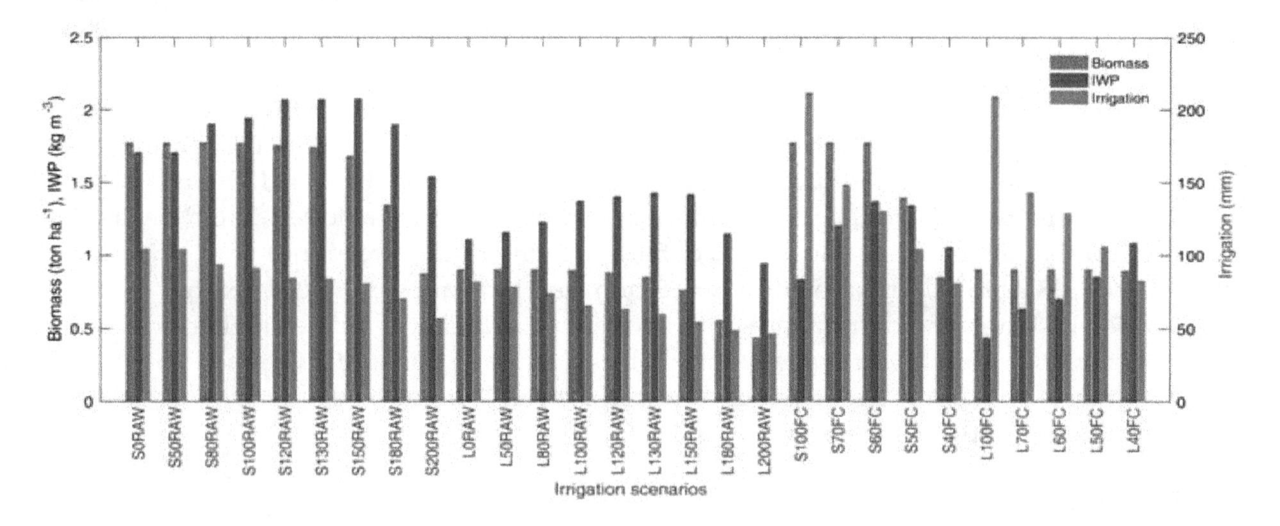

Figure 12. Comparison of biomass and water productivity response (IWP) to different irrigation scenarios. RAW is readily available water content. S0RAW-S200RAW refers to irrigation at 0–200% of RAW threshold irrigation scenarios for sand soil. L0RAW-L200RAW refers to refers to irrigation at 0–200% of RAW threshold irrigation scenarios for loam soil. S40FC-S100FC refers to deficit irrigation at 40–100% of field capacity for sand soil. L40FC-L100FC refers to deficit irrigation at 40–100% of field capacity for loam soil.

Figure 13 shows the relationship curves of biomass yield and irrigation water productivity response to irrigation scenarios. As expected, irrigation water productivity curve response to irrigation depths had parabolic relationships for both soil types in varied RAW threshold irrigation scenarios. Increasing water use efficiency can be enhanced by decreasing the irrigation to an optimum point. The optimum point, which resulted in 22% water saving for site S1, was found at the scenario with depletion of 150% of RAW (S150RAW), resulting in the irrigation water productivity = 2.07 kg m^{-3}, irrigation depth = 81 mm, and biomass yield = 1.68 ton ha^{-1}. For site S2, the optimum irrigation water productivity was at 130% of RAW scenario (L130RAW), resulting in irrigation water productivity = 1.42 kg m^{-3}, irrigation depth = 60 mm, and biomass yield = 0.85 ton ha^{-1}.

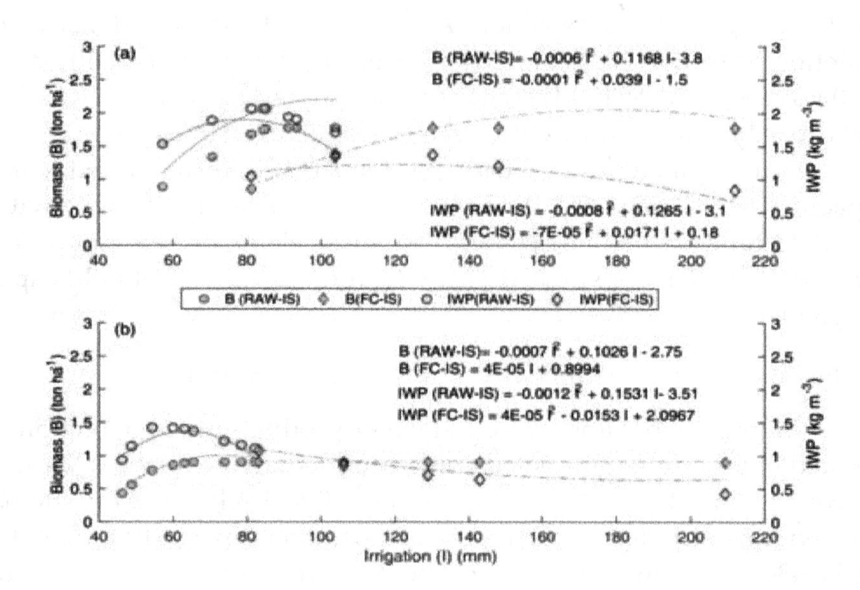

Figure 13. Relationship between biomass and irrigation water productivity responses to different scenarios: (**a**) at site S1 (sand soil) and (**b**) at site S2 (loam soil). I is irrigation, B is biomass, IWP is irrigation water productivity, RAW-IS is varied readily available water content threshold irrigation scenarios, FC-IS is varied field capacity threshold irrigation scenarios.

In varied field capacity threshold irrigation scenarios, for site S1, the optimum irrigation water productivity with 39% water saving was found at deficit irrigation at 60% of field capacity (S60FC) with irrigation water productivity = 1.36 kg m^{-3}, irrigation depth = 130 mm, and biomass yield = 1.77 ton ha^{-1}. For site S2, the optimum water productivity resulted in 60% water saving, which was found at deficit irrigation at 40% of field capacity (L40FC scenario) with irrigation water productivity = 1.08 kg m^{-3}, irrigation depth = 83 mm, and biomass yield = 0.89 ton ha^{-1}.

The varied RAW threshold irrigation scenarios resulted in higher simulated higher irrigation water productivity than the varied field capacity threshold scenarios in this study. Overall, deficit irrigation simulation scenarios in both irrigation scenario classes can provide a remarkable improvement in irrigation water productivity for water saving strategies.

3.3.4. Limitation

Crop models, like AquaCrop, are potentially valuable tools for answering questions primarily relating to research understanding, assessing crop management, and policy decision-making [49,99]. However, it is essential to test the models in diverse field environments, such as those with varied temperatures, elevation transects, or amidst latitudinal variations [99]. Particularly, AquaCrop has some limitations in terms of predicting crop yields only at the single growth cycle, single field scale, and only factoring in vertical water balance. The results of this study, obtained using climate data and field observation data relating to lettuce from a single growth cycle experiment at farm scale, allowed important information to be obtained in terms of calibrating lettuce crop parameters for sand and loam soil, and assessing limited water irrigation scenarios in the Cambodian context. However, it remains limited and the uncertainty on parameters has to be kept in mind. This study should be repeated in a contrasting range of diverse environments. Climate conditions and different cultural practices are the variables that differentiate the scenarios between different sites [99,100]. It has been emphasised that uncertainty model simulation results are themselves uncertain, due to known inadequacies of the model (residual errors in measurement) and due to unknown inadequacies of the model (by inputting new cultivars or different types of management, the model may be wrong in unsuspected ways) [101]. Despite such limitations, AquaCrop has already proven its usefulness in practical applications, and should still be tested widely in broader crop management applications, in diverse field environments [99,100].

4. Conclusions

An AquaCrop model was parameterised to simulate the canopy cover and aboveground biomass growth of lettuce under drip irrigation and plastic mulching for both sand and loam soil in the tropical monsoon climate of Cambodia. The model simulated canopy cover (RMSE < 0.8%) and aboveground biomass (RMSE < 0.01 ton ha^{-1}) in a satisfactory way after adjusting several key parameters, as mentioned in Farahani et al. [54].

Additionally, the results suggested that the incorporation of a heat stress factor affecting canopy cover and biomass growth is necessary to meet the conditions encountered in a tropical climate context.

Shortage of water in Cambodian agriculture has increased due to climate change, and this is a significant challenge facing farmers in their crop production. In this study, the AquaCrop model has helped to develop the simulation process for limited irrigation management strategies to maximise irrigation water productivity. To test the impact of different irrigation scheduling and water saving strategies, two scenario classes were explored: (i) varied readily available water (RAW) threshold irrigation and (ii) varied field capacity threshold irrigation scenarios. The irrigation scenario analysis proposed optimal irrigation strategies for lettuce.

For varied RAW threshold irrigation scenarios, the analysis proposed optimal simulated irrigation water productivity at scenarios of 150% of RAW (irrigation water productivity = 2.1 kg m^{-3}) for sand and 130% of RAW (irrigation water productivity = 1.4 kg m^{-3}) for loam soil. This can save 22% of water, and resulted in a biomass yield reduction of 5 and 2%, respectively, for sand and loam soil. For varied field capacity threshold irrigation scenarios, the optimal deficit irrigation depth was found at 60% of field capacity (irrigation water productivity of 1.4 kg m^{-3}) for sand soil, and at 40% of field capacity (irrigation water productivity of 1.0 kg m^{-3}) for loam soil. It can save water up to 39% and 60%, for sand and loam soil, respectively, maintaining biomass yields compared to full irrigation. These results suggest that deficit irrigation is worth considering as a water saving strategy for lettuce in the monsoon climate of Cambodia.

Overall, AquaCrop is a valuable tool to predict lettuce growth and to investigate different scenarios for providing irrigation scheduling strategies for water saving in Cambodia. However, further research is necessary to standardise the model parameters for lettuce in various irrigation management, environmental, and climatic conditions.

Author Contributions: P.K. performed the experiments, analysed the data, and wrote the paper. S.G. advised on the methodologies, gave comments and corrected the manuscript. C.O. supervised the research and gave comments to improve the manuscript. L.H. advised the agronomy practice during the experiments and gave comments on the manuscript. A.D. guided and supervised the research, gave comments, and corrected the manuscript.

Acknowledgments: This study was funded by the Belgian university cooperation programme, ARES-CCD (La Commission Coopération au Développement de l'Académie de Recherche et d'Enseignement supérieur).

References

1. Hoekstra, A.Y.; Wiedmann, T.O. Humanity's Unsustainable Environmental Footprint. *Science* **2014**, *344*, 1114–1117. [CrossRef] [PubMed]
2. Bae, J.; Dall'erba, S. Crop Production, Export of Virtual Water and Water-Saving Strategies in Arizona. *Ecol. Econ.* **2018**, *146*, 148–156. [CrossRef]
3. Rodríguez-Ferrero, N.; Salas-Velasco, M.; Sanchez-Martínez, M.T. Assessment of Productive Efficiency in Irrigated Areas of Andalusia. *Int. J. Water Resour. Dev.* **2010**, *26*, 365–379. [CrossRef]

4. Chartres, C. Is Water Scarcity a Constraint to Feeding Asia's Growing Population? *Int. J. Water Resour. Dev.* **2014**, *30*, 28–36. [CrossRef]
5. Jaramillo, F.; Destouni, G. Local Flow Regulation and Irrigation Raise Global Human Water Consumption and Footprint. *Science* **2015**, *350*, 1248–1251. [CrossRef] [PubMed]
6. Toumi, J.; Er-Raki, S.; Ezzahar, J.; Khabba, S.; Jarlan, L.; Chehbouni, A. Performance Assessment of AquaCrop Model for Estimating Evapotranspiration, Soil Water Content and Grain Yield of Winter Wheat in Tensift Al Haouz (Morocco): Application to Irrigation Management. *Agric. Water Manag.* **2016**, *163*, 219–235. [CrossRef]
7. Linker, R.; Ioslovich, I. Assimilation of Canopy Cover and Biomass Measurements in the Crop Model AquaCrop. *Biosyst. Eng.* **2017**, *162*, 57–66. [CrossRef]
8. Touch, V.; Martin, R.J.; Scott, J.F.; Cowie, A.; Liu, D.L. Climate Change Adaptation Options in Rainfed Upland Cropping Systems in the Wet Tropics: A Case Study of Smallholder Farms in North-West Cambodia. *J. Environ. Manag.* **2016**, *182*, 238–246. [CrossRef] [PubMed]
9. Chhinh, N.; Millington, A. Drought Monitoring for Rice Production in Cambodia. *Climate* **2015**, *3*, 792–811. [CrossRef]
10. Montgomery, S.C.; Martin, R.J.; Guppy, C.; Wright, G.C.; Tighe, M.K. Farmer Knowledge and Perception of Production Constraints in Northwest Cambodia. *J. Rural Stud.* **2017**, *56*, 12–20. [CrossRef]
11. Moreira, M.A.; Dos Santos, C.A.P.; Lucas, A.A.T.; Bianchini, F.G.; De Souza, I.M.; Viégas, P.R.A. Lettuce Production according to Different Sources of Organic Matter and Soil Cover. *Agric. Sci.* **2014**, *5*, 99–105. [CrossRef]
12. Valenzuela, H.R.; Bernard, K.; John, C. *Lettuce Production Guidelines for Hawaii*; University of Hawaii: Honolulu, HI, USA, 1996.
13. Cahn, M.; Johnson, L. New Approaches to Irrigation Scheduling of Vegetables. *Horticulturae* **2017**, *3*, 1–20. [CrossRef]
14. Domingues, D.S.; Takahashi, H.W.; Camara, C.A.P.; Nixdorf, S.L. Automated System Developed to Control pH and Concentration of Nutrient Solution Evaluated in Hydroponic Lettuce Production. *Comput. Electron. Agric.* **2012**, *84*, 53–61. [CrossRef]
15. Sokhen, C.; Kanika, D.; Moustier, P. *Vegetable Market Flows and Chains in Phnom Penh*; CIRAD-AVRDC-French MOFA: Hanoi, Vietnam, 2004.
16. De Bon, H.; Parrot, L.; Moustier, P. Sustainable Urban Agriculture in Developing Countries. A Review. *Agron. Sustain. Dev.* **2010**, *30*, 21–32. [CrossRef]
17. Morris, S.; Davies, W.; Baines, R.N. Challenges and Opportunities for Increasing Competitiveness of Vegetable Production in Cambodia. *Acta Hortic.* **2013**, *1006*, 253–260. [CrossRef]
18. Xue, J.; Huo, Z.; Wang, F.; Kang, S.; Huang, G. Untangling the Effects of Shallow Groundwater and Deficit Irrigation on Irrigation Water Productivity in Arid Region: New Conceptual Model. *Sci. Total Environ.* **2018**, *619–620*, 1170–1182. [CrossRef] [PubMed]
19. Adu, M.O.; Yawson, D.O.; Armah, F.A.; Asare, P.A.; Frimpong, K.A. Meta-Analysis of Crop Yields of Full, Deficit, and Partial Root-Zone Drying Irrigation. *Agric. Water Manag.* **2018**, *197*, 79–90. [CrossRef]
20. Liu, Y.; Luo, Y. A Consolidated Evaluation of the FAO-56 Dual Crop Coefficient Approach Using the Lysimeter Data in the North China Plain. *Agric. Water Manag.* **2010**, *97*, 31–40. [CrossRef]
21. Verstraeten, W.W.; Veroustraete, F.; Feyen, J. Assessment of Evapotranspiration and Soil Moisture Content across Different Scales of Observation. *Sensors* **2008**, *8*, 70–117. [CrossRef] [PubMed]
22. Hunsaker, D.J.; French, A.N.; Waller, P.M.; Bautista, E.; Thorp, K.R.; Bronson, K.F.; Andrade-Sanchez, P. Comparison of Traditional and ET-Based Irrigation Scheduling of Surface-Irrigated Cotton in the Arid Southwestern USA. *Agric. Water Manag.* **2015**, *159*, 209–224. [CrossRef]
23. Thompson, R.B.; Gallardo, M.; Valdez, L.C.; Fernández, M.D. Determination of Lower Limits for Irrigation Management Using in Situ Assessments of Apparent Crop Water Uptake Made with Volumetric Soil Water Content Sensors. *Agric. Water Manag.* **2007**, *92*, 13–28. [CrossRef]
24. Ferreira, M.I.; Conceição, N.; Malheiro, A.C.; Silvestre, J.M.; Silva, R.M. Water Stress Indicators and Stress Functions to Calculate Soil Water Depletion in Deficit Irrigated Grapevine and Kiwi. *Acta Hortic.* **2017**, *1150*, 119–126. [CrossRef]
25. Li, S.; Kang, S.; Li, F.; Zhang, L. Evapotranspiration and Crop Coefficient of Spring Maize with Plastic Mulch Using Eddy Covariance in Northwest China. *Agric. Water Manag.* **2008**, *95*, 1214–1222. [CrossRef]

26. Kashyap, P.S.; Panda, R.K. Evaluation of Evapotranspiration Estimation Methods and Development of Crop-Coefficients for Potato Crop in a Sub-Humid Region. *Agric. Water Manag.* **2001**, *50*, 9–25. [CrossRef]
27. Inthavong, T.; Tsubo, M.; Fukai, S. A Water Balance Model for Characterization of Length of Growing Period and Water Stress Development for Rainfed Lowland Rice. *Field Crop. Res.* **2011**, *121*, 291–301. [CrossRef]
28. Davis, S.L.; Dukes, M.D. Irrigation Scheduling Performance by Evapotranspiration-Based Controllers. *Agric. Water Manag.* **2010**, *98*, 19–28. [CrossRef]
29. Kukal, S.S.; Hira, G.S.; Sidhu, A.S. Soil Matric Potential-Based Irrigation Scheduling to Rice (*Oryza sativa*). *Irrig. Sci.* **2005**, *23*, 153–159. [CrossRef]
30. Pereira, L.S.; Paredes, P.; Sholpankulov, E.D.; Inchenkova, O.P.; Teodoro, P.R.; Horst, M.G. Irrigation Scheduling Strategies for Cotton to Cope with Water Scarcity in the Fergana Valley, Central Asia. *Agric. Water Manag.* **2009**, *96*, 723–735. [CrossRef]
31. Pereira, L.S.; Cordery, I.; Iacovides, I. Improved Indicators of Water Use Performance and Productivity for Sustainable Water Conservation and Saving. *Agric. Water Manag.* **2012**, *108*, 39–51. [CrossRef]
32. Afzal, M.; Battilani, A.; Solimando, D.; Ragab, R. Improving Water Resources Management Using Different Irrigation Strategies and Water Qualities: Field and Modelling Study. *Agric. Water Manag.* **2016**, *176*, 40–54. [CrossRef]
33. Geerts, S.; Raes, D. Deficit Irrigation as an on-Farm Strategy to Maximize Crop Water Productivity in Dry Areas. *Agric. Water Manag.* **2009**, *96*, 1275–1284. [CrossRef]
34. Chai, Q.; Gan, Y.; Zhao, C.; Xu, H.L.; Waskom, R.M.; Niu, Y.; Siddique, K.H.M. Regulated Deficit Irrigation for Crop Production under Drought Stress. A Review. *Agron. Sustain. Dev.* **2016**, *36*, 1–21. [CrossRef]
35. Lopez, J.R.; Winter, J.M.; Elliott, J.; Ruane, A.C.; Porter, C.; Hoogenboom, G. Integrating Growth Stage Deficit Irrigation into a Process Based Crop Model. *Agric. For. Meteorol.* **2017**, *243*, 84–92. [CrossRef]
36. Kögler, F.; Söffker, D. Water (Stress) Models and Deficit Irrigation: System-Theoretical Description and Causality Mapping. *Ecol. Model.* **2017**, *361*, 135–156. [CrossRef]
37. Patanè, C.; Tringali, S.; Sortino, O. Effects of Deficit Irrigation on Biomass, Yield, Water Productivity and Fruit Quality of Processing Tomato under Semi-Arid Mediterranean Climate Conditions. *Sci. Hortic. (Amsterdam)* **2011**, *129*, 590–596. [CrossRef]
38. Abd El-Wahed, M.H.; Baker, G.A.; Ali, M.M.; Abd El-Fattah, F.A. Effect of Drip Deficit Irrigation and Soil Mulching on Growth of Common Bean Plant, Water Use Efficiency and Soil Salinity. *Sci. Hortic. (Amsterdam)* **2017**, *225*, 235–242. [CrossRef]
39. Samperio, A.; Moñino, M.J.; Vivas, A.; Blanco-Cipollone, F.; Martín, A.G.; Prieto, M.H. Effect of Deficit Irrigation during Stage II and Post-Harvest on Tree Water Status, Vegetative Growth, Yield and Economic Assessment in "Angeleno" Japanese Plum. *Agric. Water Manag.* **2015**, *158*, 69–81. [CrossRef]
40. Yang, C.; Luo, Y.; Sun, L.; Wu, N. Effect of Deficit Irrigation on the Growth, Water Use Characteristics and Yield of Cotton in Arid Northwest China. *Pedosphere* **2015**, *25*, 910–924. [CrossRef]
41. Payero, J.O.; Melvin, S.R.; Irmak, S.; Tarkalson, D. Yield Response of Corn to Deficit Irrigation in a Semiarid Climate. *Agric. Water Manag.* **2006**, *84*, 101–112. [CrossRef]
42. Karam, F.; Mounzer, O.; Sarkis, F.; Lahoud, R. Yield and Nitrogen Recovery of Lettuce under Different Irrigation Regimes. *J. Appl. Hortic.* **2002**, *4*, 70–76.
43. Kuslu, Y.; Dursun, A.; Sahin, U.; Kiziloglu, F.M.; Turan, M. Short Communication. Effect of Deficit Irrigation on Curly Lettuce Grown under Semiarid Conditions. *Span. J. Agric. Res.* **2008**, *6*, 714–719. [CrossRef]
44. Geerts, S.; Raes, D.; Garcia, M. Using AquaCrop to Derive Deficit Irrigation Schedules. *Agric. Water Manag.* **2010**, *98*, 213–216. [CrossRef]
45. Hassanli, M.; Ebrahimian, H.; Mohammadi, E.; Rahimi, A.; Shokouhi, A. Simulating Maize Yields When Irrigating with Saline Water, Using the AquaCrop, SALTMED, and SWAP Models. *Agric. Water Manag.* **2016**, *176*, 91–99. [CrossRef]
46. Abderrahman, W.A.; Mohammed, N.; Al-Harazin, I.M. Computerized and Dynamic Model for Irrigation Water Management of Large Irrigation Schemes in Saudi Arabia. *Int. J. Water Resour. Dev.* **2001**, *17*, 261–270. [CrossRef]
47. Wolf, J.; Evans, L.G.; Semenov, M.A.; Eckersten, H.; Iglesias, A. Comparison of Wheat Simulation Models under Climate Change. I. Model Calibration and Sensitivity Analyses. *Clim. Res.* **1996**, *7*, 253–270. [CrossRef]

48. Ran, H.; Kang, S.; Li, F.; Tong, L.; Ding, R.; Du, T.; Li, S.; Zhang, X. Performance of AquaCrop and SIMDualKc Models in Evapotranspiration Partitioning on Full and Deficit Irrigated Maize for Seed Production under Plastic Film-Mulch in an Arid Region of China. *Agric. Syst.* **2017**, *151*, 20–32. [CrossRef]
49. Steduto, P.; Hsiao, T.C.; Raes, D.; Fereres, E. AquaCrop—The FAO Crop Model to Simulate Yield Response to Water: I. Concepts and Underlying Principles. *Agron. J.* **2009**, *101*, 426–437. [CrossRef]
50. Singh, A.; Saha, S.; Mondal, S. Modelling Irrigated Wheat Production Using the FAO AquaCrop Model in West Bengal, India, for Sustainable Agriculture. *Irrig. Drain.* **2013**, *62*, 50–56. [CrossRef]
51. Tavakoli, A.R.; Mahdavi Moghadam, M.; Sepaskhah, A.R. Evaluation of the AquaCrop Model for Barley Production under Deficit Irrigation and Rainfed Condition in Iran. *Agric. Water Manag.* **2015**, *161*, 136–146. [CrossRef]
52. Paredes, P.; Wei, Z.; Liu, Y.; Xu, D.; Xin, Y.; Zhang, B.; Pereira, L.S. Performance Assessment of the FAO AquaCrop Model for Soil Water, Soil Evaporation, Biomass and Yield of Soybeans in North China Plain. *Agric. Water Manag.* **2015**, *152*, 57–71. [CrossRef]
53. Todorovic, M.; Albrizio, R.; Zivotic, L.; Saab, M.-T.A.; Stöckle, C.; Steduto, P. Assessment of AquaCrop, CropSyst, and WOFOST Models in the Simulation of Sunflower Growth under Different Water Regimes. *Agron. J.* **2009**, *101*, 509–521. [CrossRef]
54. Farahani, H.J.; Izzi, G.; Oweis, T.Y. Parameterization and Evaluation of the AquaCrop Model for Full and Deficit Irrigated Cotton. *Agron. J.* **2009**, *101*, 469–476. [CrossRef]
55. Hussein, F.; Janat, M.; Yakoub, A. Simulating Cotton Yield Response to Deficit Irrigation with the FAO AquaCrop Model. *Span. J. Agric. Res.* **2011**, *9*, 1319–1330. [CrossRef]
56. Hsiao, T.C.; Heng, L.; Steduto, P.; Rojas-Lara, B.; Raes, D.; Fereres, E. AquaCrop—The FAO Crop Model to Simulate Yield Response to Water: III. Parameterization and Testing for Maize. *Agron. J.* **2009**, *101*, 448–459. [CrossRef]
57. Malik, A.; Shakir, A.S.; Ajmal, M.; Khan, M.J.; Khan, T.A. Assessment of AquaCrop Model in Simulating Sugar Beet Canopy Cover, Biomass and Root Yield under Different Irrigation and Field Management Practices in Semi-Arid Regions of Pakistan. *Water Resour. Manag.* **2017**, *31*, 4275–4292. [CrossRef]
58. Andarzian, B.; Bannayan, M.; Steduto, P.; Mazraeh, H.; Barati, M.E.; Barati, M.A.; Rahnama, A. Validation and Testing of the AquaCrop Model under Full and Deficit Irrigated Wheat Production in Iran. *Agric. Water Manag.* **2011**, *100*, 1–8. [CrossRef]
59. Mkhabela, M.S.; Bullock, P.R. Performance of the FAO AquaCrop Model for Wheat Grain Yield and Soil Moisture Simulation in Western Canada. *Agric. Water Manag.* **2012**, *110*, 16–24. [CrossRef]
60. Rankine, D.R.; Cohen, J.E.; Taylor, M.A.; Coy, A.D.; Simpson, L.A.; Stephenson, T.; Lawrence, J.L. Parameterizing the FAO AquaCrop Model for Rainfed and Irrigated Field-Grown Sweet Potato. *Agron. J.* **2015**, *107*, 375–387. [CrossRef]
61. Casa, A. De; Ovando, G.; Bressanini, L.; Martínez, J. Aquacrop Model Calibration in Potato and Its Use to Estimate Yield Variability under Field Conditions. *Atmos. Clim. Sci.* **2013**, *3*, 397–407. [CrossRef]
62. Wellens, J.; Raes, D.; Traore, F.; Denis, A.; Djaby, B.; Tychon, B. Performance Assessment of the FAO AquaCrop Model for Irrigated Cabbage on Farmer Plots in a Semi-Arid Environment. *Agric. Water Manag.* **2013**, *127*, 40–47. [CrossRef]
63. Deb, P.; Tran, D.A.; Udmale, P.D. Assessment of the Impacts of Climate Change and Brackish Irrigation Water on Rice Productivity and Evaluation of Adaptation Measures in Ca Mau Province, Vietnam. *Theor. Appl. Climatol.* **2016**, *125*, 641–656. [CrossRef]
64. Adeboye, O.B.; Schultz, B.; Adekalu, K.O.; Prasad, K. Modelling of Response of the Growth and Yield of Soybean to Full and Deficit Irrigation by Using Aquacrop. *Irrig. Drain.* **2017**, *66*, 192–205. [CrossRef]
65. Zeleke, K.T.; Luckett, D.; Cowley, R. Calibration and Testing of the FAO AquaCrop Model for Canola. *Agron. J.* **2011**, *103*, 1610–1618. [CrossRef]
66. Greaves, G.E.; Wang, Y.M. Assessment of Fao Aquacrop Model for Simulating Maize Growth and Productivity under Deficit Irrigation in a Tropical Environment. *Water* **2016**, *8*, 1–18. [CrossRef]
67. Montoya, F.; Camargo, D.; Ortega, J.F.; Córcoles, J.I.; Domínguez, A. Evaluation of Aquacrop Model for a Potato Crop under Different Irrigation Conditions. *Agric. Water Manag.* **2016**, *164*, 267–280. [CrossRef]
68. Klocke, N.L.; Fischbach, P.E. G84-690 Estimating Soil Moisture by Appearance and Feel. In *Historical Materials from University of Nebraska-Lincoln Extension*; University of Nebraska: Lincoln, NE, USA, 1984; pp. 1–9.

69. Allen, R.G.; Pereira, L.S.; Raes, D.; Smith, M. Crop Evapotranspiration: Guidelines for Computing Crop Water Requirements. *Irrig. Drain.* **1998**, *300*, 300.
70. Pansu, M.; Gautheyrou, J. *Handbook of Soil Analysis: Mineralogical, Organic and Inorganic Methods*; Springer Science & Business Media: Berlin, Germany, 2007.
71. Margesin, R.; Schinner, F. *Manual for Soil Analysis-Monitoring and Assessing Soil Bioremediation*; Springer Science & Business Media: Berlin, Germany, 2005.
72. Ket, P.; Garré, S.; Oeurng, C.; Degré, A. *A Comparison of Soil Water Retention Curves Obtained Using Field, Lab and Modelling Methods in Monsoon Context of Cambodia*; ARES-CCD: Brussels, Belgium, 2018.
73. Gallardo, M.; Jackson, L.E.E.; Schulbach, K.; Snyder, R.L.L.; Thompson, R.B.B.; Wyland, L.J.J. Production and Water Use in Lettuces under Variable Water Supply. *Irrig. Sci.* **1996**, *16*, 125–137. [CrossRef]
74. Razzaghi, F.; Zho, Z.; Andersen, M.; Plauborg, F. Simulation of Potato Yield in Temperate Condition by the AquaCrop Model. *Agric. Water Manag.* **2017**, *191*, 113–123. [CrossRef]
75. Raes, D.; Steduto, P.; Hsiao, T.C.; Fereres, E. AquaCrop—The FAO Crop Model to Simulate Yield Response to Water: II. Main Algorithms and Software Description. *Agron. J.* **2009**, *101*, 438–447. [CrossRef]
76. Steduto, P.; Raes, D.; Hsiao, T.; Fereres, E. AquaCrop: A New Model for Crop Prediction under Water Deficit Conditions. *Options Méditerr.* **2009**, *33*, 285–292.
77. Steduto, P.; Hsiao, T.C.; Fereres, E.; Raes, D. *Crop Yield Response to Water*; The Food and Agriculture Organization (FAO): Rome, Italy, 2012.
78. Paredes, P.; de Melo-Abreu, J.P.; Alves, I.; Pereira, L.S. Assessing the Performance of the FAO AquaCrop Model to Estimate Maize Yields and Water Use under Full and Deficit Irrigation with Focus on Model Parameterization. *Agric. Water Manag.* **2014**, *144*, 81–97. [CrossRef]
79. Morris, M.D. Factorial Plans for Preliminary Computational Experiments. *Technometrics* **1991**, *33*, 161–174. [CrossRef]
80. Stott, L.D. The Influence of Diet on the δ13C of Shell Carbon in the Pulmonate Snail Helix Aspersa. *Earth Planet. Sci. Lett.* **2002**, *195*, 249–259. [CrossRef]
81. Krause, P.; Boyle, D.P. Comparison of Different Efficiency Criteria for Hydrological Model Assessment. *Adv. Geosci.* **2005**, *5*, 89–97. [CrossRef]
82. Wellens, J.; Raes, D.; Tychon, B. On the Use of Decision-Support Tools for Improved Irrigation Management: AquaCrop-Based Applications. In *Current Perspective on Irrigation and Drainage*; INTECH: London, UK, 2017; pp. 53–67.
83. Zhuo, L.; Hoekstra, A. The Effect of Different Agricultural Management Practices on Irrigation Efficiency, Water Use Efficiency and Green and Blue Water Footprint. *Front. Agric. Sci. Eng.* **2017**, *4*, 185–194. [CrossRef]
84. Raes, D.; Steduto, P.; Hsiao, T.C.; Fereres, E. Calculation Procedures. In *AquaCrop-Reference Manual*; The Food and Agriculture Organization (FAO): Rome, Italy, 2017.
85. Raes, D.; Geerts, S.; Kipkorir, E.; Wellens, J.; Sahli, A. Simulation of Yield Decline as a Result of Water Stress with a Robust Soil Water Balance Model. *Agric. Water Manag.* **2006**, *81*, 335–357. [CrossRef]
86. Navarro-Hellín, H.; Martínez-del-Rincon, J.; Domingo-Miguel, R.; Soto-Valles, F.; Torres-Sánchez, R. A Decision Support System for Managing Irrigation in Agriculture. *Comput. Electron. Agric.* **2016**, *124*, 121–131. [CrossRef]
87. Raes, P.D.; Steduto, T.C.; Hsiao, E.F. *FAO Crop.-Water Productivity Model to Simulate Yield Response to Water. Reference Manual*; Food and Agriculture Organization of the United Nations: Rome, Italy, 2017.
88. Allen, R.G.; Pereira, L.S.; Raes, D.; Smith, M.; Ab, W. Crop Evapotranspiration-Guidelines for Computing Crop Water Requirements. In *FAO Irrigation and Drainage Paper 56*; Food and Agriculture Organization: Rome, Italy, 1998; pp. 1–15.
89. Lamn, F.; Ayars, J.; Nakayama, F. Irrigation Scheduling. In *Microirrigation for Crop Production*; Developments in Agricultural Engineering 13; Freddie, R., Lamm James, E., Ayars Francis, S., Nakayama, Eds.; Elsevier: Houston, TX, 2015; pp. 61–128.
90. Sutton, B.; Merit, N. Maintenance of Lettuce Root Zone at Field Capacity Gives Best Yields with Drip Irrigation. *Sci. Hortic. (Amsterdam)* **1993**, *56*, 1–11. [CrossRef]
91. Tan, S.; Wang, Q.; Zhang, J.; Chen, Y.; Shan, Y.; Xu, D. Performance of AquaCrop Model for Cotton Growth Simulation under Film-Mulched Drip Irrigation in Southern Xinjiang, China. *Agric. Water Manag.* **2018**, *196*, 99–113. [CrossRef]

92. Fazilah, W.; Ilahi, F.; Ahmad, D.; Husain, M.C. Effects of Root Zone Cooling on Butterhead Lettuce Grown in Tropical Conditions in a Coir-Perlite Mixture. *Hortic. Environ. Biotechnol.* **2017**, *58*, 1–4. [CrossRef]
93. Zhang, G.; Johkan, M.; Hohjo, M.; Tsukagoshi, S.; Maruo, T. Plant Growth and Photosynthesis Response to Low Potassium Conditions in Three Lettuce (*Lactuca sativa*) Types. *Hortic. J.* **2017**, *86*, 229–237. [CrossRef]
94. Dufault, R.J.; Ward, B.; Hassell, R.L. Dynamic Relationships between Field Temperatures and Romaine Lettuce Yield and Head Quality. *Sci. Hortic. (Amsterdam)* **2009**, *120*, 452–459. [CrossRef]
95. Parker, R.O. *Plant. & Soil Science: Fundamentals and Applications*, 1st ed.; Delmar Cengage Learning: Independence, KY, USA, 2009.
96. Wheeler, T.R.; Hadley, P.; Morison, J.I.; Ellis, R.H. Effects of Temperature on the Growth of Lettuce (*Lactuca sativa* L.) and the Implications for Assessing the Impacts of Potential Climate Change. *Eur. J. Agron.* **1993**, *2*, 305–311. [CrossRef]
97. Abdullah, K.; Ismail, T.G.; Yusuf, U.; Belgin, C. Effects of Mulch and Irrigation Water Amounts on Lettuce's Yield, Evapotranspiration, Transpiration and Soil Evaporation in Isparta Location, Turkey. *J. Biol. Sci.* **2004**, *4*, 751–755.
98. Gallardo, M.; Snyder, R.L.; Schulbach, K.; Jackson, L.E. Crop Growth and Water Use Model for Lettuce. *Irrig. Drain. Eng.* **1996**, *122*, 354–359. [CrossRef]
99. Boote, K.J.; Jones, J.W.; Pickering, N.B. Potential Uses and Limitations of Crop Models. *Agron. J.* **1996**, *88*, 704–716. [CrossRef]
100. Silvestro, P.C.; Pignatti, S.; Yang, H.; Yang, G.; Pascucci, S.; Castaldi, F.; Casa, R. Sensitivity Analysis of the Aquacrop and SAFYE Crop Models for the Assessment of Water Limited Winter Wheat Yield in Regional Scale Applications. *PLoS ONE* **2017**, *12*, 1–30. [CrossRef] [PubMed]
101. Wallach, D.; Keussayan, N.; Brun, F.; Lacroix, B.; Bergez, J. Assessing the Uncertainty When Using a Model to Compare Irrigation Strategies. *Agron. J.* **2008**, *104*, 1274–1283. [CrossRef]

9

Sustainable Irrigation in Agriculture: An Analysis of Global Research

Juan F. Velasco-Muñoz [1], José A. Aznar-Sánchez [1,*], Ana Batlles-delaFuente [1] and Maria Dolores Fidelibus [2]

[1] Department of Economy and Business, Research Centre CAESCG and CIAIMBITAL, University of Almería, 04120 Almería, Spain
[2] Department of Civil, Environmental, Land, Building Engineering and Chemistry, Polytechnic University of Bari, 70126 Bari, Italy
* Correspondence: jaznar@ual.es

Abstract: Irrigated agriculture plays a fundamental role as a supplier of food and raw materials. However, it is also the world's largest water user. In recent years, there has been an increase in the number of studies analyzing agricultural irrigation from the perspective of sustainability with a focus on its environmental, economic, and social impacts. This study seeks to analyze the dynamics of global research in sustainable irrigation in agriculture between 1999 and 2018, including the main agents promoting it and the topics that have received the most attention. To do this, a review and a bibliometric analysis were carried out on a sample of 713 articles. The results show that sustainability is a line of study that is becoming increasingly more prominent within research in irrigation. The study also reveals the existence of substantial differences and preferred topics in the research undertaken by different countries. The priority issues addressed in the research were climatic change, environmental impact, and natural resources conservation; unconventional water resources; irrigation technology and innovation; and water use efficiency. Finally, the findings indicate a series of areas related to sustainable irrigation in agriculture in which research should be promoted.

Keywords: sustainable irrigation; bibliometric analysis; climate change; innovation and technology; water use efficiency; unconventional water resources

---

1. Introduction

The current global context is conditioned by the growth of the world's population and the progressive and continuous deterioration of the environment. This creates the challenge of ensuring the supply of basic resources, such as food and water, and sustainable development [1], where water plays an essential role in the survival of human society [2] and contributes to the provision of a wide range of services on which the wellbeing of society is based [3–5]. However, water resources are subject to severe degradation due to many factors, such as the consequences of global climate change, alterations in the use of land, agricultural and urban expansion, and overexploitation due to economic development [6–8]. In parallel with this degradation and overexploitation of ecosystems and water resources, the demand for the services supplied by these resources is expected to increase.

Agricultural ecosystems are the principal suppliers of food, but they are also the main users of water resources on a global level [9,10]. These ecosystems use between 60% and 90% of the available water, depending on the climate and economic development of the region [11,12]. The global area dedicated to irrigated crops is estimated to be 275 million hectares, with an upward growth trend of 1.3% per year [13]. This accounts for just 23% of farmed land; however, 45% of total food production is obtained through these types of crops [14,15]. It has been estimated that in order to satisfy the food demand in 2050, world production must increase by 70% [16]. In a scenario of low production, in order to fulfil this objective, it will be necessary to increase the use of water resources on a global level by 53% [17]—around 50% in developing countries and 16% in developed countries [18]—keeping the current values of variables like productivity and technology.

Currently, different approaches are being used to address the challenges of food provision and the supply of water for different uses and to maintain an environmental balance. Some works point to the development of measures to control demand so that irrigation water sustainability can be reached. The development of efficient water markets can be an optimal measure in underdeveloped areas and with a high level of water scarcity, like in South Africa [19,20]. The implementation of joint restrictions based on the establishment of quotas and the payment of fees can be an effective control system for the use of agriculture water in developed regions specialized in the production of high-quality crops and where overexploitation of water resources is currently taking place [21]. Regarding water supply, many authors recommend the joint use of different water resources and the development of infrastructures as nonconventional water sources [22,23]. Another line of research is focused on the improvement of the efficiency of water use and the development of clean production models that guarantee sustainability from social and economic perspectives [24,25]. In order to achieve this objective, the whole irrigation process must be analyzed. This process covers different phases beginning with the water source and ending with its use for agriculture. Zhang et al. [26] identified three phases in irrigation: The first includes the extraction of water from the source and its transfer through channels to the point of use; the second consists of the distribution of the water to the root system to facilitate its absorption by crops (this includes both traditional irrigation using floods and furrows and modern irrigation through drip systems and microsprinklers); and the third covers the whole crop-growing process, whereby the water is transported from the roots to the rest of the plant. The goal is to save resources through minimizing water losses during these three phases and to improve the efficiency in the use of water resources.

The so-called "Science of Sustainability" also studies how to address these challenges. It is defined as "a discipline that points the way towards a sustainable society" and is "aimed at understanding the fundamental character of interactions between natural, human, and social systems, covers a wide range of academic disciplines", for the development of agricultural systems and the sustainable use of water [27–29]. At the end of the 1990s, sustainability was used as a characteristic to describe ecosystems, referring to the capacity to maintain the flow of services in different environmental, economic, and social contexts [30]. When it is applied to the management of water resources in agriculture, sustainability is considered to be a series of practices that increase crop yield and minimize water losses [31]. The objectives of the sustainable management of water resources in agriculture consider the continuity of the agricultural system from physical and biological perspectives, as well as the economic efficiency of the use of the resources and social participation in the decision-making processes [32]. An evaluation of a change in water use requires, therefore, a multidisciplinary approach that includes an analysis of the body of water under study in order to understand the possible impacts on the quantity and quality of the water and the timetable of the different uses. A comprehensive evaluation of the marginal productivity of water is also required, together with an analysis of its nonmarketable value, such as that derived from ecosystem services [33].

In recent years, there has been an increase in the number of studies analyzing agricultural irrigation from the perspective of sustainability with a focus on its environmental, economic, and social impacts. The objective of this study is to analyze the dynamics of the research on sustainable irrigation in agriculture over the last twenty years. In order to fulfil this objective, a two-fold analysis was undertaken: quantitatively through a bibliometric analysis; and qualitatively through a systemic review based on keyword analysis. The study analyzes the evolution of the number of articles published, the main authors, institutions and countries that promote this research field, the disciplines involved in the research, the main lines of research, the differences in academic approach and the countries considered, and the main issues that affect the research in this field.

Bibliometric analysis was introduced by Garfield in the 1950s [34], and its objective is to identify, classify, and evaluate the principal components within a specific research field [35]. Bibliometry combines tools of quantitative analysis to study the trends of a research topic and identify the main driving agents and the relevance of their publications [36,37]. In bibliometric analyses, three types of indicators can be distinguished, which were defined by Durieux and Gevenois [38]: productivity indicators, relevance indicators, and structural indicators. In addition to these indicators, different approaches exist in bibliometric analysis. Co-occurrence, co-citation, and bibliographic coupling analysis are among the traditional approaches. This extended methodology can be considered as a new one in some research areas. This has also continuously been developing. In this sense, this work introduces some new methodological aspects which provide a contribution regarding previous works—in fact, the sample search process, a mixed quantitative and qualitative review, and the production of keyword networks to identify main trends per country. The results of this study provide a basis on which to establish priorities and to develop new projects in future research on this topic.

2. Methodology

In order to conduct this study, a traditional approach based on co-occurrence was selected, which included the assessment of productivity, quality, and structural indicators. In this approach, first, the agents with the highest number of publications were identified, and second, the impact of the publications of these authors was analyzed. This type of analysis, particularly with respect to journals, is highly interesting for researchers, given that it constitutes a way to assess the relevance of the journals in which authors publish their studies [39]. Finally, we used mapping techniques to analyze the structure of the network between different agents. The Scopus database was used to select the sample of studies to analyze. This database has proven to be the most suitable for our area of study, enabling us to ensure the selection of a representative sample of the studies carried out on sustainable irrigation (SI). Furthermore, it is easy to access, allows the visualization and analysis of data, and allows data to be downloaded in different formats for subsequent processing using software applications [40]. Nevertheless, if some works on SI are not indexed in the Scopus database, they have not been considered in our sample.

The term used to carry out the search was "sustainable irrigation", and this selection was based on previous studies on the same topic [41–43]. This term was searched for under authors' keywords and titles. The study period selected was 1999 to 2018. Research activity in this topic peaked during these years. Furthermore, this period immediately followed the 1st World Water Forum held in Marrakesh in 1997, which is considered to be one of the main landmarks in this field. Only documents until 2018 were included so that complete annual periods could be compared. In order to avoid duplication, the sample only included original articles [44]. It is worth pointing out that a different search query could give rise to different results. The search was carried out in January 2019. The sample of this study was composed of 713 articles. In addition, a search of articles on "irrigation" was also carried out with the same restrictions in order to analyze the relative importance of sustainability within this general theme. Figure 1 shows an outline of the methodology on which this study was based.

The analyzed variables were the number of articles, their years of publication, all of the authors of the articles, the institutions and countries of all of the authors, the subject areas in which Scopus classifies the studies, the name of the journals in which they were published, and the keywords. After downloading this information, the first task was to eliminate duplications. The names of authors and institutions can be found in different formats. This can lead to errors when counting these records. Therefore, these two variables were analyzed, and the different records were regrouped so that the same author and institution were not counted more than once. Once the information had been refined, different tables and figures were drawn up, and the analysis of the data was conducted. The programs used were Excel (version 2016) and SciMAT (v1.1.04) (University of Granada, Granada, Spain). The tool used to create the network maps was VOSviewer, which is widely used in this type of study [42]. Finally, keyword analysis was used to extract the principal research trends [45]. The terms were regrouped in order to eliminate duplications due to plurals, hyphens, words in upper case letters, etc. For the grouping of keywords by topics, standardized grouping algorithms were used with the following tools: Vosviewer (Association strength) and SciMAT (network analysis).

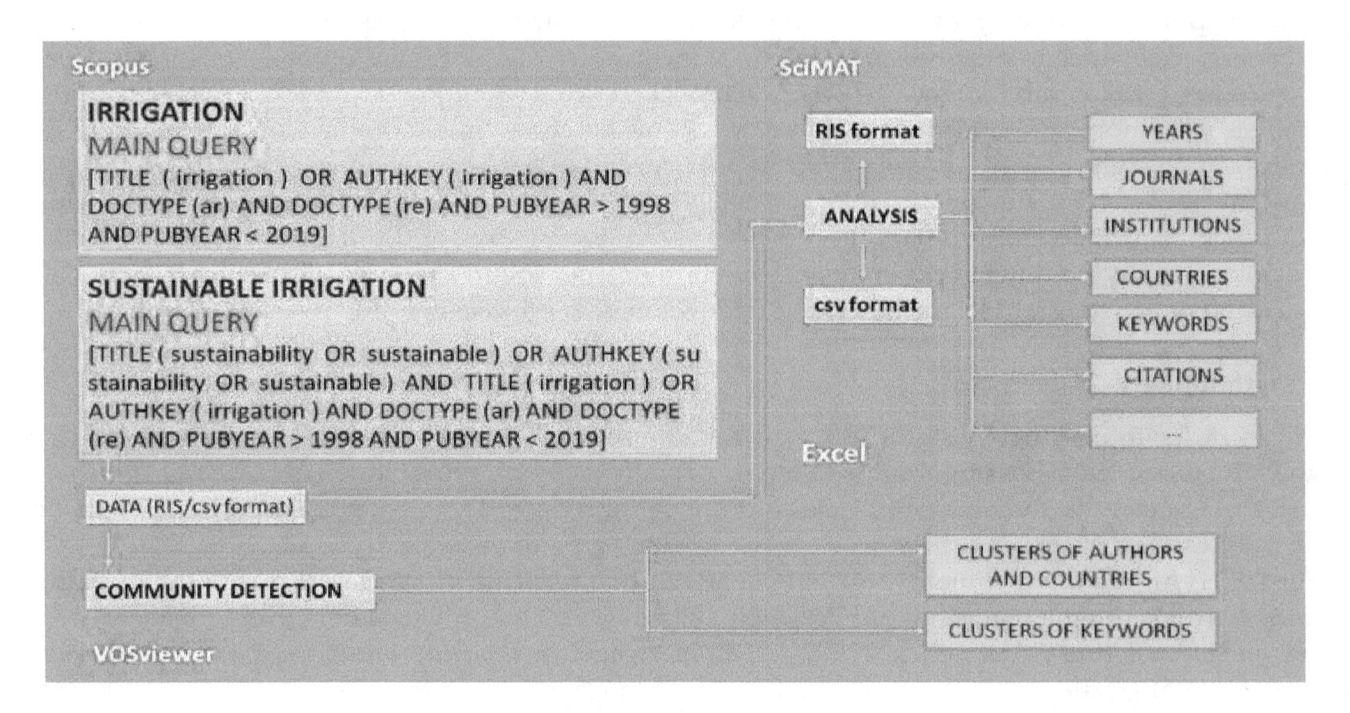

Figure 1. Outline of the methodological development of this study.

As for the methodology, this work includes some novel aspects compared to previous studies dealing with a similar topic. Firstly, regarding the sample selection of articles to be analyzed, some previous studies made a search based on titles, abstracts, and keywords [41–43,46,47]. In this work, the search was conducted in the fields of title and authors' keywords. Furthermore, before getting the final sample, it was checked that all included articles were related to the actual SI research. Secondly, most works on bibliometric reviews include the analysis of keywords. Nevertheless, this is the first search analysis which aims at detecting SI research trends based on the disciplinary approach of the study and the country where the research was conducted. Finally, the work includes a quantitative review based on a bibliometric analysis, as well as a qualitative one based on the traditional review.

3. Results

3.1. *Evolution of the General Characteristics of Research on Sustainable Irrigation (SI)*

Table 1 shows the evolution of the main variables related to research on SI during the period 1999–2018. During the studied period, relevant events like international declarations and congresses decisively influenced on the sustainability research. The Kyoto Protocol (UNFCCC, 2008), which commits world countries to reduce greenhouse gas emissions, should be highlighted, as well as the Economics of Ecosystems and Biodiversity of 2010; the Rio +20 of 2012; the Millennium Development Goals of the United Nations (UN, 2015), which provides guidelines for improving livelihoods and the environment globally; or the Paris Agreement on Climate Change of 2016; among others. These happenings additionally stimulate research on this topic [48]. This could also explain the existence of peaks regarding the publication of articles on SI research, like in 2017. A further reason explaining the higher number of published articles in 2017 compared to 2018 is that the sample selection was conducted in January 2019. The Scopus database updates itself continuously and, at the time of the sample search, not all published articles in 2018 had been registered. If the sample selection were to be performed at the end of 2019, the number of published and indexed articles on SI in Scopus in 2018 would increase.

Table 1. Main characteristics of sustainable irrigation (SI) research.

Year	Articles	Authors	Journals	Countries	Citation	Average Citation [1]
1999	6	13	6	5	0	0.0
2000	11	22	9	7	14	0.8
2001	15	30	11	10	5	0.6
2002	15	43	11	14	14	0.7
2003	12	24	12	11	35	1.2
2004	15	33	15	13	47	1.6
2005	20	57	16	15	76	2.0
2006	35	112	23	25	130	2.5
2007	22	68	19	15	181	3.3
2008	31	89	24	20	211	3.9
2009	25	56	18	21	289	4.8
2010	47	135	39	29	331	5.2
2011	37	118	31	29	442	6.1
2012	37	107	26	22	517	7.0
2013	45	149	38	27	650	7.9
2014	63	214	47	38	743	8.5
2015	53	202	46	30	901	9.4
2016	68	244	55	34	1268	10.5
2017	88	325	58	37	1515	11.4
2018	68	292	45	42	1707	12.7

[1] Total number of citations accumulated to date divided by the total number of articles published to date.

In general terms, we observed a growth trend in all of the variables analyzed, which indicates the development of this line of research. More than 45% of the total number of studies in the sample are concentrated in the last five years of the period analyzed. In order to confirm the growth of this field of study, the evolution of the number of articles on SI during the period of analysis was compared with all of the articles published on irrigation and all of the articles published on sustainability. Figure 2 shows the percentage of annual variation in the number of articles published in these lines of research. The average annual growth of the articles on irrigation was 1.6%, the one of articles on sustainability 3.8%, while that of articles on SI was 5.2%. This enabled us to confirm that SI is a line of study that is becoming increasingly more prominent within research in irrigation and in sustainability in general. These results agree with other works on water and sustainability [1,39,49].

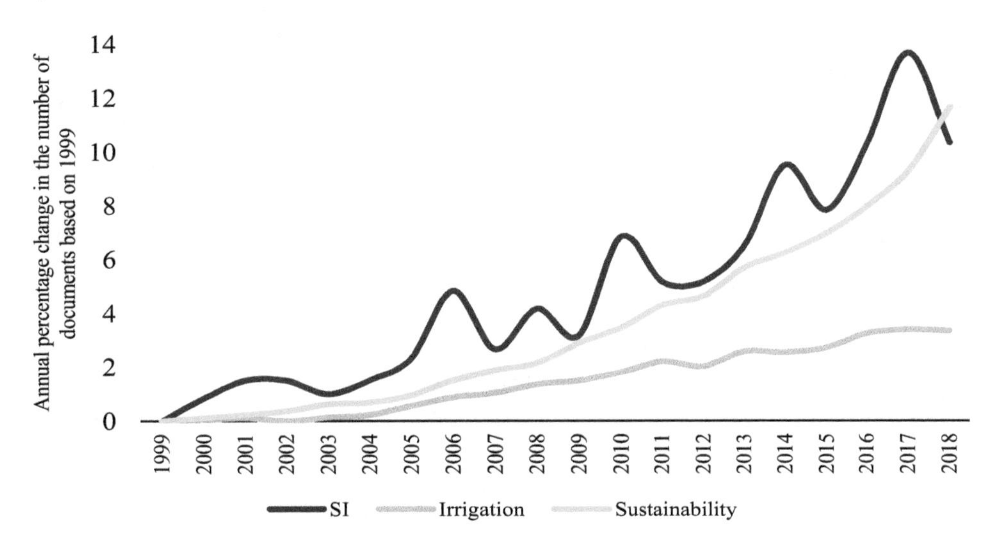

Figure 2. Comparative trends in irrigation, sustainability, and SI research.

With respect to the rest of the variables included in Table 1, the average number of authors per article doubled from two at the beginning of the period to four at the end. The number of journals in which articles on SI were published increased from six in 1999 to 45 in 2018. The number of countries also grew during the period analyzed (from five in 1999 to 42 in 2018). The annual number of references increased from 0.8 in 2000 to 12.7 in 2018.

3.2. Evolution of Research in SI by Subject Area

Figure 3 shows the evolution of the main subject areas into which the articles on SI included in the Scopus database were classified. It should be noted that an article may belong to more than one category. From the beginning of the period, the category in which the highest number of studies were classified was Environmental Sciences, which accounted for almost 65% of the total sample. The second largest block of studies was classified in the Agricultural and Biological Sciences category, with 44.3% of the total sample. In third place was the Social Sciences category with 21.1% of the articles. These three categories have dominated research on SI since the beginning of the studied period. However, in contrast to some previous works [37,39,48], our results revealed that over the last five years, the Earth and Planetary Sciences, Engineering, Energy, and Economics categories have begun to gain relevance, although none of them include more than 15% of the total articles in the sample. The Scopus classification distinguishes between the following categories: Business, Management, and Accounting; and Economics, Econometrics, and Finance, which also differ from Social Sciences. For the purpose of simplification, we grouped these two categories into only one and termed it "Economics".

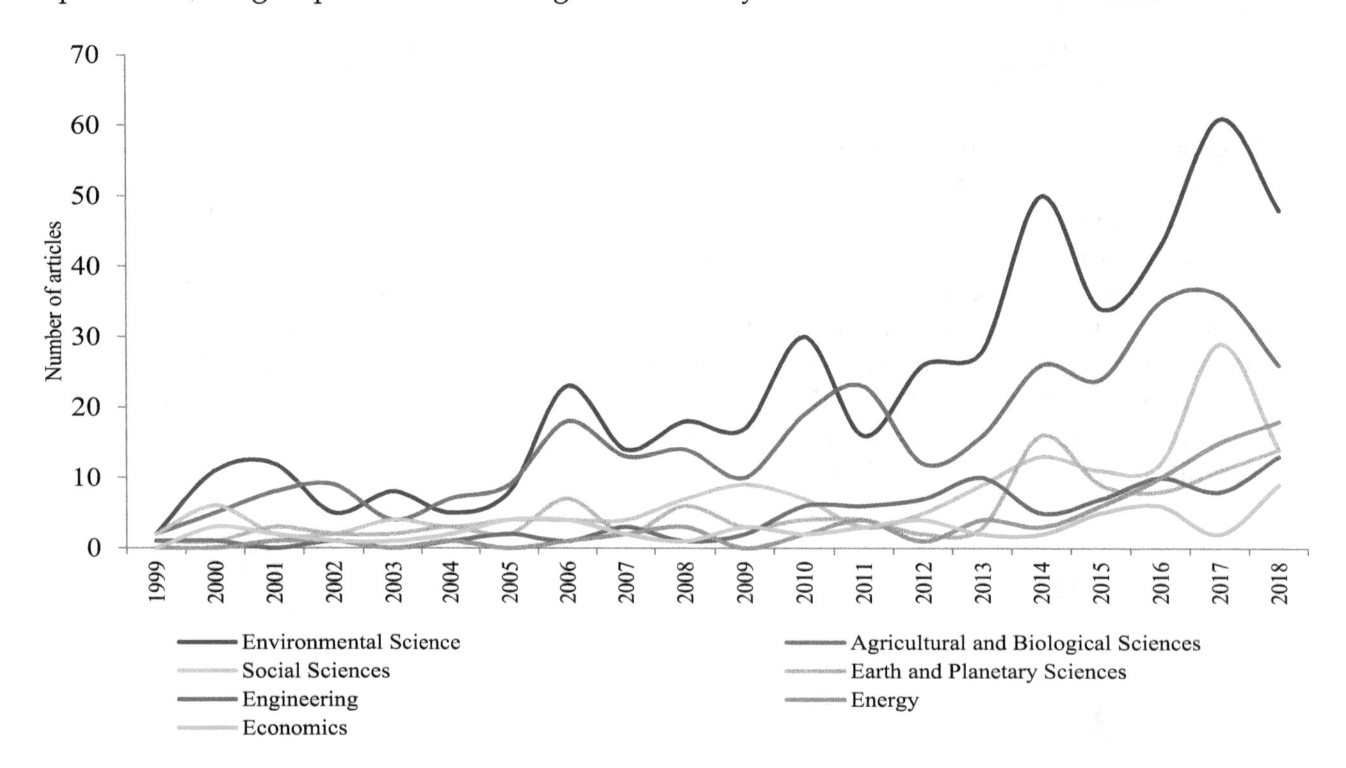

Figure 3. Comparative trends of subject categories related to SI research.

The keyword analysis revealed that there is a series of commonly used terms in research on SI, irrespective of the approach of the study. These terms include, among others, agriculture, alternative agriculture, climate change, crops, groundwater, irrigation system, salinity, sustainable development, water conservation, water management, water resources, water supply, water use, water use efficiency, and water quality. When we took the analysis beyond these terms, we identified a group of keywords used specifically by each discipline.

In the studies classified in the category of Environmental Sciences, there was an emphasis on the state of the soil (soils, soil moisture), aquifers, and surface water. With respect to processes, from the environmental approach, recycling and wastewater reclamation were prominent. In terms of methodology, the keywords that stood out were numerical model and decision making, particularly related to the management of the available water resources (water budget, water availability). With regard to the geographical dimension, China, the United States, and India were particularly prominent, as were the regions of Eurasia and Asia.

In the studies classified within the category of Agricultural and Biological Sciences, technical terms were predominant. Studies in this category mostly focused on soils and groundwater. There was particular emphasis on different types of crops (Triticum aestivum, Zea mays, fruit, Gossypium hirsutum, and rice) and irrigation processes (deficit irrigation, drainage, drip irrigation, leaching, waterlogging, agricultural irrigation). Furthermore, from the agronomic perspective, the environmental dimension was also considered (Environmental Impact). In these studies, China and the United States stood out, together with the regions of Africa and Asia.

The studies carried out from a Social Sciences approach had a more multidisciplinary perspective. They focused primarily on the stakeholders, water demands, and food security. However, technical concepts were also prominent, particularly those related to irrigation and water management (drip irrigation, sustainable water management), crops (Triticum aestivum), and economics (water economics). Unlike the over categories, land use was found to be one of the prominent subjects of this group of studies. A focus on management and decision making at different levels was also characteristic of these studies (governance approach, water planning, policy making, resource management, decision support system). With respect to geographical distribution, the United States, China, India, Spain, and Australia were among the most cited countries, and Asia, Europe and Africa stood out on a regional level.

Finally, the studies carried out based on an economic approach (Economic Sciences) were the most multidisciplinary, including technical, social, and environmental aspects. Among the main themes analyzed in these studies, we found food supply, food security, the development and innovation of irrigation systems (agricultural technology, irrigation performance) and management processes (integrated resource management, managed change, project management, strategic change, strategic management, strategic planning, decision making), and issues related to economic and social management (efficiency, investment, performance, economic and social effects) and the environment (environmental impact, environmental sustainability).

3.3. Most Relevant Journals in the Research on SI

Table 2 shows the main characteristics of the most prolific journals in the field of SI. The group of journals with the highest number of articles published on SI accounted for 25.7% of the total articles in the sample. This indicates that there is a high level of dispersion in terms of the journals that publish articles on this subject area. The leading journal in terms of the total number of articles published during the whole period analyzed was *Agricultural Water Management*, with a total of 52 articles on SI. This journal has the highest H index and the most citations of the journals with articles published in this area, and a Scimago journal rank (SJR) factor of 1.272. It published its first issue on this subject in 2001. Since then, it has remained among the top positions in terms of the number of articles published on SI, becoming the leader in 2014. The journal in second place was *Irrigation and Drainage*, with a total of 30 articles on SI. This journal published its first article on SI in 2001 and was the most prolific journal until 2005. It has the second highest H index, an average of 10.4 citations per article and an SJR index of 0.342. The third journal was *Sustainability*, with 17 articles on SI. This journal is among the most recently incorporated journals, as its first article on SI was published in 2013. However, in only five years, it rose to third position in terms of the number of articles for the whole of the period of study. It has an average of 3.8 citations per article, an H index of 6 and an SJR of 0.537. The journal with the highest average number of citations per article was *Science of the Total Environment* with 33.8; followed by the *Journal of Hydrology* with 31.5 and *Water Resources Management* with 29.4.

Table 2. Main characteristics of the most active journals related to SI research.

Journal	Articles	SJR [1]	H index [2]	Country	Citation	Average Citation [3]	1st Article	Last Article
Agricultural Water Management	52	1.272 (Q1)	21	Netherlands	1128	21.7	2001	2018
Irrigation and Drainage	30	0.342 (Q2)	10	USA	313	10.4	2001	2018
Sustainability	17	0.537 (Q2)	6	Switzerland	65	3.8	2013	2018
Water	15	0.634 (Q1)	8	Switzerland	129	8.6	2009	2018
Water Policy	11	0.461 (Q2)	6	UK	74	6.7	2005	2018
Water Resources Management	11	1.185 (Q1)	9	Netherlands	323	29.4	2000	2015
Acta Horticulturae	10	0.198 (Q3)	2	Belgium	17	1.7	2011	2018
Journal of Hydrology	10	1.832 (Q1)	8	Netherlands	315	31.5	2010	2017
Journal of Cleaner Production	9	1.467 (Q1)	5	Netherlands	49	5.4	2015	2018
Journal of Irrigation and Drainage Engineering	9	0.521 (Q2)	4	USA	47	5.2	2007	2017
Science of The Total Environment	9	1.546 (Q1)	6	Netherlands	304	33.8	2004	2018

[1] Scimago Journal Rank 2017; [2] only sample documents; [3] total number of citations divided by the total number of articles.

3.4. *Most Relevant Countries in Research on SI*

Table 3 shows the principle characteristics of the articles on SI from the most prolific countries. During the analyzed period, the United States was the leading country in research on SI in terms of the number of articles, with a total of 143. The country with the second highest number of articles was India, with a total of 74. This was followed by Australia with 67, Spain with 61, and Italy with 55. Due to the differences in terms of the size and economic development of the different countries, these data were analyzed to determine the number of articles per capita, measured as the number of articles per million inhabitants. Based on this variable, Australia was shown to be the most productive country with 2.7 articles per million inhabitants. This was followed by the Netherlands with 1.5, Spain with 1.3, Italy with 0.9, and the United Kingdom with 0.8. France was shown to be the country with the most citations per article, with 23.9, followed by the United Kingdom with 22.5, Iran with 21.9, the Netherlands with 18.6, and the United States with 18.1.

Table 3. Main characteristics of the most active countries related to SI research.

Country	Articles	Average per Capita Articles [1]	Citation	Average Citation [2]	H Index [3]	1st Article	Last Article	% of Cultivated Area Equipped for Irrigation (Ranking Countries) [4]
USA	143	0.439	2585	18.1	25	1999	2018	16.94 (72)
India	74	0.055	688	9.3	14	1999	2018	41.54 (38)
Australia	67	2.724	941	14.0	17	2000	2018	5.72 (110)
Spain	61	1.310	815	13.4	14	2004	2018	21.61 (64)
Italy	55	0.908	359	6.5	9	2002	2018	44.22 (35)
China	52	0.038	791	15.2	15	2004	2018	51.48 (28)
UK	51	0.772	1147	22.5	16	1999	2018	3.41 (126)
Germany	36	0.435	509	14.1	12	2004	2018	5.65 (111)
France	29	0.432	692	23.9	10	2000	2018	14.53 (82)
Japan	26	0.205	347	13.3	8	2001	2017	54.96 (25)
Netherlands	26	1.518	483	18.6	11	2001	2018	46.85 (31)
Brazil	22	0.105	163	7.4	6	2006	2018	5.79 (108)
Canada	22	0.599	179	8.1	8	2005	2018	2.44 (134)
Iran	19	0.234	416	21.9	8	2009	2018	51.88 (27)
South Africa	19	0.335	109	5.7	7	2002	2018	12.93 (83)

[1] Total number of articles per million inhabitants; [2] total number of citations divided by the total number of articles; [3] only sample documents; [4] FAO Aquastat (2019), last available data.

The percentage of cultivated area ready for irrigation per country has been included in the last column of the table, as well as the position they have in the world ranking regarding this variable. It can be stated that these countries do not occupy leading positions as far as irrigation-equipped cultivated land is concerned. From the available information about 177 countries, Japan is the country with the highest percentage of irrigation-equipped land surface—54.96%—, reaching the 25th place—followed by Iran with 51.88% (27th place), China with 51.48% (28th place), and Netherlands with 46.85% (31st place). However, some countries leading research on SI place themselves on lower positions within the irrigation-equipped cultivated surface ranking. This is the case for the USA with 16.94% (72nd place), Australia with 5.72% (110th place), or the UK with 3.41% (126th place).

Table 4 shows the principal variables related to the international collaboration of countries with the highest numbers of articles. The average percentage of articles carried out through international collaboration was 50.3%. The countries with the highest percentage of studies carried out in collaboration were Canada with 81.8%, France with 79.3%, Germany with 75.1%, the Netherlands with 65.4%, and China with 55.8%. The United States was found to have the largest collaboration network, with 33 different collaborators. In addition, similarly to Australia, this country forms part of the group of the main collaborators of 10 of the 15 countries in the table. These data reveal the global nature of research in this subject area, with very high percentages and extensive collaboration networks on a global level. The majority of the countries obtained a higher average number of citations per article when they worked in collaboration with other countries. The articles produced through collaboration obtained an average of 14.6 citations as opposed to 13.5 citations of noncollaborative articles. When comparing these results to those of related works on irrigation and water [37,39,48], it can be observed that studies on sustainability trigger a higher level of international cooperation.

Table 4. International collaboration of the most active countries related to SI research.

Country	Percentage of Collaboration [1]	Number of Collaborators	Main Collaborators	Average Citation	
				Collaboration [2]	Noncollaboration [3]
USA	45.5	33	China, Italy, Mexico, Canada, Sweden	23.1	13.9
India	21.6	15	USA, Ethiopia, France, UK, Australia	11.6	8.7
Australia	52.2	22	China, Canada, Spain, USA, Bangladesh	18.6	9.1
Spain	40.9	18	Portugal, Australia, Germany, Italy, Belgium	22.1	7.3
Italy	40.0	19	USA, Netherlands, Spain, France, South Africa	5.0	7.5
China	55.8	16	USA, Australia, Canada, Bangladesh, Germany	15.7	14.6
UK	47.1	28	India, Italy, Netherlands, Philippines, Spain	22.3	22.6
Germany	75.1	30	Uzbekistan, USA, Spain, Switzerland, Australia	15.6	9.9
France	79.3	21	India, USA, Australia, Belgium, Italy	19.1	42.2
Japan	38.5	9	Thailand, USA, Vietnam, Australia, Egypt	9.9	15.5
Netherlands	65.4	16	Italy, Australia, China, Germany, Pakistan	23.6	9.1
Brazil	31.8	7	Spain, USA, Argentina, Germany, Italy	14.7	4.0
Canada	81.8	13	Australia, China, USA, Denmark, India	9.2	3.3
Iran	26.3	4	Australia, USA, Germany, Netherlands	2.2	28.9
South Africa	52.6	13	Australia, Italy, Belgium, Bolivia, Denmark	6.2	5.2

[1] Number of articles made through international collaboration divided by the total number of articles; [2] number of citations obtained by articles made through international collaboration divided by the number of articles; [3] number of citations obtained for articles not made through international collaboration divided by the number of articles.

Figure 4 shows a network map of the collaborations carried out between countries, where the size of the circle represents the number of documents per country and the color corresponds to the cluster formed by the different groups of countries. Three clusters can be distinguished, led by the United States, Australia, and Spain in terms of the number of articles. The first (shown in blue) includes some of the most prolific countries, such as India, Italy, China, France, Japan, and the Netherlands, and others, such as Mexico, Egypt, and Bangladesh. Together with Australia, the second cluster (shown in red) includes Canada, Iran, South Africa, Sweden, Switzerland, Pakistan, Sri Lanka, and Uzbekistan. The group led by Spain (shown in green) includes some European countries, such as the United Kingdom, Germany, Belgium, Portugal, and Greece, as well as countries in the Mediterranean basin, such as Israel, Jordan, Morocco, and Turkey, and others, such as Brazil, Thailand, and New Zealand.

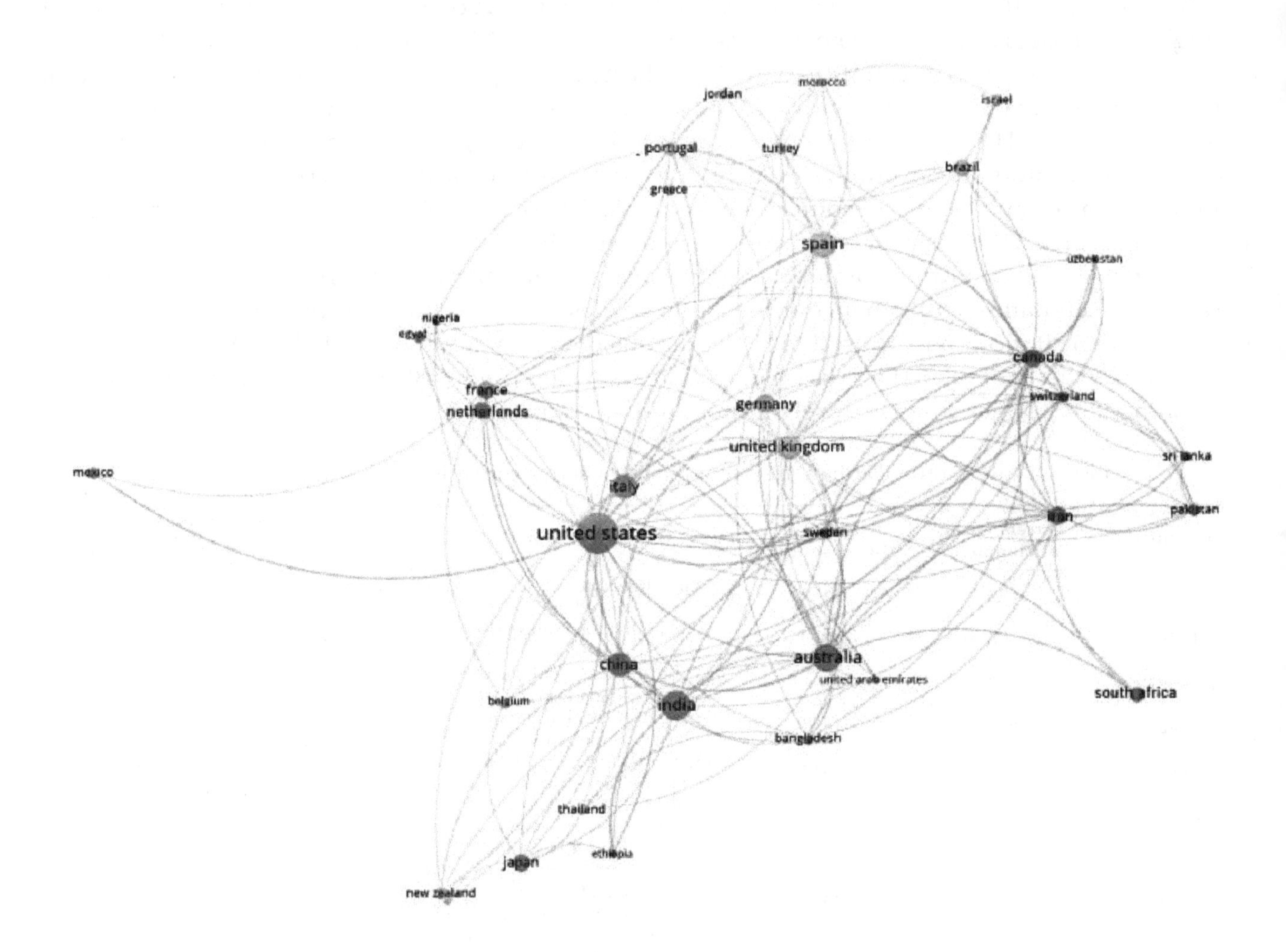

Figure 4. Main relationships between countries in SI research.

A keyword analysis was used to detect the preferences in the research conducted by the countries included in Table 4 (Table 5). We established that there is a group of terms that make up a general line from which the different specific topics are derived.

In the studies conducted in the United States, the central themes included food supply and food security (food-supply, food-security), the conservation of natural resources and environmental impacts (conservation-of-natural-resources, environmental-impact), and land use (land-use). Agronomic issues, such as crop evapotranspiration and productivity, were prominent. Both surface water and groundwater were studied with special emphasis on the availability of the resource, the water budget, the water table, and water stress (surface-water, water-budget, water-availability, water-table, water-stress). The term most used in relation to methodology was numerical model (numerical-model), and the term most used with regard to crops was Zea mays (Zea-mays). No noteworthy geographical terms other than the United States were identified.

Table 5. Main keywords of the most active countries related to SI research.

Country	Keywords
USA	crop-yield, surface-water, alternative-agriculture, numerical-model, food-security, water-budget, wastewater-reclamation, water-availability, food-supply, evapotranspiration, conservation-of-natural-resources, environmental-impact, water-table, water-stress, Zea-mays, land-use
India	optimization, irrigation-planning, water-table, waterlogging, drainage, Maharashtra, irrigation-projects, rice, arid-regions, ecosystems, fertilizers, nutrient, vegetables, waste-water, South-Asia, surface-water
Australia	Australasia, wastewater, irrigation-efficiency, Murray-Darling-basin, environmental-protection, evapotranspiration, food-supply, hydrogeology, runoff, water-availability, controlled-study, hydrology, water-treatment, rice, recycling, Canada
Spain	energy-efficiency, fruit, irrigation-networks, semiarid-region, profitability, southern-Europe, soil-moisture, deficit-irrigation, carbon-dioxide, drip-irrigation, water-economics, economic-analysis, water-productivity, decision-support-systems, energy-resources, stakeholder
Italy	deficit-irrigation, decision-support-system, southern-Europe, farms, fruit, orchard, Mediterranean-environment, forestry, stem-water-potential, crop-yield, dicotyledon, stomatal-conductance, water-stress, drainage, decision-making, environmental-impact
China	alternative-agriculture, Zea-mays, Triticum-aestivum, Xinjiang-Uygur, North-China-plain, decision-making, irrigation-district, evapotranspiration, environmental-protection, hydrological-modelling, ecology, landforms, soil-moisture, integrated-approach, water-availability, uncertainty
UK	alternative-agriculture, environmental-impact, water-treatment, drainage, Africa, environment, wastewater, cost-benefit-analysis, rice, wetland, recycling, runoff, drainage-and-irrigation, arid-regions, crop-yield, hydrocarbon
Germany	food-supply, Triticum-aestivum, Uzbekistan, alternative-agriculture, arid-region, fertilizer, oasis, climate-models, economic-and-social-effects, drip-irrigation, agricultural-intensification, food-security, leaching, common-pool-resource, cropping-system, greenhouse-gas
France	farming-system, stakeholder, groundwater-overexploitation, evapotranspiration, deficit-irrigation, decision-making, environmental-policy, governance-approach, surface-water, public-private-partnership, pricing-policy, water-economics, chemical-composition, dynamic-model, linear-programming, ecophysiological-responses
Japan	irrigation-development, water-users'-organization, rainfall, sustainable-rice-production, institutional-development, Triticum-aestivum, Zea-mays, water-policy, participatory-irrigation-management, saline-water-irrigation, soil-water-salinity, sorghum, electrical-conductivity, semiarid-region, sorghum-bicolour, drought
Netherlands	drainage, Triticum-aestivum, food-production, recirculations, well, groundwater-abstraction, agricultural-extension, smallholder, agricultural-development, food-supply, agricultural-management, decision-making, rain, alternative-agriculture, catchments, networking-system
Brazil	biofuel, expansion, water-availability, environmental-impact, bioenergy, biomass-power, Cerrado, sugar-cane, glycine-max, sustainable-production, Saccharum-officinarum, carbon-dioxide, chemistry, evapotranspiration, metabolism, rainwater
Canada	alternative-agriculture, water-policies, sensitivity-analysis, stochastic-programming, decision-making, environmental-protection, uncertainty-analysis, water-availability, water-stress, Gossypium-hirsutum, food-security, global-perspective, food-production, Triticum-aestivum, water-sharing, yield-response
Iran	crop-yield, cropping-pattern, economic-and-social-effects, rainfall, food-supply, Zea-mays, FAO, optimum-decision, surface-water-resources, recycling, genetic-algorithm, GHG-emission, irrigation-district, farmers-motivation, untreated-wastewater-irrigation, Bayesian-networks
South Africa	Sub-Saharan-Africa, wastewater, evapotranspiration, water-economics, drought, simulation, controlled-study, electric-conductivity, economic-and-social-effects, GIS, water-balance, computer-simulation, food-supply, project-management, mine-water, environmental-impact

The articles from India were found to place more emphasis on new alternative agriculture systems (alternative-agriculture) and the agricultural activity in arid areas (arid-regions) on an ecosystem level (ecosystems). Agronomic aspects such as the nutrition and fertilization of crops (fertilizers, nutrient) and their productivity (crop-yield, productivity) were also prominent. Both surface and groundwater were studied (surface-water), as was the planning and development of irrigation projects for optimizing the resource (irrigation-planning, irrigation-project, optimization), with particular emphasis on irrigation processes (drainage, waterlogging, water-table) and alternative water sources (wastewater). The term most used in relation to methodology was numerical model (numerical-model), and with regard to crops, the most used terms were rice and vegetables (rice, vegetables). Prominent geographical terms, such as the State of Maharashtra or the region of South Asia, were identified.

The studies conducted in Australia covered food supply (food-supply), runoff (runoff), and environmental protection (environmental-protection). Agronomic issues, such as crop evapotranspiration (evapotranspiration) and irrigation efficiency (irrigation-efficiency), were prominent. With respect to water resources, particular emphasis was placed on the availability of the resources, the water budget (water-availability, water-budget), and the use of alternative sources through recycling and wastewater treatment (wastewater, water-treatment, recycling). From a methodological point of view, the hydrological and geological approaches were prominent (hydrology, hydrogeology), as were controlled studies (controlled-study), and in terms of crops, the prominent term was rice (rice). Noteworthy geographical terms, such as the basin of the river Murray–Darling (Murray-Darling-basin), the region of Australasia and the country of Canada, were identified.

In the case of Spain, relevant topics were the management of the use of energy resources (energy-efficiency, energy-resources), the productivity of water (water-productivity), efficiency (efficiency), semiarid regions (semiarid-regions), and environmental protection (environmental-protection). Agronomic issues such as soil moisture (soil-moisture) were prominent, as were innovations in agricultural and irrigation systems (alternative-agriculture, irrigation-networks, drip-irrigation, deficit-irrigation) and irrigation efficiency (irrigation-efficiency). In terms of methodology, the economic approach was prominent (water-economics, economic-analysis, profitability), as was the social approach (stakeholders, decision-support-systems). With respect to crops, fruit was the most relevant term (fruit). Noteworthy geographical terms, such as the regions of Southern Europe and Eurasia (Southern-Europe, Europe, Eurasia), were identified.

The studies conducted in Italy were similar to those conducted in Spain. The main differences were found in certain agronomic terms related to crops (crop-yield, dicotyledon, stomatal-conductance, stem-water-potential) or irrigation (drainage, soil-moisture). The studies were carried out on a farm level (farms, orchard) and focused on the Mediterranean environment. In terms of methodology, the numerical models focused on decision making (decision-making) were noteworthy. With respect to crops, as well as fruit (fruit), Zea mays was also prominent. The central themes included water stress (water-stress), the environmental impact (environmental-impact), and forestry (forestry).

The articles conducted in China placed greater emphasis on environmental protection (environmental-protection), agricultural activity in arid regions (arid-regions), decision-making processes (decision-making), and issues at the district level (irrigation-district). Agronomic issues such as crop evapotranspiration (evapotranspiration) and crop productivity (crop-yield) were also prominent. With respect to water, the central theme was the level of water resources (water-level), while in the methodological area, hydrogeological models were prominent (hydrogeological-modelling). With respect to crops, rice, corn maize and wheat were found to be noteworthy (rice, Zea-mays, Triticum-aestivum, maize). The most prominent geographical terms were the regions of the North China Plain and Xinjiang Uygur.

The central themes in the studies conducted in the United Kingdom were crop yield (crop-yield); the environment and the assessment of the environmental impact (environment, environmental-impact), particularly in arid regions (arid-regions); and hydrocarbon (hydrocarbon). Processes related to water and irrigation, such as drainage (drainage), runoff (runoff), and the recycling of water in wetlands and wastewater treatment (water-treatment, wastewater, wetlands-recycling), stood out. The most prominent term related to methodology was cost benefit analysis (cost-benefit-analysis), and in terms of crops, the study of rice was predominant (rice). This may be due to the country's connection with Asian countries (Asia, Eurasia, and South Asia).

The articles conducted in Germany, as in the case of other countries, were conditioned by its collaborative ties with other nations. The main themes included food supply and food security (food-supply, food security); the new alternative agricultural systems and the intensification of agricultural activity (alternative-agriculture, agricultural-intensification); agricultural activity in arid regions (arid-regions, oasis); and the effects of agricultural activity on economic and social levels and environmental pollution (economic-and-social-effects, greenhouse-gas, particle-size). The prominent agronomic aspects were fertilizers (fertilizers), leaching (leaching), and drip irrigation (drip-irrigation). The most commonly used term with respect to methodology was related to the study of the effects of climate change: climate models (climate-models). Further fields of interest are common pool resources; and, with respect to crops, studies on wheat (Triticum-aestivum). One of Germany's principal collaborators was China, which is why this country appeared prominently among the keywords of German studies. Similarly, Uzbekistan is one of Germany's main trading partners and also appears among the keywords.

The studies carried out in France particularly focused on the policy and institutional dimension (environmental-policy, decision-making, governance-approach, public-private-partnership, pricing-policy). French studies analyzed the level of exploitation (farming-system) and contemplated the different agents involved (stakeholders). The priority issues included the overexploitation of groundwater and surface water (groundwater-overexploitation, surface-water); the chemical composition (chemical-composition); crop evapotranspiration (evapotranspiration) and the response to possible alterations (ecophysiological-responses); and deficit irrigation (deficit-irrigation). In terms of methodology, dynamic models and linear programming were prominent (dynamic-model, linear-programming), together with economic issues related to water (water-economics).

Studies conducted in Japan considered issues such as the development of irrigation, particularly through participative processes on both institutional and irrigation water user levels (irrigation-development, water-users'-organization, participatory-irrigation-management, institutional-development, water-policy). The Japanese studies analyzed aspects related to the salinity of irrigation water and soil (saline-water-irrigation, soil-water-salinity) and the electrical conductivity of water (electrical-conductivity). Water shortages due to drought, particularly in semiarid regions, were also found to be relevant issues (drought, semi-arid-region). The studies from this country analyzed the use of rainwater as a source for irrigation (rainfall). It is the country with the highest number of crops, and the study of the sustainable production of rice was found to be particularly relevant (sustainable-rice-production, Triticum-aestivum, Zea-mays, sorghum, sorghum-bicolour).

In the Netherlands, the primary topics identified were the management and development of agriculture, particularly towards the use of new alternative systems (agricultural-development, agricultural-management, alternative-agriculture), and there was special emphasis on the extension of agricultural practices (agricultural-extension). Other subject areas of many of this country's studies were the security and production of food (food-supply, food-production). The crop that is most studied was wheat (Triticum-aestivum). The articles focused on groundwater and rainwater with respect to water resources (well, groundwater-abstraction, rain), recirculation processes and network development (drainage, recirculations, networking-system) and the studies on the level of the basins were predominant (catchments). On a social level, the point of view of the small farmers (smallholders) was given special attention, particularly in relation to the decision-making processes (decision-making).

The studies conducted in Brazil contemplated the environmental impacts (environmental-impact), water availability (water-availability), and the expansion of the agricultural activity (agricultural-expansion). With respect to crop processes, evapotranspiration, metabolism, sustainable production, and carbon dioxide were prominent (carbon-dioxide, sustainable-production, evapotranspiration, metabolism). This country has published a large number of studies on the use of rainwater for irrigation (rainwater). Particularly noteworthy is the research on crops related to the use of biomass for different purposes (biofuel, bioenergy, biomass-power, sugar-cane, glycine-max, saccharum-officinarum). On a geographic level, studies in the region of Cerrado were predominant. Additionally, the most prominent methodological approach was found to be chemistry (chemistry).

In Canada, the most relevant themes were the new forms of agriculture (alternative-agriculture), environmental protection (environmental-protection), food security (food-security, food-production), and decision making (decision-making, water-policies). The global perspective of this line of research (global-perspective) was found to be noteworthy. From a methodological point of view, the sensitivity models, stochastic programming, and uncertainty analysis were prominent (sensitivity-analysis, stochastic-programming, uncertainty-analysis). The predominant terms with respect to water were water availability, water stress, and water sharing. With regard to crops, wheat and cotton stood out (Gossypium-hirsutum, Triticum-aestivum).

In the case of Iran, concern for the food supply was found to be prominent, despite its relationship with the FAO—Food and Agriculture Organization of the United Nations (food-supply, FAO, crop-yield). On a methodological level, genetic algorithms stood out (genetic-algorithm), together with processes for optimizing decisions (optimum-decision), and Bayesian networks (Bayesian-networks). The economic and social levels were represented through the motivation of farmers and the assessment of the economic and social effects (economic-and-social-effects, farmers-motivation). In the agronomic field, cropping patterns and the use of untreated wastewater were priority areas (cropping-pattern, untreated-wastewater-irrigation). In addition to wastewater, the combined integral use of surface water, recycled water, and rainwater for irrigation was prominent (rainfall, surface-water-resources, recycling).

Finally, the articles from South Africa were based on controlled studies, geographical information systems, and computer simulations (computer-simulation, simulation, controlled-study, GIS). The environmental impacts (environmental-impact), particularly those related to the use of mine water and wastewater (mine-water, wastewater); food supply (food-supply); project management (project-management); and the economic and social effects related to irrigation (economic-and-social-effects) were priority themes. With respect to the agronomic dimension, the evapotranspiration processes, drought, electric conductivity, and water balance stood out (evapotranspiration, drought, electric-conductivity, water-balance).

3.5. Most Relevant Institutions in Research on SI

Table 6 shows the main characteristics of the institutions with the highest number of articles on SI. The Chinese Academy of Sciences holds the first position with 14 articles. This institution has accumulated a total of 376 citations in these articles, with an average of 26.9 citations per article and an H index of 7. The institution with the second largest number of articles is the Commonwealth Scientific and Industrial Research Organisation—Land and Water (CSIRO Land and Water), with a total of 12 studies published. It has 212 citations, an average of 17.7 citations per article, and an H index of 8. In third place is the University of South Australia, with 10 articles. This institution has accumulated a total of 24 citations, an average of 2.4 citations per article, and an H index of 3. The institution with the largest number of citations and the highest average citations per article in its studies on SI is the University of Texas, with a total of 562 citations and 93.7 citations per article.

With respect to the international collaboration of institutions, the average percentage of articles carried out jointly was 46.6%. The institutions with the highest percentage of articles carried out in collaboration were the IHE Delft Institute for Water Education and the Universidade de Lisboa (University of Lisbon) with 83.3%. These two institutions were followed by the University of South Australia with 80.1%, Columbia University with 75.1%, China Agricultural University and Texas A and M University with 66.7%, and the University of California with 62.5%. The average number of citations of the jointly-written articles of the group of 22 institutions was 18.3 as opposed to 16.7 citations for the rest. The institutions with the highest number of citations in articles written in collaboration were the University of Texas, the University of California, and China Agricultural University.

Table 6. Main characteristics of the most active institutions related to SI research.

Institution	Country	Articles	Citation	Average Citation [1]	H Index [2]	Percentage of Collaboration [3]	Average Citation	
							Collaboration [4]	Noncollaboration [5]
Chinese Academy of Sciences	China	14	376	26.9	7	35.7	22.2	29.4
CSIRO Land and Water	Australia	12	212	17.7	8	41.7	21.2	15.1
University of South Australia	Australia	10	24	2.4	3	80.1	2.5	2.0
USDA Agricultural Research Service, Washington DC	USA	9	86	9.6	5	22.2	8.5	9.9
Wageningen University and Research Centre	Netherlands	9	117	13.0	6	55.6	16.2	9.0
Indian Institute of Technology Kharagpur	India	9	163	18.1	7	33.3	6.0	24.2
University of California, Riverside	USA	8	406	50.8	5	62.5	49.6	52.7
Columbia University in the City of New York	USA	8	97	12.1	5	75.1	13.2	9.0
Universidad de Cordoba	Spain	7	73	10.4	7	28.6	14.5	8.8
University of California, Davis	USA	7	87	12.4	3	42.9	26.0	2.3
Università degli Studi della Basilicata	Italy	7	16	2.3	3	14.3	0.0	2.7
Harran Üniversitesi	Turkey	6	77	12.8	3	0.0	0.0	12.8
University of Texas at Austin	USA	6	562	93.7	5	50.0	110.7	76.7
China Agricultural University	China	6	104	17.3	3	66.7	26.0	0.0
Universidad de Almeria	Spain	6	40	6.7	4	50.0	2.7	10.7
Texas A and M University	USA	6	42	7.0	3	66.7	5.5	10.0
IHE Delft Institute for Water Education	Netherlands	6	142	23.7	5	83.3	22.2	31.0
Ben-Gurion University of the Negev	Israel	6	66	11.0	6	16.7	6.0	12.0
Alma Mater Studiorum Università di Bologna	Italy	6	155	25.8	3	16.7	13.0	28.4
Commonwealth Scientific and Industrial Research Organization	Australia	6	92	15.3	3	50.0	16.7	14.0
Northwest A & F University	China	6	42	7.0	5	50.0	7.0	7.0
Universidade de Lisboa	Portugal	6	65	10.8	5	83.3	13.0	0.0

[1] The total number of citations divided by the total number of articles; [2] only sample documents; [3] the number of articles produced through international collaboration divided by the total number of articles; [4] the number of citations obtained by articles produced through international collaboration divided by the number of articles; [5] the number of citations obtained for articles not made through international collaboration divided by the number of articles.

3.6. Most Relevant Authors in Research on SI

Table 7 shows the main characteristics of the authors who have produced the highest numbers of articles on SI. A large number of authors published articles on SI within the study period, but the number of publications per author was small. The three authors with the most articles were Henning Bjornlund from the University of South Australia, Bartolomeo Dichio from the Università degli Studi della Basilicata, and Ajay Kumar R. Singh from the Indian Institute of Technology Kharagpur. Although they published the same number of articles, there were large differences in terms of the relevance of the publications of the different authors. Bjornlund has a total of 20 citations for his articles, Dichio 11, and Singh 145. James D. Oster of the University of California was the author with the largest number of citations with a total of 369; and the highest average number of citations per article with 73.8. He was followed by Mohammad Valipour from the Islamic Azad University, with a total of 366 citations and 73.2 per article. In third place was Dennis Wichelns from the Stockholm Environment Institute with 189 total citations and 37.8 citations per article. The author with the oldest publication was Bart Schultz of the IHE Delft Institute for Water Education, who published his first article in 2001. This author, who has written four articles on SI, has accumulated a total of 122 citations and published his last article on this subject in 2005. The authors who have made more recent contributions to this line of research are P. Amparo López-Jiménez and Modesto Pérez-Sánchez from the Universitat Politècnica de València.

Table 7. Major characteristics of the most active authors related to SI research.

Author	Articles	Citation	Average Citations [1]	H Index [2]	Country	Affiliation [3]	1st Article	Last Article
Bjornlund, Henning	6	20	3.3	2	Australia	University of South Australia	2010	2017
Dichio, Bartolomeo	6	11	1.8	2	Italy	Università degli Studi della Basilicata	2010	2018
Singh, Ajay Kumar R.	6	145	24.2	5	India	Indian Institute of Technology Kharagpur	2010	2017
Oster, James D.	5	369	73.8	3	USA	University of California	2003	2013
Scholz, Miklas	5	49	9.8	4	UK	University of Salford	2006	2018
Valipour, Mohammad	5	366	73.2	5	Iran	Islamic Azad University	2015	2017
Wichelns, Dennis	5	189	37.8	4	Sweden	Stockholm Environment Institute	2002	2014
Xiloyannis, Cristos	5	11	2.2	2	Italy	Università degli Studi della Basilicata	2010	2018
Annandale, John George	4	28	7.0	3	South Africa	Universiteit van Pretoria	2002	2017
Aydogdu, Mustafa Hakki	4	15	3.8	3	Turkey	Harran Üniversitesi	2015	2017
López-Jiménez, P. Amparo	4	50	12.5	3	Spain	Universitat Politècnica de València	2016	2018
Montanaro, Giuseppe	4	9	2.3	1	Italy	Università degli Studi della Basilicata	2010	2018
Pérez-Sánchez, Modesto	4	50	12.5	3	Spain	Universitat Politècnica de València	2016	2018
Schultz, Bart	4	122	30.5	4	Netherlands	IHE Delft Institute for Water Education	2001	2005
Vico, Giulia	4	46	11.5	3	Sweden	Swedish University of Agricultural Sciences	2011	2018

[1] Total number of citations divided by the total number of articles; [2] only sample documents; [3] last verified affiliation.

3.7. *Main Issues in SI Research*

The most relevant issues in SI research were determined based on the analysis of keywords. These issues included the concern for the state of natural resources, including water and soil and their conservation; the impact of agriculture on the environment and the consequences of climate change; the use of nonconventional water resources as an alternative for irrigation, including desalinated seawater, reused water, and harvested rainwater; the developments in innovation and technology for irrigation systems, particularly drip irrigation and deficit irrigation; and, finally, the improvements in the efficiency of the use of irrigation water. Below is an overview of the research carried out on these four priority issues. The weight of each priority research line has been established through the repetition number of the main keywords within each research line as the article average of the sample. It has to be taken into account that an article can be classified under different topics. For example, an article which analyzes water efficiency regarding nonconventional water resources shows as the most weighted research line climatic change, environmental impact, and natural resource conservation, with a 56.8% out of the total sample works; followed by water use efficiency with 42.5%, irrigation technology and innovation with 37.6%, and unconventional water resources with 31.2%.

3.7.1. Climatic Change, Environmental Impact, and Natural Resource Conservation

It is predicted that the consequences derived from global climate change will be alterations in precipitation cycles, triggering long-term droughts, more frequent and more intense extreme phenomena, and water supply imbalances [39,50]. Furthermore, these consequences will be reflected in agriculture by way of variations in soil humidity and in the evapotranspiration and runoff flows [14,51]. The United Nations report on the development of global water resources of 2015 estimates that there will be a drinking water shortage of 40% on a global level by the year 2030 [52].

Bad practice in agriculture produces a series of impacts that can have consequences on environmental, economic, and social levels. As well as water use, current irrigation agriculture requires the addition of fertilizers and other chemical products [53]. When the use of chemical products is incomplete or inefficient or when excessive water is applied, the resulting filtration ends up in drainage systems or in the groundwater recharge areas under the cultivated land [54]. The most deteriorated ecosystems currently include the majority of the groundwater bodies on a global scale. These water resources have enabled the development of agricultural activity in arid and semiarid regions and also in more humid regions where there are mismatches between precipitation and the needs of the crops [55]. In recent decades, the intensification of agriculture has given rise to a fall in piezometric levels, the development of salinization processes, seawater intrusion, and pollution by agricultural nitrates, among other effects [23].

Due to the estimated increase in the amount of fresh water required to meet the future irrigation demands, a drastic reduction in biodiversity is expected to take place, together with an increase in the salinity or flooding of soil, a loss in the flow of complementary services provided by the ecosystems, and the degradation of water sources and ecosystems in general [56,57]. On a social level, an increase in the vulnerability and inequality between users is expected [58,59].

In order to mitigate these adverse effects and to contribute to the conservation of the ecosystems, important legislation is currently being developed on a global level. Among the objectives established by the United Nations for the Horizon 2030 on Sustainable Development is one specifically related to water and sanitation (ODS 6), which addresses aspects ranging from water shortage to water use efficiency [60]. The Horizon 2020 Plan of the European Parliament includes the requirement for sustainable production in agricultural systems [61]. Many countries, including the United States, China, India, and Costa Rica, have consolidated payment systems for environmental services provided by agricultural ecosystems with the objective of conserving water resources in good condition.

3.7.2. Unconventional Water Resources

The current scenario is one in which so-called conventional water sources are being exhausted and degraded in large parts of the world. These water sources include both surface water (rivers, lakes, reservoirs) and groundwater (aquifers). The principal option for increasing the water supply for irrigation consists of using alternative sources, also called nonconventional water sources [23]. These other sources include the reuse of urban and industrial water, the desalination of seawater, and rainwater harvesting. In recent years, water from nontraditional sources has become a competitive option in the supply of quality water for irrigation, particularly in arid and semiarid regions [62] The use of these types of resources has a series of advantages—two in particular. First, the contribution of these new water sources represents an increase in the supply of the resource, which is capable of satisfying the growing demand of the different sectors (urban supply, agricultural activity, tourism sector, industrial sector, and environmental requirements) [63]. Second, the use of water from alternative sources should serve to diminish the use of traditional water sources so that the state of deterioration of the wetlands, rivers, and aquifers can be restored or at least alleviated [64]. If these two advantages are to be efficient, they have to be accompanied by demand control. In addition to these two principal functions, the use of nonconventional water resources gives rise to other advantages. They provide a greater reliability in the supply, they supply higher quality water, they can generate increases in crop yields, they contribute to ensuring the stability of agricultural incomes, and they can have positive effects on seawater intrusion processes in the aquifers [65–67].

The use of each of these alternatives also gives rise to a series of limitations and disadvantages. Evidently, the construction of seawater desalination plants is only feasible in coastal areas. A wastewater treatment plant requires a volume of a large enough size for the facility to be viable [64]. Therefore, the use of this resource is not possible in areas where the activities generating wastewater (population nuclei, industrial facilities, livestock farms, etc.) do not have sufficient water use [63]. Thus, despite water reuse being the ideal way of maintaining continuous use of the resource, this type of facility is not appropriate for many rural areas where the population is dispersed. The main problem of rainwater harvesting systems is the low volume of water that can be supplied in comparison with the demand [67]. Furthermore, the seasonality of rain in many regions means that the water must be stored for long periods of time for use when needed. To these limitations we must also add the high cost of water derived through these systems. The installation costs are usually very high, and we must also take into account the cost of production. In the case of desalination and reuse, these costs usually establish a price for water that is much higher than the price of conventional resources [68]. This means that many farmers throughout the world are not willing to pay the price of the water unless there is no alternative available. Studies have been carried out on experience with the use of desalinated seawater, reused water or harvested rainwater all over the world [69–75].

3.7.3. Irrigation Technology and Innovation

Irrigation technology has evolved continuously over the last few decades. Flood irrigation, sprinkler irrigation, furrow irrigation, and drip irrigation are some of the methods that have emerged, and their advantages and disadvantages have been studied with respect to different types of crops, soils, and climatic conditions. New technologies have given rise to the development of comprehensive automated systems that combine the use of tensiometers, lysimeters, software applications, and even geographical information systems. However, drip irrigation and deficit irrigation are the terms that appear among the most used keywords.

De Wrachienb et al. [76] date the beginning of drip irrigation systems to the 1940s in Australia. The development of this system came about after the emergence of polypropylene tubes. It was not until two decades later that this system was improved in Israel, from where it was exported all over the world. Currently, thanks to automation and the use of microcontrollers, sensors, and integrated systems, this method has been perfected, and the drip irrigation system is now considerably more advantageous than traditional systems such as flood irrigation or sprinklers [77–79]. The main contribution of this system is that it enables a substantial saving in the use of water for irrigation, which enables the development or expansion of agricultural activity in arid and semiarid regions, where it would not be possible otherwise [80,81]. Another advantage is that it can prevent evaporation, as it supplies water directly to the roots of plants [82]. Different studies show that the use of drip irrigation increases the marketable yield and quality of crops and stabilizes production when deficit irrigation is used and that fertigation through drip irrigation helps to reduce the use of fertilizers and, therefore, the risk of pollution due to leachate [80,83]. Salvador and Aragüés [84] analyzed the advantages and disadvantages of the use of underground drip irrigation systems. They demonstrated their usefulness, profitability, and sustainability and indicated that the design, handling, and maintenance of this system, together with the quality of the irrigation water and type of soil, are fundamental aspects that determine their sustainability. On the other hand, Puy et al. [85] indicated that this type of system can have harmful consequences in terms of the degradation of the soil or the production of greenhouse gas emissions.

Deficit irrigation was introduced as a measure to limit the vegetative growth of crops [86]. This irrigation technique has been fully developed, and it is used extensively [87]. This method has been used with both drip irrigation and microsprinkling on different crops and can be combined with remote sensing technology or infrared techniques to produce significant water savings while crop yields remain unaffected. Du et al. [88] analyzed the use of deficit irrigation as a sustainable strategy for managing water resources in agriculture for food security in China. These authors concluded that the current understanding of physiological processes enables the deficit irrigation methods to be adjusted to different crops and environments in order to increase water use efficiency and the yield and quality of crops. Many studies have been carried out on this subject area [89–93].

Though many authors support drip irrigation as a sustainability measure, some recent studies question it. Perry et al. [94] confirm the "zero-sum game" hypothesis which argues that the impact of high-technology watering in a farm increases the demand of local water and land production at the expense of water availability and production in other places. Furthermore, due to the advantageous effects of drip irrigation, it makes water more affordable and, at the time, it allows irrigating larger areas, obtaining greater profits, and shift to more valuable crops. The most foreseeable impact of water efficiency improvement will be the increase of current water demands. In this sense, water scarcity would remain difficult to manage. Paul et al. [95], in their review of the rebound effects on the management of land and cultivation soils, found evidence for the presence of rebound effects and the Jevons paradox, together with productivity increases and efficiency of irrigation water due to technological innovations. Further studies agree with these results [96,97].

3.7.4. Water Use Efficiency

All of these innovations have the objective of improving water use efficiency for irrigation. In the year 2000, Kofi Annan, the Secretary General of the United Nations, proposed a "Blue Revolution in Agriculture" that was proposed to be capable of increasing productivity per unit of water. This strategy became known by the slogan "more crop per drop" [24]. According to Yang [98], obtaining the ideal water efficiency for irrigating crops involves the reduction of losses caused by evaporation, runoff, and underground draining while increasing production. Zhang et al. [26] indicated that the use of technology to save irrigation water not only saves water and increases production but also improves the nutritional value of agricultural products and guarantees food safety by improving the environmental conditions. Water use efficiency in agriculture generally implies a reduction in water use to meet a specific production objective or to increase the production of a specific water supply [99]. The aim of improving water use efficiency is to increase food production, boost financial gains, and guarantee the supply of ecosystem services at lower social and environmental costs per unit of water used [100,101]. The practices used to achieve this objective include rainwater harvesting, complementary irrigation, deficit irrigation, and the use of precision irrigation techniques and practices to conserve groundwater [24,102]. The priority areas where it is possible to significantly increase the productivity of water include areas with a high level of poverty and a low level of water productivity; areas with physical water shortage, where competition for water is high; areas with limited development of water resources, where the high yields of additional water have a considerable impact; and areas with degraded ecosystems driven by water, such as depleting water tables and dried-up rivers [103,104].

Among the different improvements developed over the last few decades, the use of drip irrigation has been fundamental in the improvement of water use efficiency and saving. Different studies have shown that drip irrigation has a water-saving potential of between 18% and 75%. According to Narayanamoorthy [82], drip irrigation saves an average of 25 to 75% of water compared to flood irrigation. Similar results were found, although with different percentages, in studies by Ibragimov et al. [105], Maisiri et al. [106], Yazar et al. [107], or Peterson and Ding [108], Abdulai et al. [109], Cremades et al. [110], and Jalota et al. [111].

4. Conclusions

This study presented the dynamics of global research in sustainable irrigation in agriculture over the last two decades, the main agents promoting it, and the topics that have received the most attention. The main concerns stated in the Introduction section related to the improvement of irrigation water use in order to increase food production, the world overexploitation of water resources, and the effects of global climate change. Our analysis verified how these questions are addressed by countries taking into account interdisciplinary approaches, and it also proved how these questions are mirrored in the main research lines on SI.

The results of the analysis of the principal variables revealed that the study of sustainable irrigation has grown in recent years in all of the variables considered: articles, authors, journals, institutions, and countries. Despite the fact that the growth trend in this topic is higher than that of general research in irrigation, an even greater research effort using a sustainability-based approach is required to further knowledge in this area. Traditionally, studies on sustainability have focused on one of the areas of which it is composed, namely, the environmental, social or economic dimensions. In the study of irrigation, the dominant area has been the environmental dimension, far more than the social or economic perspectives. The studies that analyzed just one of these dimensions provide highly useful information, but this information is only partial. It is necessary to integrate the three aspects of sustainability in order to gain full knowledge of the feasibility of certain practices, not only in terms of their environmental impacts but also with respect to income generation for farmers and the wellbeing of the community.

The keyword study revealed the existence of diversity between studies carried out using specific approaches and in different countries. In general, the study of environmental impacts and climate change, water availability, the improvement in efficiency, sustainable development, food supply, and the conservation of water bodies, particularly aquifers that have deteriorated, are common themes. However, certain practices, such as deficit irrigation or drip irrigation and aspects related to energy consumption and certain crops, are priority issues for particular countries. The methodological approaches used and the tools applied are other points of differentiation of the research carried out by each country. The keyword analysis showed four main research lines on SI: climatic change, environmental impact, and natural resource conservation; unconventional water resources; irrigation technology and innovation; and water use efficiency. Due to the large number of analyzed documents and the scope of this work, an in-depth content analysis per topic has not been undertaken. It will be highly interesting for future studies in order to provide more detailed information of these four specific topics.

As a final conclusion, we believe that certain aspects of the research on sustainable irrigation in agriculture in each of the dimensions of sustainability should be promoted. From a technical point of view, innovation and technology have furthered the development of irrigation systems and new available water sources that can contribute to improving the efficiency of water use and the sustainability of rural areas, particularly agricultural activity in arid regions. However, effort should be made to make this technology accessible, as its cost is economically unfeasible for small-scale agriculture in many countries. New water sources, such as those derived from desalination, reuse and rainwater harvesting systems, are very expensive for farmers compared to traditional sources. The production processes for desalination and reuse should be improved, particularly with respect to energy consumption in order to bring down the final price of the water. Furthermore, although the use of these nonconventional water resources has proved to have a series of advantages for the crops and the soil, this knowledge has not been transmitted to the farmers, and therefore, they are still reluctant to use it for irrigation. Greater effort should be made to communicate the results of the research to society. Finally, greater knowledge of the environmental impacts of irrigation-related practices in different areas on plot, district, basin and regional levels is needed. Water bodies are connected to each other, so certain practices that generate a small impact on river source areas can have a multiplying effect and be experienced in the underground bodies of coastal areas.

Author Contributions: The four authors have equally contributed to this paper. All authors have revised and approved the final manuscript.

Acknowledgments: This work was partially supported by the Spanish Ministry of Economy and Competitiveness and the European Regional Development Fund by means of the research project ECO2017-82347-P, and by the Research Plan of the University of Almería through a Postdoctoral Contract to Juan F. Velasco Muñoz.

References

1. Hossain, M.S.; Pogue, S.J.; Trenchard, L.; Van Oudenhoven, A.P.E.; Washbourne, C.L.; Muiruri, E.W.; Tomczyk, A.M.; García-Llorente, M.; Hale, R.; Hevia, V.; et al. Identifying future research directions for biodiversity, ecosystem services and sustainability: Perspectives from early-career researchers. *Int. J. Sustain. Dev. World Ecol.* **2018**, *25*, 249–261. [CrossRef]
2. Manju, S.; Sagar, N. Renewable energy integrated desalination: A sustainable solution to overcome future fresh-water scarcity in India. *Sustain. Energy. Rev.* **2017**, *73*, 594–609. [CrossRef]
3. Millennnium Ecosystem Assessment (MA). *Ecosystems and Human Well-Being: Biodiversity Synthesis*; World Resources Institute: Washington, DC, USA, 2005.
4. Wang, M.H.; Li, J.; Ho, Y.S. Research articles published in water resources journals: A bibliometric analysis. *Desalin. Water Treat.* **2011**, *28*, 353–365. [CrossRef]
5. Flávio, H.M.; Ferreira, P.; Formigo, N.; Svendsen, J.C. Reconciling agriculture and stream restoration in Europe: A review relating to the EU Water Framework Directive. *Sci. Total Environ.* **2017**, *596–597*, 378–395. [CrossRef] [PubMed]
6. Zhang, Y.; Chen, H.; Lu, J.; Zhang, G. Detecting and predicting the topic change of Knowledge-based Systems: A topic-based bibliometric analysis from 1991 to 2016. *Knowl. Based Syst.* **2017**, *133*, 255–268. [CrossRef]
7. Damkjaer, S.; Taylor, R. The measurement of water scarcity: Defining a meaningful indicator. *Ambio* **2017**, *46*, 513–531. [CrossRef] [PubMed]
8. Liu, J.; Wang, Y.; Yu, Z.; Cao, X.; Tian, L.; Sun, S.; Wu, P. A comprehensive analysis of blue water scarcity from the production, consumption, and water transfer perspectives. *Ecol. Indic.* **2017**, *72*, 870–880. [CrossRef]
9. Forouzani, M.; Karami, E. Agricultural water poverty index and sustainability. *Agron. Sustain. Dev.* **2011**, *31*, 415–432. [CrossRef]
10. Fu, H.Z.; Wang, M.H.; Ho, Y.S. Mapping of drinking water research: A bibliometric analysis of research output during 1992–2011. *Sci. Total Environ.* **2013**, *443*, 757–765. [CrossRef]
11. Pedro-Monzonís, M.; Solera, A.; Ferrer, J.; Estrela, T.; Paredes-Arquiola, J. A review of water scarcity and drought indexes in water resources planning and management. *J. Hydrol.* **2015**, *527*, 482–493. [CrossRef]
12. Adeyemi, O.; Grove, I.; Peets, S.; Norton, T. Advanced monitoring and management systems for improving sustainability in precision irrigation. *Sustainability* **2017**, *9*, 353. [CrossRef]
13. Hedley, C.B.; Knox, J.W.; Raine, S.R.; Smith, R. Water: Advanced irrigation technologies. In *Encyclopedia of Agriculture and Food Systems*, 2nd ed.; Elsevier Academic Press: San Diego, CA, USA, 2014; pp. 378–406. ISBN 978-0-444-52512-3.

14. Zhang, Y.; Zhang, Y.; Shi, K.; Yao, X. Research development, current hotspots, and future directions of water research based on MODIS images: A critical review with a bibliometric analysis. *Environ. Sci. Pollut. Res. Int.* **2017**, *24*, 15226–15239. [CrossRef] [PubMed]
15. Gago, J.; Douthe, C.; Coopman, R.E.; Gallego, P.P.; Ribas-Carbo, M.; Flexas, J.; Escalona, J.; Medrano, H. UAVs challenge to assess water stress for sustainable agriculture. *Agric. Water Manag.* **2015**, *153*, 9–19. [CrossRef]
16. Wu, W.; Ma, B. Integrated nutrient management (INM) for sustaining crop productivity and reducing environmental impact: A review. *Sci. Total Environ.* **2015**, *512–513*, 415–427. [CrossRef] [PubMed]
17. De Fraiture, C.; Wichelns, D. Satisfying future water demands for agriculture. *Agric. Water Manag.* **2010**, *97*, 502–511. [CrossRef]
18. Fischer, G.; Tubiello, F.N.; van Velthuizen, H.; Wiberg, D.A. Climate change impacts on irrigation water requirements: Effects of mitigation, 1990–2080. *Technol. Forecast. Soc.* **2007**, *74*, 1083–1107. [CrossRef]
19. Matchaya, G.; Nhamo, L.; Nhlengethwa, S.; Nhemachena, C. An Overview of Water Markets in Southern Africa: An Option for Water Management in Times of Scarcity. *Water* **2019**, *11*, 1006. [CrossRef]
20. Graveline, N. Combining flexible regulatory and economic instruments for agriculture water demand control under climate change in Beauce. *Water Resour. Econ.* **2019**, 100143. [CrossRef]
21. Jothibasu, A.; Anbazhagan, S. Hydrogeological assessment of the groundwater aquifers for sustainability state and development planning. *Environ. Earth Sci.* **2018**, *77*, 88. [CrossRef]
22. Singh, A. Conjunctive use of water resources for sustainable irrigated agriculture. *J. Hydrol.* **2014**, *519*, 1688–1697. [CrossRef]
23. Aznar-Sánchez, J.A.; Belmonte-Ureña, L.J.; Velasco-Muñoz, J.V.; Valera, D.L. Aquifer Sustainability and the Use of Desalinated Seawater for Greenhouse Irrigation in the Campo de Níjar, Southeast Spain. *Int. J. Environ. Res. Public Health* **2019**, *16*, 898. [CrossRef] [PubMed]
24. Morison, J.I.L.; Baker, N.R.; Mullineaux, P.M.; Davies, W.J. Improving water use in crop production. *Philos. Trans. R. Soc. Lond. B Biol. Sci.* **2017**, *363*, 639–658. [CrossRef] [PubMed]
25. Melo-Zurita, M.L.; Thomsen, D.C.; Holbrook, N.J.; Smith, T.F.; Lyth, A.; Munro, P.G.; de Bruin, A.; Seddaiu, G.; Roggero, P.P.; Baird, J.; et al. Global water governance and climate change: Identifying innovative arrangements for adaptive transformation. *Water* **2018**, *10*, 29. [CrossRef]
26. Zhang, B.; Fu, Z.; Wang, J.; Zhang, L. Farmers' adoption of water-saving irrigation technology alleviates water scarcity in metropolis suburbs: A case study of Beijing, China. *Agric. Water Manag.* **2019**, *212*, 349–357. [CrossRef]
27. Komiyama, H.; Takeuchi, K. Sustainability science: Building a new discipline. *Sustain. Sci.* **2006**, *1*, 1–6. [CrossRef]
28. Yarime, M.; Takeda, Y.; Kajikawa, Y. Towards institutional analysis of sustainability science: A quantitative examination of the patterns of research collaboration. *Sustain. Sci.* **2010**, *5*, 115–125. [CrossRef]
29. Juwana, I.; Muttil, N.; Perera, B.J.C. Indicator-based water sustainability assessment—A review. *Sci. Total Environ.* **2012**, *438*, 357–371. [CrossRef]
30. Becker, B. *Sustainability Assessment: A Review of Values, Concepts and Methodological Approaches*; Issues in Agriculture 10; World Bank-Consultative Group on International Agriculture Research (CGIAR): Washington, DC, USA, 1997.
31. Mancosu, N.; Snyder, R.L.; Kyriakakis, G.; Spano, D. Water scarcity and future challenges for food production. *Water* **2015**, *7*, 975–992. [CrossRef]
32. Ioris, A.A.R.; Hunter, C.; Walker, S. The development and application of water management sustainability indicators in Brazil and Scotland. *J. Environ. Manag.* **2008**, *88*, 1190–1201. [CrossRef]
33. Ward, F.A.; Michelsen, A. The economic value of water in agriculture: Concepts and policy applications. *Water Policy* **2002**, *4*, 423–446. [CrossRef]
34. Huang, L.; Zhang, Y.; Guo, Y.; Zhu, D.; Porter, A.L. Four dimensional science and technology planning: A new approach based on bibliometrics and technology roadmapping. *Technol. Forecast. Soc. Chang.* **2014**, *81*, 39–48. [CrossRef]
35. Aznar-Sánchez, J.A.; Belmonte-Ureña, L.J.; López-Serrano, M.J.; Velasco-Muñoz, J.F. Forest ecosystem services: An analysis of worldwide research. *Forests* **2018**, *9*, 453. [CrossRef]
36. Li, W.; Zhao, Y. Bibliometric analysis of global environmental assessment research in a 20-year period. *Environ. Impact Assess. Rev.* **2015**, *50*, 158–166. [CrossRef]

37. Aznar-Sánchez, J.A.; Belmonte-Ureña, L.J.; Velasco-Muñoz, J.F.; Manzano-Agugliaro, F. Economic analysis of sustainable water use: A review of worldwide research. *J. Clean Prod.* **2018**, *198*, 1120–1132. [CrossRef]
38. Durieux, V.; Gevenois, P.A. Bibliometric Indicators: Quality Measurements of Scientific Publication. *Radiology* **2010**, *255*, 342. [CrossRef] [PubMed]
39. Velasco-Muñoz, J.F.; Aznar-Sánchez, J.A.; Belmonte-Ureña, L.J.; López-Serrano, M.J. Advances in water use efficiency in agriculture: A bibliometric analysis. *Water* **2018**, *10*, 377. [CrossRef]
40. Aznar-Sánchez, J.A.; Velasco-Muñoz, J.F.; Belmonte-Ureña, L.J.; Manzano-Agugliaro, F. The worldwide research trends on water ecosystem services. *Ecol. Indic.* **2019**, *99*, 310–323. [CrossRef]
41. Tancoigne, E.; Barbier, M.; Cointet, J.P.; Richard, G. The place of agricultural sciences in the literature on ecosystem services. *Ecosyst. Serv.* **2014**, *10*, 35–48. [CrossRef]
42. Velasco-Muñoz, J.V.; Aznar-Sánchez, J.A.; Belmonte-Ureña, L.J.; Román-Sánchez, I.M. Sustainable water use in agriculture: A review of worldwide research. *Sustainability* **2018**, *10*, 1084. [CrossRef]
43. Hassan, S.U.; Haddawy, P.; Zhu, J. A bibliometric study of the world's research activity in sustainable development and its sub-areas using scientific literature. *Scientometrics* **2014**, *99*, 549–579. [CrossRef]
44. Cossarini, D.M.; MacDonald, B.H.; Wells, P.G. Communicating marine environmental information to decision makers: Enablers and barriers to use of publications (grey literature) of the Gulf of Maine Council on the Marine Environment. *Ocean Coastal Manag.* **2014**, *96*, 163–172. [CrossRef]
45. Aznar-Sánchez, J.A.; Velasco-Muñoz, J.F.; Belmonte-Ureña, L.J.; Manzano-Agugliaro, F. Innovation and technology for sustainable mining activity: A worldwide research assessment. *J. Clean Prod.* **2019**, *221*, 38–54. [CrossRef]
46. Cogato, A.; Meggio, F.; Migliorati, M.; Marinello, F. Extreme Weather Events in Agriculture: A Systematic Review. *Sustainability* **2019**, *11*, 2547. [CrossRef]
47. Cui, X.; Guo, X.; Wang, Y.; Wang, X.; Zhu, W.; Shi, J.; Lin, C.; Gao, X. Application of remote sensing to water environmental processes under a changing climate. *J. Hydrol.* **2019**, *574*, 892–902. [CrossRef]
48. Aznar-Sánchez, J.A.; Piquer-Rodríguez, M.; Velasco-Muñoz, J.F.; Manzano-Agugliaro, F. Worldwide research trends on sustainable land use in agriculture. *Land Use Pol.* **2019**, *87*, 104069. [CrossRef]
49. Zhou, X.Y. Spatial explicit management for the water sustainability of coupled human and natural systems. *Environ. Pollut.* **2019**, *251*, 292–301. [CrossRef] [PubMed]
50. Sillmann, J.; Roeckner, E. Indices for extreme events in projections of anthropogenic climate change. *Clim. Chang.* **2008**, *86*, 83–104. [CrossRef]
51. Mitrică, B.; Mitrică, E.; Enciu, P.; Mocanu, I. An approach for forecasting of public water scarcity at the end of the 21st century, in the Timiş Plain of Romania. *Technol. Forecast. Soc. Chang.* **2017**, *118*, 258–269. [CrossRef]
52. United Nations World Water Assessment Programme (WWAP). *Water for a Sustainable World*; The United Nations World Water Development Report; UNESCO: Paris, France, 2015.
53. Wichelns, D.; Oster, J.D. Sustainable irrigation is necessary and achievable, but direct costs and environmental impacts can be substantial. *Agric. Water Manag.* **2006**, *86*, 114–127. [CrossRef]
54. Hadas, A.; Hadas, A.; Sagiv, B.; Haruvy, N. Agricultural practices, soil fertility management modes and resultant nitrogen leaching rates under semi-arid conditions. *Agric. Water Manag.* **1999**, *42*, 81–95. [CrossRef]
55. Sears, L.; Caparelli, J.; Lee, C.; Pan, D.; Strandberg, G.; Vuu, L.; Lawell, C.Y. Jevons' Paradox and Efficient Irrigation Technology. *Sustainability* **2018**, *10*, 1590. [CrossRef]
56. Singh, A. Decision support for on-farm water management and long-term agricultural sustainability in a semi-arid region of India. *J. Hydrol.* **2010**, *391*, 63–76. [CrossRef]
57. Kögler, F.; Söffker, D. Water (stress) models and deficit irrigation: System-theoretical description and causality mapping. *Ecol. Model.* **2017**, *361*, 135–156. [CrossRef]
58. Richard-Ferroudji, A.; Faysse, N.; Bouzidi, Z.; Menon, R.T.P.; Rinaudo, J.D. The DIALAQ project on sustainable groundwater management: A transdisciplinary and transcultural approach to participatory foresight. *Curr. Opin. Environ. Sustain.* **2016**, *20*, 56–60. [CrossRef]
59. García-Caparrós, P.; Contreras, J.I.; Baeza, R.; Segura, M.L.; Lao, M.T. Integral management of irrigation water in intensive horticultural systems of Almería. *Sustainability* **2017**, *9*, 2271. [CrossRef]
60. Vanham, D.; Hoekstra, A.Y.; Wada, Y.; Bouraoui, F.; de Roo, A.; Mekonnen, M.M.; van de Bund, W.J.; Batelaan, O.; Pavelic, P.; Bastiaanssen, W.G.M.; et al. Physical water scarcity metrics for monitoring progress towards SDG target 6.4: An evaluation of indicator 6.4.2. "Level of water stress". *Sci. Total Environ.* **2018**, *613–614*, 218–232. [CrossRef] [PubMed]

61. Geoghegan-Quin, M. Role of Research & Innovation in Agriculture. 2013. European Commission-SPEECH/13/505. Available online: http://europa.eu/rapid/press-release_SPEECH-13-505_en.htm (accessed on 20 January 2019).
62. Aznar-Sánchez, J.A.; Belmonte-Ureña, L.J.; Valera, D.L. Perceptions and Acceptance of Desalinated Seawater for Irrigation: A Case Study in the Níjar District (Southeast Spain). *Water* **2017**, *9*, 408. [CrossRef]
63. Ghaffour, N.; Missimer, T.M.; Amy, G.L. Technical review and evaluation of the economics of water desalination: Current and future challenges for better water supply sustainability. *Desalination* **2013**, *309*, 197–207. [CrossRef]
64. Zepeda-Quintana, D.S.; Loeza-Rentería, C.M.; Munguía-Vega, N.E.; Esquer-Peralta, J.; Velazquez-Contreras, L.E. Sustainability strategies for coastal aquifers: A case study of the Hermosillo Coast aquifer. *J. Clean Prod.* **2018**, *195*, 1170–1182. [CrossRef]
65. Duarte, T.K.; Minciardi, R.; Robba, M.; Sacile, R. Optimal control of coastal aquifer pumping towards the sustainability of water supply and salinity. *Sustain. Water Qual. Ecol.* **2015**, *6*, 88–100. [CrossRef]
66. Martínez-Granados, D.; Calatrava, J. Combining economic policy instruments with desalinisation to reduce overdraft in the Spanish Alto Guadalentín aquifer. *Water Policy* **2017**, *19*, 341–357. [CrossRef]
67. Jorreto, S.; Sola, F.; Vallejos, A.; Sánchez-Martos, F.; Gisbert, J.; Molina, L.; Rigol, J.P.; Pulido-Bosch, A. Evolution of the geometry of the freshwater-seawater interface in a coastal aquifer affected by an intense pumping of seawater. *Geogaceta* **2017**, *62*, 87–90.
68. Quintana, J.; Tovar, J. Evaluación del acuífero de Lima (Perú) y medidas correctoras para contrarrestar la sobreexplotación. *Bol. Geológico Y Min.* **2002**, *113*, 303–312.
69. Assouline, S.; Shavit, U. Effects of management policies, including artificial recharge, on salinization in a sloping aquifer: The Israeli Coastal Aquifer case. *Water Resour. Res.* **2004**, *40*, W04101. [CrossRef]
70. Boisson, A.; Villesseche, D.; Baisset, M.; Perrin, J.; Viossanges, M.; Kloppmann, W.; Chandra, S.; Dewandel, B.; Picot-Colbeaux, G.; Rangarajan, R.; et al. Questioning the impact and sustainability of percolation tanks as aquifer recharge structures in semi-arid crystalline context. *Environ. Earth. Sci.* **2015**, *73*, 7711–7721. [CrossRef]
71. Khezzani, B.; Bouchemal, S. Variations in groundwater levels and quality due to agricultural over-exploitation in an arid environment: The phreatic aquifer of the Souf oasis (Algerian Sahara). *Environ. Earth Sci.* **2018**, *77*, 142. [CrossRef]
72. Rupérez-Moreno, C.; Senent-Aparicio, J.; Martinez-Vicente, D.; García-Aróstegui, J.L.; Cabezas-Calvo-Rubio, F.; Pérez-Sánchez, J. Sustainability of irrigated agriculture with overexploited aquifers: The case of Segura basin (SE, Spain). *Agric. Water Manag.* **2017**, *182*, 67–76. [CrossRef]
73. Salcedo-Sánchez, E.R.; Esteller, M.V.; Garrido-Hoyos, S.E.; Martínez-Morales, M. Groundwater optimization model for sustainable management of the Valley of Puebla aquifer, Mexico. *Environ. Earth Sci.* **2013**, *70*, 337–351. [CrossRef]
74. Palacios-Vélez, O.L.; Escobar-Villagrán, B.S. La sustentabilidad de la agricultura de riego ante la sobreexplotación de acuíferos. *Tecnol. Y Cienc. del Agua* **2016**, *7*, 5–16.
75. Reca, J.; Trillo, C.; Sánchez, J.A.; Martínez, J.; Valera, D. Optimization model for on-farm irrigation management of Meditarranean greenhouse crops using desalinated and saline water from different sources. *Agric. Syst.* **2018**, *166*, 173–183. [CrossRef]
76. De Wrachien, W.; Medicia, M.; Lorenzini, G. The Great Potential of Micro-Irrigation Technology for Poor-Rural Communities. *Irrigat. Drain. Syst. Eng.* **2014**, *3*, e124. [CrossRef]
77. Guerbaoui, M.; Afou, Y.; Ed-Dahhak, A.; Lachhab, A.; Bouchikhi, B. Pc-based automated drip irrigation system. *Int. J. Eng. Sci. Technol.* **2013**, *5*, 221–225.
78. De Wrachien, W.; Lorenzini, G.; Medici, M. Sprinkler irrigation systems: State-of-the-art of kinematic analysis and quantum mechanics applied to water jets. *Irrig. Drain.* **2013**, *62*, 407–413. [CrossRef]
79. Lorenzini, G.; Saro, O. Thermal fluid dynamic modelling of a water droplet evaporating in air. *Int. J. Heat Mass Transf.* **2013**, *62*, 323–335. [CrossRef]
80. Surendran, U.; Jayakumar, M.; Marimuthu, S. Low cost drip irrigation: Impact on sugarcane yield, water and energy saving in semiarid tropical agro ecosystem in India. *Sci. Total Environ.* **2016**, *573*, 1430–1440. [CrossRef] [PubMed]

81. Kalpakian, J.; Legrouri, A.; Ejekki, F.; Doudou, K.; Berrada, F.; Ouardaoui, A.; Kettani, D. Obstacles facing the diffusion of drip irrigation technology in the Middle Atlas region of Morocco. *Int. J. Environ. Stud.* **2014**, *71*, 63–75. [CrossRef]
82. Narayanamoorthy, A. Economic Viability of Drip Irrigation: An Empirical Analysis from Maharashtra. *Indian J. Agric. Econ.* **1997**, *52*, 728–739.
83. Lamm, F.R. Cotton, tomato, corn, and onion production with subsurface drip irrigation: A review. *Trans. ASABE* **2016**, *59*, 263–278. [CrossRef]
84. Salvador, R.; Aragüés, R. Estado de la cuestión del riego por goteo enterrado: Diseño, manejo, mantenimiento y control de la salinidad del suelo. *ITEA-Inf. Tec. Econ. Agrar.* **2013**, *109*, 395–407. [CrossRef]
85. Puy, A.; García-Avilés, J.M.; Balbo, A.L.; Keller, M.; Riedesel, S.; Blum, D.; Bubenzer, O. Drip irrigation uptake in traditional irrigated fields: The edaphological impact. *J. Environ. Manag.* **2017**, *202*, 550–561. [CrossRef]
86. Holzapfel, E.A.; Pannunzio, A.; Lorite, I.; Silva de Oliveira, A.; Farkas, I. Design and Management of Irrigation Systems. *Chil. J. Agric. Res.* **2009**, *69*, 17–25. [CrossRef]
87. Fereres, E.; Soriano, M.A. Deficit irrigation for reducing agricultural water use. *J. Exp. Bot.* **2007**, *58*, 147–159. [CrossRef] [PubMed]
88. Du, T.; Kang, S.; Zhang, J.; Davies, W.J. Deficit irrigation and sustainable water-resource strategies in agriculture for China's food security. *J. Exp. Bot.* **2015**, *66*, 2253–2269. [CrossRef] [PubMed]
89. Girona, J.; Mata, M.; Marsal, J. Regulated deficit irrigation during the kernel-filling period and optimal irrigation rates in almond. *Agric. Water Manag.* **2005**, *75*, 152–167. [CrossRef]
90. Hutmacher, R.B.; Nightingale, H.I.; Rolston, D.E.; Biggar, J.W.; Dale, F.; Vail, S.S.; Peters, D. Growth and yield responses of almond (Prunus amygdalus) to trickle irrigation. *Irrig. Sci.* **1994**, *14*, 117–126. [CrossRef]
91. Bassoi, L.H.; Hopmans, J.W.; de C. Jorge, L.A.; de Alencar, C.M.; Silva, J. Grapevine root distribution in drip and microsprinkler irrigation. *Sci. Agric.* **2003**, *60*, 377–387. [CrossRef]
92. Sepulcre-Cantó, G.; Zarco-Tejada, P.J.; Jiménez-Muñoz, J.C.; Sobrino, J.A.; de Miguel, E.; Villalobos, F.J. Detection of water stress in an olive orchard with thermal remote sensing imagery. *Agric. For. Meteorol.* **2006**, *136*, 31–44. [CrossRef]
93. Falkenberg, N.; Piccinni, G.; Cothren, J.T.; Leskovar, D.I.; Rush, C.M. Remote sensing of biotic and abiotic stress for irrigation management of cotton. *Agric. Water Manag.* **2007**, *87*, 23–31. [CrossRef]
94. Perry, C.; Steduto, P.; Karajeh, F. *Does Improved Irrigation Technology Save Water? A Review of the Evidence*; Food and Agriculture Organization of the United Nations: Cairo, Egypt, 2017.
95. Paul, C.; Techen, A.; Robinson, J.; Helming, K. Rebound effects in agricultural land and soil management: Review and analytical framework. *J. Clean. Prod.* **2019**, *227*, 1054–1067. [CrossRef]
96. Grafton, R.Q.; Williams, J.; Perry, C.J.; Molle, F.; Ringler, C.; Steduto, P.; Udall, B.; Wheeler, S.A.; Wang, Y.; Garrick, D.; et al. The paradox of irrigation efficiency. *Science* **2018**, *361*, 748–750. [CrossRef]
97. Berbel, J.; Pedraza, V.; Giannoccaro, G. The trajectory towards basin closure of a European river: Guadalquivir. *Int. J. River Basin Manag.* **2013**, *11*, 111–119. [CrossRef]
98. Yang, C. Technologies to improve water management for rice cultivation to cope with climate change. *Crop Environ. Bioinform.* **2012**, *9*, 193–207.
99. Ma, H.; Shi, C.; Chou, N. China's water utilization efficiency: An analysis with environmental considerations. *Sustainability* **2016**, *8*, 516. [CrossRef]
100. Boutraa, T. Improvement of water use efficiency in irrigated agriculture: A review. *J. Agron.* **2010**, *9*, 1–8. [CrossRef]
101. Xue, J.; Guan, H.; Huo, Z.; Wang, F.; Huang, G.; Boll, J. Water saving practices enhance regional efficiency of water consumption and water productivity in an arid agricultural area with shallow groundwater. *Agric. Water Manag.* **2017**, *194*, 78–89. [CrossRef]
102. Attwater, R.; Derry, C. Achieving resilience through water recycling in peri-urban agriculture. *Water* **2017**, *9*, 223. [CrossRef]
103. Molden, D.; Oweis, T.; Steduto, P.; Bindraban, P.; Hanjra, M.A.; Kijne, J. Improving agricultural water productivity: Between optimism and caution. *Agric. Water Manag.* **2010**, *97*, 528–535. [CrossRef]
104. Fang, S.; Jia, R.; Tu, W.; Sun, Z. Assessing factors driving the change of irrigation water-use efficiency in China based on geographical features. *Water* **2017**, *9*, 759. [CrossRef]

105. Ibragimov, N.; Evett, S.R.; Esanbekov, Y.; Kamilov, B.S.; Mirzaev, L.; Lamers, J.P.A. Water use efficiency of irrigated cotton in Uzbekistan under drip and furrow irrigation. *Agric. Water Manag.* **2007**, *90*, 112–120. [CrossRef]
106. Maisiri, N.; Sanzanje, A.; Rockstrom, J.; Twomlow, S.J. On farm evaluation of the effect of low cost drip irrigation on water and crop productivity compared to conventional surface irrigation system. *Phys. Chem. Earth* **2005**, *30*, 783–791. [CrossRef]
107. Yazar, A.; Sezen, S.M.; Sesveren, S. LEPA and trickle irrigation of cotton in the Southeast Anatolia Project (GAP) area in Turkey. *Agric. Water Manag.* **2002**, *54*, 189–203. [CrossRef]
108. Peterson, J.M.; Ding, Y. Economic adjustments to groundwater depletion in the high plains: Do water-saving irrigation systems save water? *Am. J. Agric. Econ.* **2005**, *87*, 147–159. [CrossRef]
109. Abdulai, A.; Owusu, V.; Bakang, J.E.A. Adoption of safer irrigation technologies and cropping patterns: Evidence from Southern Ghana. *Ecol. Econ.* **2011**, *70*, 1415–1423. [CrossRef]
110. Cremades, R.; Wang, J.; Morris, J. Policies, economic incentives and the adoption of modern irrigation technology in China. *Earth Syst. Dyn.* **2015**, *6*, 399–410. [CrossRef]
111. Jalota, S.K.; Singh, K.B.; Chahal, G.B.S.; Gupta, R.K.; Chakraborty, S.; Sood, A.; Ray, S.S.; Panigrahy, S. Integrated effect of transplanting date, cultivar and irrigation on yield, water saving and water productivity of rice (Oryza sativa L.) in Indian Punjab: Field and simulation study. *Agric. Water Manag.* **2009**, *96*, 1096–1104. [CrossRef]

10

Analysis of the Effects of High Precipitation in Texas on Rainfed Sorghum Yields

Om Prakash Sharma [1], Narayanan Kannan [2,*], Scott Cook [3], Bijay Kumar Pokhrel [2] and Cameron McKenzie [4]

1 Department of Chemistry, Geosciences and Physics, Tarleton State University, Stephenville, TX 76402, USA; omprakash.sharma@go.tarleton.edu
2 Texas Institute for Applied Environmental Research (TIAER), Tarleton State University, Stephenville, TX 76402, USA; pokhrel@tarleton.edu
3 Department of Mathematics, Tarleton State University, Stephenville, TX 76402, USA; scook@tarleton.edu
4 Department of History, Sociology, Geography and GIS, Tarleton State University, Stephenville, TX 76402, USA; cameron.mckenzie@tarleton.edu
* Correspondence: kannan@tarleton.edu

Abstract: Most of the recent studies on the consequences of extreme weather events on crop yields are focused on droughts and warming climate. The knowledge of the consequences of excess precipitation on the crop yield is lacking. We attempted to fill this gap by estimating reductions in rainfed grain sorghum yields for excess precipitation. The historical grain sorghum yield and corresponding historical precipitation data are collected by county. These data are sorted based on length of the record and missing values and arranged for the period 1973–2003. Grain sorghum growing periods in the different parts of Texas is estimated based on the east-west precipitation gradient, north-south temperature gradient, and typical planting and harvesting dates in Texas. We estimated the growing season total precipitation and maximum 4-day total precipitation for each county growing rainfed grain sorghum. These two parameters were used as independent variables, and crop yields of sorghum was used as the dependent variable. We tried to find the relationships between excess precipitation and decreases in crop yields using both graphical and mathematical relationships. The result were analyzed in four different levels; 1. Storm by storm consequences on the crop yield; 2. Growing season total precipitation and crop yield; 3. Maximum 4-day precipitation and crop yield; and 4. Multiple linear regression of independent variables with and without a principal component analysis (to remove the correlations between independent variables) and the dependent variable. The graphical and mathematical results show decreases in rainfed sorghum yields in Texas for excess precipitation could be between 18% and 38%.

Keywords: grain sorghum; precipitation; rainfed; multiple linear regression; crop yield; principal component analysis

1. Introduction

Sorghum is a crop that can be grown as either a grain or cash crop. It is one of the top five cereal crops in the world. Sorghum is also required for the survival of humankind in different parts of the world, especially in Africa and Asia. The United States (US) is the largest producer of sorghum in the world [1]. In the US, sorghum usually grows throughout the sorghum belt from South Dakota to southern Texas [2]. The top five sorghum producing states are Kansas, Texas, Colorado, Oklahoma, and South Dakota. In the US, sorghum grain is primarily used for feeding of livestock and ethanol production, but it is becoming popular in the consumer food industry and other markets [3]. The livestock industry is one of the oldest standing marketplaces for sorghum grain in the US. Sorghum is utilized in feed rations for poultry, beef, dairy, and swine [3]. Also, a large portion of sorghum is used for biofuel production. It is also exported to the different parts of the world, including Mexico, China, and Japan.

Sorghum grain is highly resistant to drought and can withstand waterlogging better than any other cereal crop. Sorghum has a special fibrous root system, which can extend to a depth of 1.2 to 1.8 m (4 to 6 feet) deep in the soil. More than 75% of water and nutrients taken by root system are from the top 0.9 m (3 feet). Therefore, the deep extension of the root system helps sorghum withstand drought conditions better than any other cash crops [4]. Grain sorghum exhibits yield stability greater than maize. Drought resistance and heat tolerance make it a popular choice for marginal rainfall areas of semiarid zones of Africa where food shortages are common.

Total water use by a sorghum crop depends on the variety, maturation, planting date, and geographical and environmental conditions. It is estimated that the total use of 1750 mm/ha (28 inches of water/acre) water is needed for good sorghum yield of 783 kg/ha (700 lb/acre) [5]. The water use of sorghum depends on the growth stage of the sorghum plant (Table 1). During the early part of plant development, water use is relatively low but water stress during this time can affect plant growth and yield. Rainfall of 25 to 50 mm (1 to 2 inches) in the second week following sorghum pollination would result in the best yield if the period of pollination had adequate soil moisture [5,6]. The period from sorghum pollination to maturity is about 60 days. At the time of growth, a dry spell in the field from 14 to 60 days after pollination may have a small effect on the final harvesting yield of the sorghum crop. If no rain were to occur during the final period of 46 days, the yield of the sorghum crop would be greatly reduced [5,6]. Therefore, rainfall and its timing are important factors for the growth and yield of sorghum.

Table 1. Estimated grain sorghum water use by growth stage [5].

Days after Sorghum Planting	Water Requirement (Inches/Day)	Water Requirement (mm/day)
0–30 (early plant growth)	0.05–0.10	1.3–2.5
30–60 (rapid plant growth)	0.10–0.20	2.5–5.0
60–80 (boot and flowering)	0.25–0.30	6.3–7.5
80–120 (grain fill to maturity)	0.10–0.25	2.5–6.3

Although sorghum is tolerant of some waterlogging, it suffers damage under prolonged wetting of soil under very high rainfall [6]. Researchers from Australia, Germany, and the US have quantified the overall of extremes climate effects like drought, heat wave problems and precipitation on the crop yield variability of different staple crops around the world [7]. The year-to-year overall changes in the climatic factor in the growing season of maize, rice, sorghum, and wheat accounted the fluctuations of 20% to 49% of total yields [8]. Climatic extremes like hot and cold climates, drought, and heavy rainfall accounted for 18% to 43% of inter-annual variations in different crops yields [9]. Therefore, it is important to understand the consequences of climate extremes on crop yields to secure our food supply. A large body of literature already exists for drought. However, studies on the consequences of extremely high precipitation on crop yield are sparse, especially for grain sorghum. Therefore, an attempt is made in this study to analyze the consequences of high precipitation on rainfed grain sorghum yields.

Extreme precipitation events are producing more and more rain, and are now becoming one of the most common events since the beginning of the 1950s in many regions of the world, including the US. Scientists expect heavy rainfall as a consequence of a warming planet [10,11]. Warmer air mass can hold more water vapor content than cold air mass. For each degree of warming in the earth, the air mass capacity for holding water vapor goes up by about 7%. An atmosphere with more moist air can produce more heavy and continuous rainfall events, which is what has been observed all over the world since the 1950s [10,11].

An increase in continuous heavy rainfall events may not always show the increases in total rainfall over a season or year. Some studies show a small decrease in rainfall and show an increase of dry periods, which offsets rainfall increases falling during heavy events. The most immediate effect of heavy rainfall is the flooding. There are several recent examples of heavy rainfall events. In August 2017, Hurricane Harvey produced 1220 mm (48 inches) of heavy rainfall on Houston, Texas from a single event and was the biggest threat from tropical cyclones. In July 2016, more than 150 mm (6 inches) of heavy rainfall occurred in less than two hours in Ellicott City, Maryland, the estimated cost of the damage is more than $22 million dollars. In summary, we incur a huge economic loss because of heavy precipitation events, including some losses coming from a reduction in crop yields.

Precipitation is generally useful in recharging the soil profile, which is very important for crop growth. The precipitation efficiency in recharging the soil profile depends on intensity and rate at which precipitation occurs. Precipitation that falls on the soil at rates greater than 127 cm/h (0.5 inches an hour) are less efficient compared to lighter rain, because the water that runs off from the surface carries the fertile soil to the streams, lakes, and rivers which decreases future yields. The timing of rainfall while crops are growing is critical. During germination and stand establishment, either heavy rainfall or little rainfall can substantially affect the yield.

In general, the more precipitation during the crop growing season, the better the crop growth. However, too much precipitation will damage the crop by saturating the soil profile and removing air, which is also important for healthy plant growth. The majority of the previous studies relating extreme climate events and food production are focused on increasing temperatures and drought. The consequences of high precipitation on probable reduction in crop yields are often ignored. There is a big knowledge gap of understanding the consequences of extremely high precipitation on the yield of food crops and relating it to subsequent consequences in food production scenarios at different spatial scales. Addressing the knowledge gap and exploring the less-studied relationship between excess precipitation and rainfed food crop yields are the novelties of this study. Detailed analysis of the above-mentioned relationship using a combination of established mathematical principles and graphical tools are some of the unique aspects of this study. The results from our study and other similar studies have applications in crop insurance, parameterization of computer models (estimating crop yield reductions based on aeration stress), policy level decisions on rainfed crop selection, yield forecasting, estimating food production, and water footprint analysis.

The specific objectives of the study are to: (1) Identify historic extreme high precipitation events during the crop growth of rainfed sorghum in Texas, (2) Extract continuous serially complete crop yield information for rainfed sorghum by county, (3) Collect continuous records of daily average precipitation corresponding to the sorghum crop yield data, (4) Estimate the growing season total precipitation and 4-day maximum precipitation using the precipitation data, and (5) Relate items 3 and 4 above using visual patterns and statistical principles to quantify the consequences of high precipitation on crop yields.

2. Materials and Methods

2.1. Data Collection and Arrangement

2.1.1. GIS Data

The map of county boundaries was downloaded from the Texas Natural Resources Information System (TNRIS) website [12]. The cultivated area map of Texas was downloaded from the National Land Cover Dataset (NLCD) [13] and overlaid with county boundaries. A map showing the location of meteorological stations in Texas was developed using the latitude and longitude information that came with the precipitation data. It was overlaid with county boundaries to identify the list of weather stations within each county. Continuous records of Sorghum yield data (without gaps) are required for the analysis. In addition, the data availability period had to be consistent for different counties in Texas. The period from 1973 to 2000 satisfied the criteria of no data gaps and consistent availability of data for many counties. Therefore, only those counties with rainfed sorghum yield data (Figure 1, Table A1 in Appendix A) for the period 1973–2000 are included in the analysis and 26 United State Geological Survey (USGS) precipitation gaging data satisfied these criteria are considered for further analysis. Twenty-six meteorological stations (precipitation data from USGS) correspond to the counties having rainfed sorghum yield.

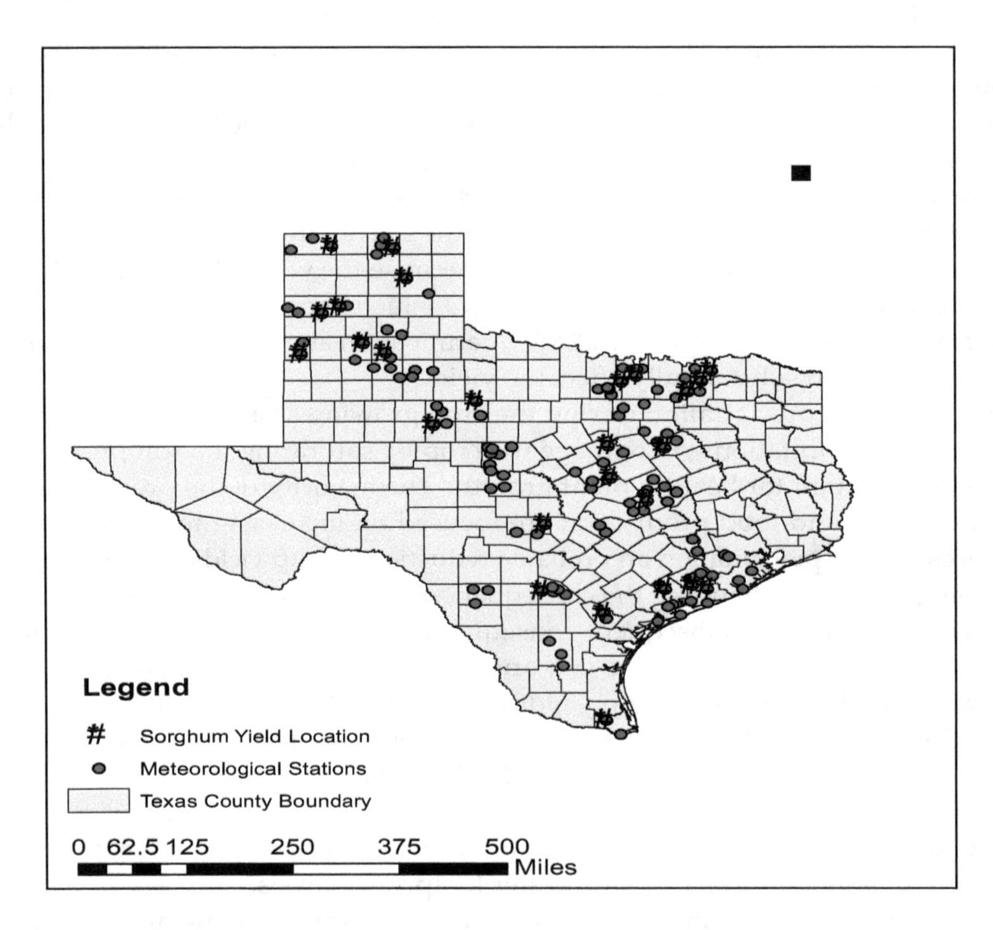

Figure 1. Map showing the location of Meteorological stations and rainfed sorghum cultivated location.

2.1.2. Estimation of Sorghum Growing Season for Different Counties in Texas

Grain sorghum is a hot season crop grown in most arid plain states that do not have enough moisture to grow other crops. Sorghum is planted once the soil temperature is consistent at about 15.5 °C (60 °F). This sometimes depends on the local condition so it can occur as early as late February in warmer climates or May in colder climates. This crop has longer maturity stages than other corn and cereal crops.

The planting dates of sorghum were estimated from USDA-ARS [14], taking into consideration the north-south temperature gradient. The harvest dates were estimated based on the planting date and the crop duration of 120 days (assumption). The detail of dates of planting and harvesting estimated for different counties in Texas are shown below in Table 2.

There is a north-south temperature gradient in Texas. Therefore, planting starts from the south and moves toward the northern region of Texas. Sorghum is planted in the southern region of Texas first around the last week of March and then towards the south-central region followed by the far eastern and eastern regions and finally ends toward the north in the last week of May. We selected a date from the range of dates in between the early and late planting dates of each county listed in the table above. That day is taken as a base for analysis with the precipitation data (Figure 2).

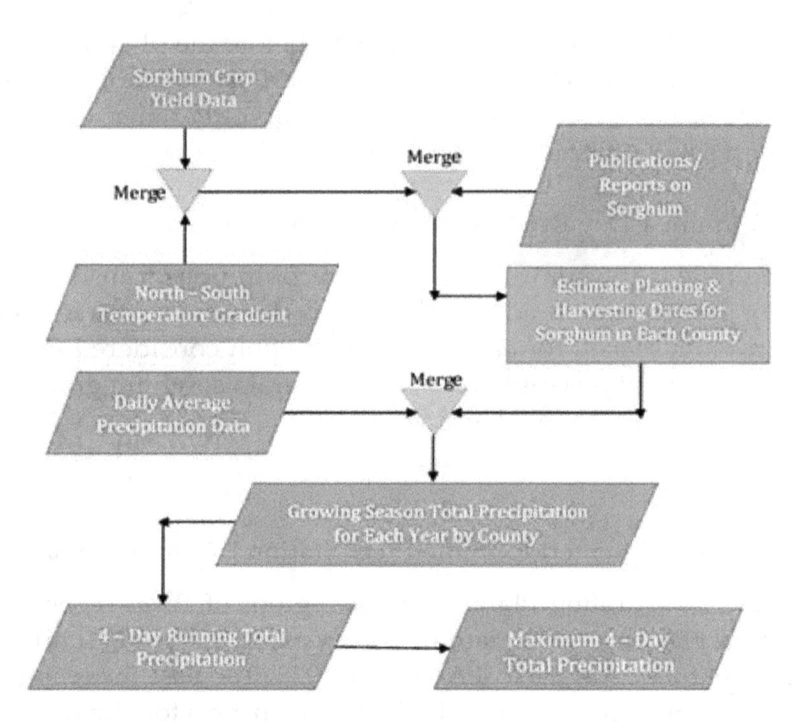

Figure 2. Estimation of growing season total precipitation and maximum 4-day total precipitation.

Table 2. Plant date of sorghum in different counties of Texas [14].

County	Planting Date			Harvest Date	Precipitation Start	Precipitation End
	Early	Late	Used			
Atascosa	3/10–3/15	3/15–3/25	15-March	3-July	5-March	25-June
Bailey	3/5–3/10	3/10–3/20	10-March	28-June	1-March	18-June
Bee	1/21–1/30	1/31–2/10	30-January	20-May	20-Jan	10-May
Bosque	3/15–3/25	3/26–4/5	25-March	13-July	15-March	3-July
Cameron	1/21–1/30	1/31–2/10	30-January	20-May	20-Jan	10-May
Collin	3/25–4/4	4/5–4/15	4-April	23-July	26-March	13-July
Cooke	3/15–3/25	3/26–4/5	25-March	13-July	15-March	3-July
Coryell	3/15–3/25	3/25–4/5	25-March	13-July	15-March	3-July
Dallam	3/5–3/10	3/10–3/20	10-March	28-June	1-March	18-June
Fannin	3/25–4/4	4/5–4/15	4-April	23-July	26-March	13-July
Floyd	3/5–3/10	3/10–3/20	10-March	28-June	1-March	18-June
Gillespie	3/10–3/15	3/15–3/25	15-March	3-July	5-March	25-June
Gray	3/5–3/10	3/10–3/20	10-March	28-June	1-March	18-June
Hale	3/5–3/10	3/10–3/20	10-March	28-June	1-March	18-June
Hansford	3/5–3/10	3/10–3/20	10-March	28-June	1-March	18-June
Hunt	3/25–4/4	4/5–4/15	4-April	23-July	26-March	13-July
Jackson	2/15–2/21	2/22–3/5	21-February	11-June	11-February	1-June
Jones	3/5–3/10	3/10–3/20	10-March	28-June	1-March	18-June
Matagorda	2/15–2/21	2/22–3/5	21-February	11-June	11-February	1-June
Milam	3/15–3/25	3/26–4/5	25-March	13-July	15-March	3-July
Navarro	3/25–4/4	4/5–4/15	4-April	23-July	26-March	13-July
Nolan	3/5–3/10	3/10–3/20	10-March	28-June	1-March	18-June
Randall	3/5–3/10	3/10–3/20	10-March	28-June	1-March	18-June
Wharton	2/15–2/21	2/22–3/5	21-February	11-June	11-February	1-June
Wise	3/15–3/25	3/26–4/5	25-March	13-July	15-March	3-July

2.1.3. Estimation of Growing Season Precipitation by County

The growing season is the number of consecutive days from the beginning of planting date to the harvesting date. It is calculated for every county. To obtain the growing season total precipitation, the precipitation of all daily values within the growing season is added together. The precipitation data used for analysis for each county was taken from 10 days before the planting and harvesting dates of each station from the base date. This is because farmers would use soil moisture from any precipitation event before planting the seeds. Also, they harvest the crop only when the crop is adequately dry, avoiding days for harvest soon after precipitation (Figure 2).

2.1.4. Estimation of Maximum 4-Day Running Total Precipitation

The 4-day running total is the cumulative value of continuous four days of precipitation data. Continuous four days of precipitation is added to get one value, and so on. In this way, it is calculated for every day in the growing season for each year and station considered for the analysis. Finally, the maximum of four days of total precipitation within the grain sorghum growing season each year is calculated for every station for further graphical analysis.

2.2. Data Analysis

2.2.1. Level 1: Historically Documented Extreme Precipitation Events and Sorghum Yield in Texas

The High Plains and Low Rolling Plains climatic regions of Texas received an extreme rainfall of 508 mm (20 inches) over 26 km^2 (10 square miles) area and 254 mm (10 inches) over 26,000 km^2 (10 thousand square miles) from 1 August to 4 August in 1978. The East Texas and Upper Coast climatic regions of Texas received an extreme rainfall of 1000 mm (40 inches) for about 26 km^2 (10 square miles) area and 254 mm (10 inches) for 26,000 km^2 (10 thousand square miles) from 24 July to 28 July in 1979. Randall County had a storm during 26 May to 27 May 1978. The rainfall amount during the period averaged 100 mm to 254 mm (4 in. to 10 in.) on the High Plains. Out of all the extreme precipitation events documented, only the May 1978 storm in Randall County fell within the sorghum-growing season. Therefore, only the details of the May 1978 storm will be included for further analysis under this category [15,16].

2.2.2. Level 2: Growing Season Precipitation and Rainfed Sorghum Crop Yields

The growing season's total precipitation and rainfed sorghum crop yield for different years is plotted to identify graphical relationships (Figure A1). The trends in data for every county were analyzed.

2.2.3. Level 3: Maximum 4-Day Running Total Precipitation and Crop Yield

The maximum 4-day running total precipitation and rainfed sorghum crop yield for different years is plotted to identify graphical relationships (Figure A2). The trends in data were studied for every county considered for the analysis.

2.2.4. Level 4: Generation of Mathematical Relationships between Rainfed Sorghum Yield and Excess Precipitation

Principal component analysis (PCA) is a commonly used mathematical tool used to display patterns in multivariate data. It removes correlation within a large set of variables and sorts them according to importance (explained variance) [17]. While PCA is commonly used for dimensionality reduction, it was not used for that purpose in this study. Total precipitation and max 4-day precipitation are somewhat correlated, which could affect the regression relationships. PCA transforms the input variables to remove such correlation. A downside of PCA is that while the original variables have clear interpretations (total growing season precipitation and max 4-day precipitation), the PCA-transformed variables do not. They are called "principal components" 1 and 2.

In our regression analysis, the dependent variable was taken as the rainfed grain sorghum yield data, and the independent variables were growing season total precipitation and maximum 4-day total precipitation (Figure 3). Multiple linear regression (MLR) analysis (Equation (1)) [18] was performed with the data analysis tool available in Microsoft Excel.

$$Y = A + B_1X_1 + B_2X_2, \quad (1)$$

where Y is crop yield, A is an intercept, X_1 and X_2 are growing season total precipitation and maximum 4-day total precipitation respectively, and B_1 and B_2 are partial regression coefficients [18].

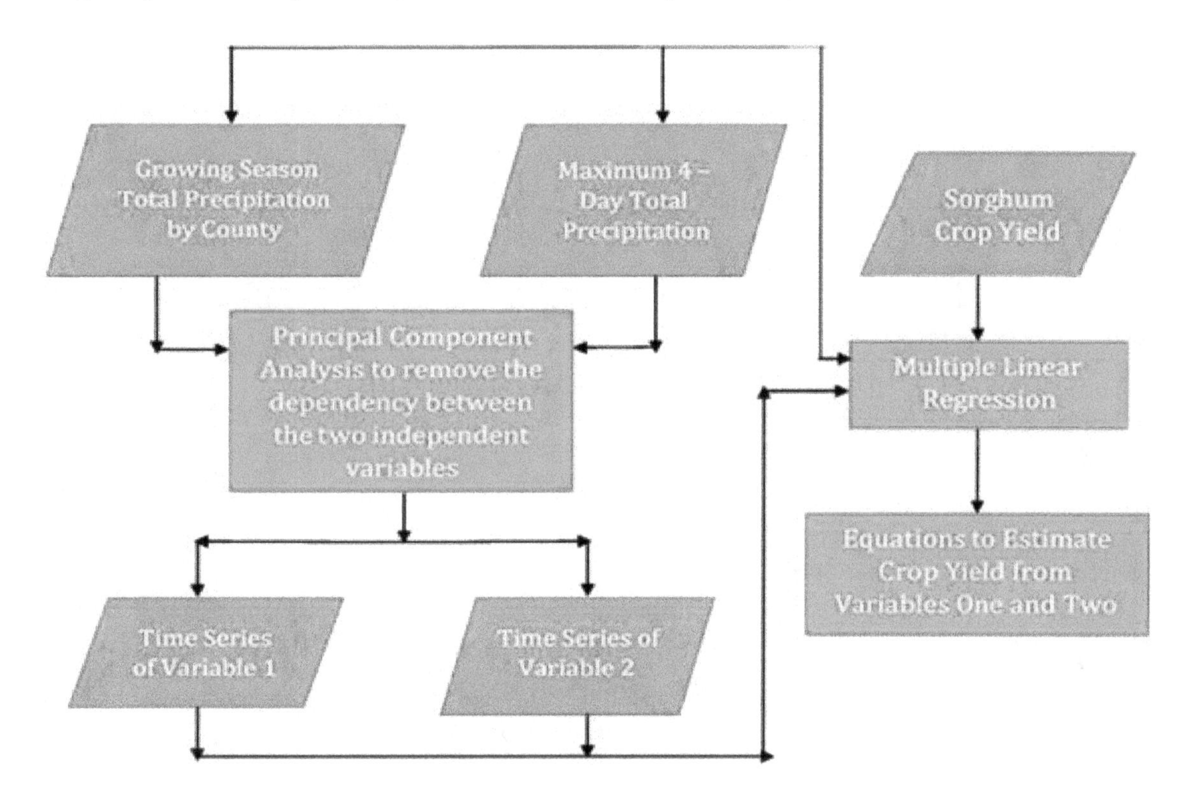

Figure 3. Schematic for the multiple linear regression (MLR) with and without a principal component analysis (PCA) (level 4 results).

3. Results

3.1. Level 1 Results

In the year 1978, Randall County encountered a storm event during the grain sorghum crop-growing period (26–27 May) (Figure 4). The 4-day maximum precipitation during the crop growing period was 182.9 mm (7.2 inches) which is 206% more than the average 4-day maximum precipitation (60.9 mm [2.4 inches]) that occurred during the sorghum crop growing period between 1973 and 2000. Also, the growing season total precipitation during the 1978 grain sorghum crop growing period was 271.8 mm (10.7 inches) which is 72.3% more than the average of the growing season total precipitation (157.5 mm (6.2 inches)) that occurred during the sorghum crop growing period between 1973 and 2000. The storm event could have brought down the rainfed sorghum yield by 27.5% (corresponding to the year 1978) when compared to the average rainfed sorghum yield from 1973 to 2000. This is evident from Figures 5 and 6, which show the sharp declines in crop yields based on 4-day maximum precipitation and growing season total precipitation separately.

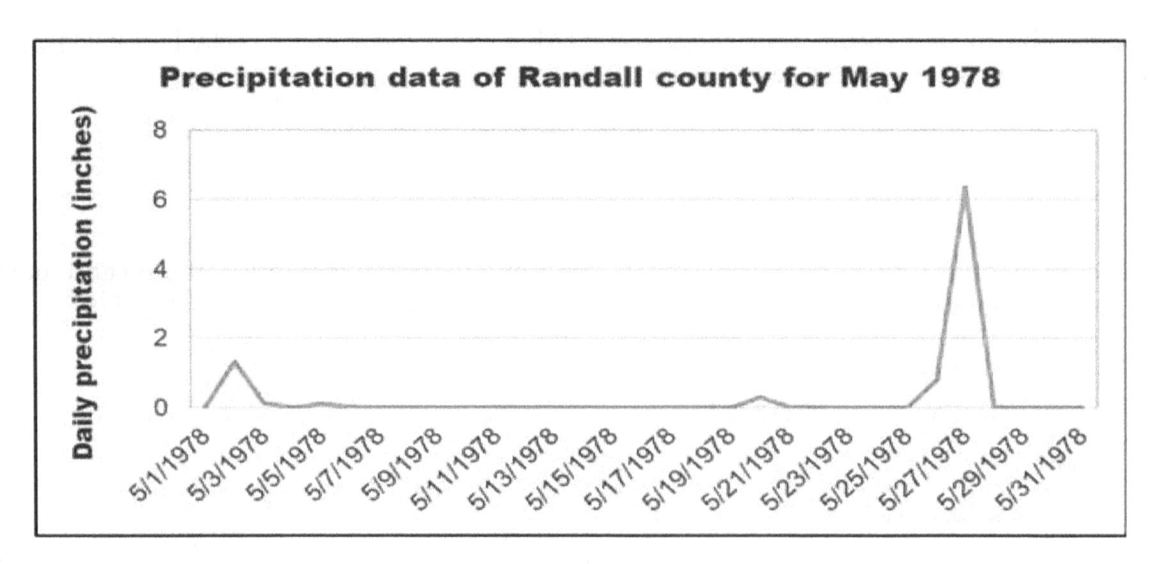

Figure 4. Storm event of May 1978 in Randall County.

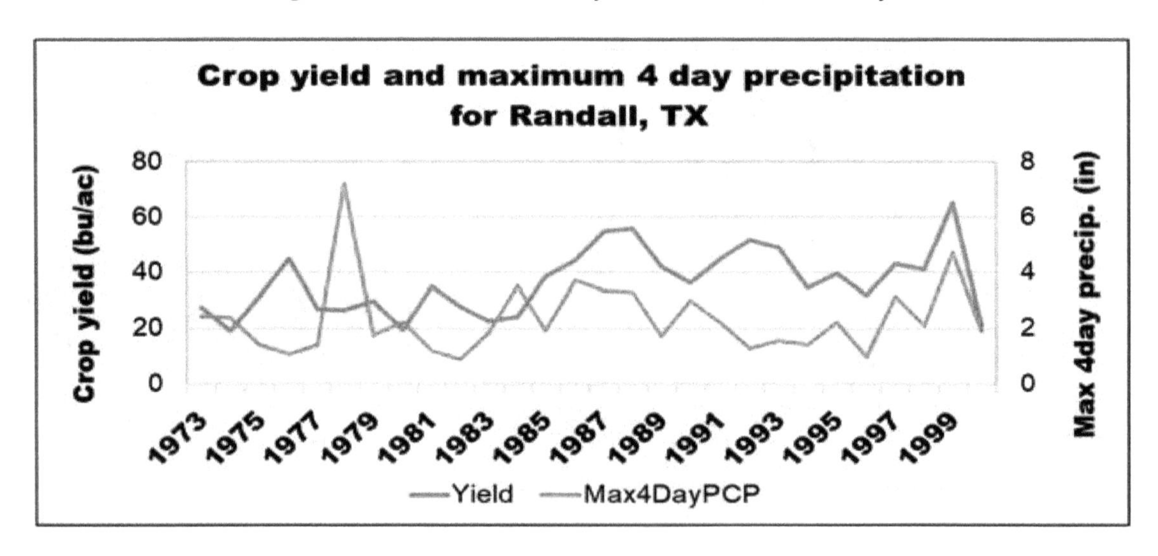

Figure 5. Sorghum yield reductions for Randall County in 1978 coming from maximum 4-day total precipitation.

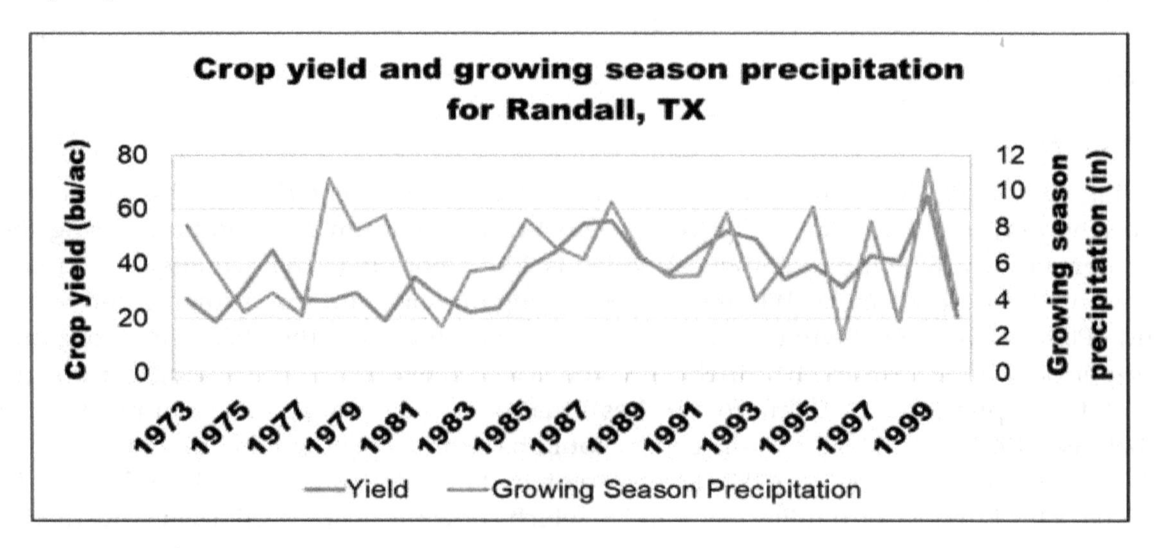

Figure 6. Sorghum yield reductions for Randall County in 1978 coming from excess growing season precipitation.

3.2. Level 2 Results

The graphical relationship between growing season total precipitation and rainfed sorghum crop yield was studied. Crop yield trends closely followed the growing season total precipitation for Texas counties Bailey, Bee, Cameron, Collin, Cooke, Dallam, Fannin, Hansford, Hunt, Jackson, and Wharton. When there was an increase in precipitation, there was a corresponding increase in the crop yield and vice versa (Figure 7). However, for some counties (e.g., Figure 8) there were declines in crop yield for excess precipitation. For Bosque County in 1976, growing season precipitation increased to 635 mm (25 inches) which resulted in a sharp decrease of crop yield. For Coryell County, when the annual growing season rainfall increased to 381 mm (15 inches) in 1976, it showed a decrease in crop yield. For Milam County in 1976, 1978, and 1994, increases in growing season total precipitation brought decreases in crop yield. Similar noticeable yield declines for excess precipitation results were observed for Atascosa, Gillespie, Hansford, Navarro, Randall, and Wise counties in Texas (Table 3) (graphs not shown in the manuscript for the sake of brevity).

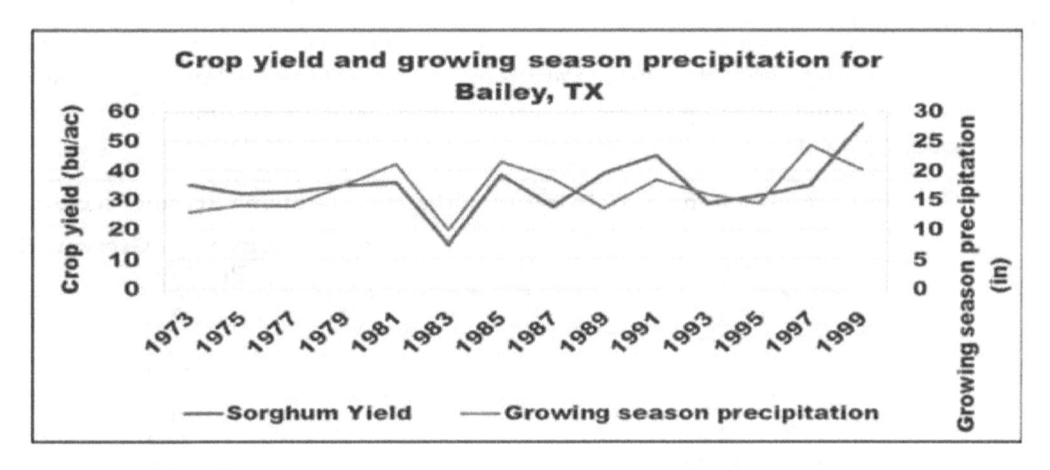

Figure 7. Example for crop yield trends closely following growing season total precipitation.

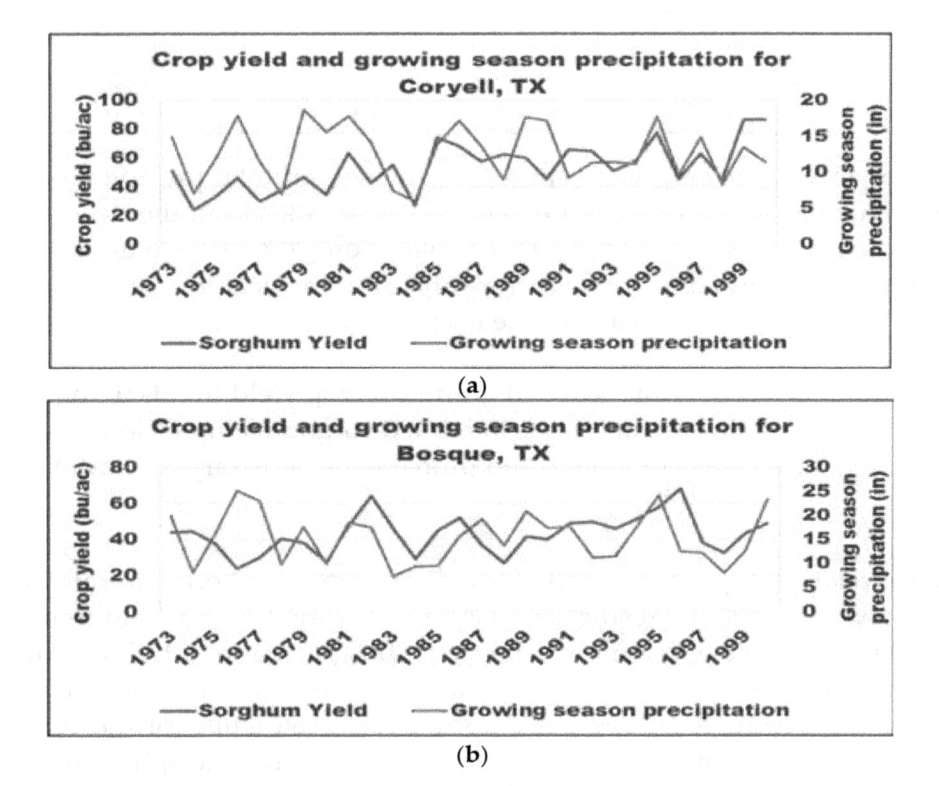

Figure 8. *Cont.*

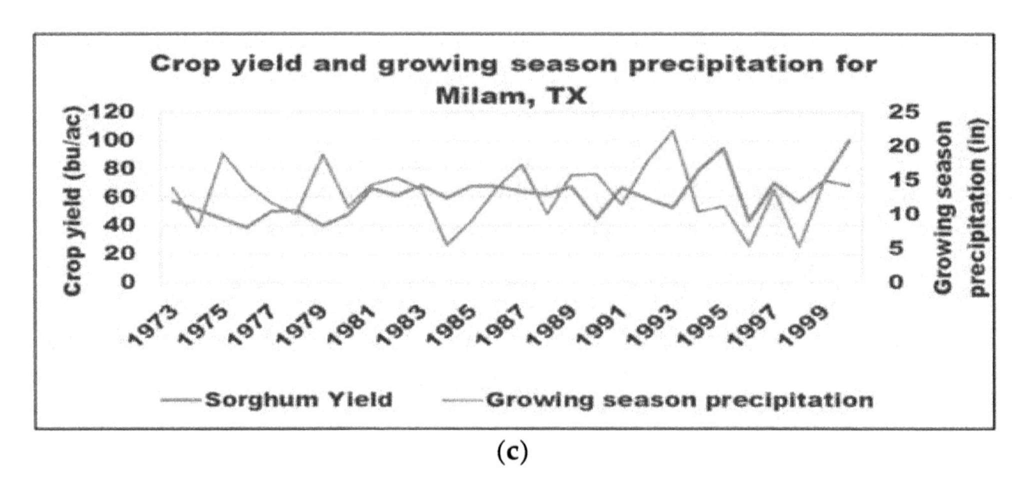

(c)

Figure 8. Relationship between growing season precipitation and crop yield for (**a**) Coryell County, (**b**) Bosque County, and (**c**) Milam County.

Table 3. Differences in sorghum yield between average growing season total precipitation (for years 1973–2000) and years showing high growing season total precipitation (column 2) and years nearby high growing season total precipitation (column 3).

County	% Differences in Sorghum Yield between the High Growing Season Precipitation and	
	Growing Season Precipitation for 1973–2000	Years Nearby High Growing Season Precipitation
Atascosa	40.5	9.5
Bosque	37.5	31.1
Coryell	33.4	27.3
Gillespie	4.88	−22.7
Hansford	34.59	28.5
Milam	27.87	21.9
Navarro	34.72	23.3
Randall	37.34	24.5
Wise	20.71	14.3
Average	30.2	17.5
95% CI	23 to 37	7 to 28

The numerical analysis of crop yields and growing season total precipitation are provided in Table 3. When compared to the average for the period 1973 to 2000, the decreases in crop yield corresponding to the year(s) with excess precipitation is about 30% (95% confidence intervals 23% to 37%). When compared to the nearby years (before and after the year with excess precipitation), the years with excess precipitation showed a decrease in crop yield of 17% (95% confidence intervals 7% to 28%) (Table 3).

In summary, the analysis of numerical and graphical crop yield trends with respect to growing season total precipitation highlighted decreases in rainfed sorghum crop yield when the precipitation received is higher than the average or what could probably be necessary for healthy crop growth.

3.3. Level 3 Results

The graphical relationships of maximum 4-day total precipitation with rainfed sorghum crop yields were analyzed. Some of the results are shown in Figure 9. Crop yield trends closely follow the maximum 4-day total precipitation for Bailey, Bee, Bosque, Fannin, Dallam, Hale, Hunt, Jones, Matagorda, Nolan, and Wise counties. Atascosa County shows four days maximum total of 8 inches and results in the sharp decrease in crop yield for the year 1980 while for the other years the crop yield trends follow precipitation. For Hunt County, the four days precipitation go above 254 mm (10 inches) and result in a decrease in crop yield comparing to other years. The decrease in crop yield was observed for Milam County as well when the maximum 4-day total precipitation reached 254 mm (10 inches).

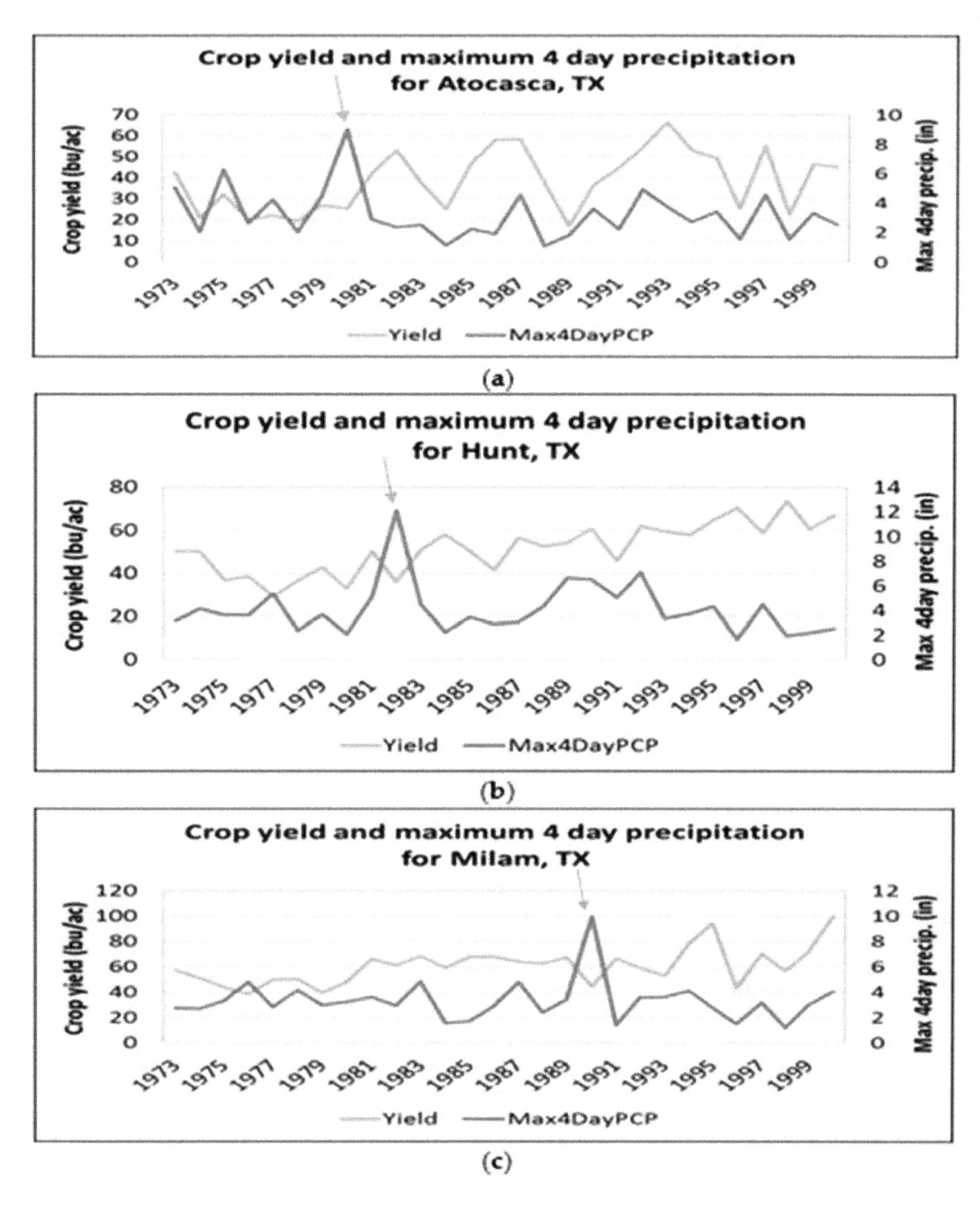

Figure 9. Relationship between maximum 4-day total precipitation and rainfed sorghum yield for (**a**) Atascosa (**b**) Hunt, and (**c**) Milam counties.

Similar rainfed sorghum yield declines were observed for high values of maximum 4-day running total precipitation for Coryell, Gillespie, Grey, Hansford, Matagorda, Navarro, Nolan, Randall, Wharton, and Wise counties. In summary, whenever the maximum four days running total precipitation is higher, that results in a decrease in crop yield of rainfed grain sorghum.

The numerical analysis of crop yields and maximum 4-day total precipitation are provided in Table 4. When compared to the average for the period 1973 to 2000 the decreases in crop yield corresponding to the year(s) with excess four-day precipitation is about 25% (95% confidence intervals 18% to 31%). When compared to the nearby years, the years with excess 4-day maximum precipitation showed a decrease in crop yield of 22% (95% confidence intervals 14% to 31%) (Table 4). In summary, the analysis of graphical and numerical crop yield trends with respect to maximum 4-day total precipitation pointed out decreases in rainfed sorghum crop yield when the precipitation received was much higher than the average or what could be necessary for healthy crop growth.

Table 4. Differences in sorghum yield between maximum 4-day precipitation (for years 1973–2000) and years showing high growing season total precipitation and years nearby high growing season total precipitation (Level 3 results).

County	% Differences in Sorghum Yield between the High Growing Season Precipitation and	
	Growing Season Precipitation for 1973–2000	Years Nearby High Growing Season Precipitation
Atascosa	35.0	26.8
Coryell	31.7	31.5
Hunt	30.6	29.0
Gillespie	10.2	−1.3
Gray	19.31	24.39
Hansford	22.29	24.95
Matagorda	7.16	5.80
Milam	26.52	33.23
Navarro	53.88	53.75
Nolan	20.14	32.52
Randall	27.54	10.13
Wharton	13.37	5.14
Wise	25.21	16.12
Average	24.8	22.5
95% CI	18 to 31	14 to 31

3.4. Level 4 Results

A multiple linear regression (MLR) analysis was performed with growing season total precipitation and maximum 4-day total precipitation as independent variables and rainfed sorghum yield as dependent variable the results of which are presented in Table 5. Although the R^2 values (Column 5 of Table 5) appear smaller, the regression relationships are significant, as evidenced by the F values of regression relationships presented in Table 6. Looking at the regression relationships by county, negative coefficients appear for growing season total precipitation for counties Bosque, Dallam, Hansford, and Milam only. Twenty-three out of 27 counties analyzed mathematically did not show declines in crop yield for excess precipitation when analyzed by growing season total precipitation. However, when analyzed by the maximum 4-day total precipitation, 21 out of 27 counties show negative coefficients substantiating the declines in crop yield for excess precipitation. The counties that do not show negative coefficients (with maximum 4-day total precipitation) are Bee, Bosque, Dallam, Deaf Smith, Floyd, and Hansford. Majority of the counties analyzed mathematically exhibit declining crop yields for excess precipitation showing negative coefficients mostly in maximum 4-day total precipitation and some in growing season total precipitation. Milam is the only county showing a negative coefficient for both the independent variables. Although Deaf Smith and Floyd showed some graphical relationships, they were the only counties that did not mathematically exhibit the regression relationship between the independent variables and the dependent variable.

Table 5. Results of multiple linear regression analysis (without principal component analysis) using annual growing season precipitation, 4-day maximum precipitation, and crop yield.

County	Coefficients for Independent Variables		Intercept	Regression Analysis without PCA
	Growing Season Precipitation	Maximum 4-Day Precipitation		Calculated (R^2)
Atascosa	1.922	−2.499	27.274	0.204
Bailey	1.595	−1.923	9.759	0.239
Bee	0.364	5.311	35.281	0.363
Bosque	−0.070	0.970	40.030	0.031
Cameron	1.346	−1.337	49.332	0.072
Collin	0.710	−0.481	46.576	0.086
Cooke	0.938	−1.453	48.932	0.151
Coryell	2.119	−2.830	36.120	0.119
Dallam	−0.476	1.762	31.160	0.024
Deaf Smith	0.506	2.239	31.146	0.053
Fannin	0.478	−2.411	56.630	0.094
Floyd	0.109	1.000	36.176	0.018
Gillespie	2.597	−3.674	26.381	0.350
Gray	1.568	−4.507	34.518	0.058
Hale	1.974	−3.905	32.998	0.056
Hansford	−1.161	3.661	42.566	0.028
Hunt	0.274	−1.941	55.071	0.064
Jackson	0.270	−2.128	78.370	0.103
Jones	2.003	−1.146	16.200	0.317
Matagorda	0.620	−0.750	69.784	0.075
Milam	−0.066	−0.977	64.811	0.014
Navarro	1.175	−4.179	50.711	0.070
Nolan	1.621	−1.206	21.611	0.268
Randall	1.730	−0.666	27.504	0.105
Wharton	0.177	−0.652	77.381	0.008
Wise	0.464	−1.952	41.285	0.054
All stations	2.523	−5.651	36.647	0.371

An MLR analysis like the one described above was performed with a PCA. The PCA was carried out to remove the relationship between the two independent variables. The results of the MLR are presented in Table 6; although the R^2 values (Column 5 of Table 7) appear smaller, the regression relationships are significant as evidenced by the F values of regression relationships presented in Table 6. Looking at the regression relationships (with PCA) by county, negative coefficients appear for growing season total precipitation for Fannin, Hansford, Hunt, Jackson, and Milam counties only. Twenty-two out of 27 counties analyzed did not show declines in crop yield for excess precipitation when analyzed mathematically using regression relationships with growing season total precipitation and crop yields. However, when analyzed by the maximum 4-day total precipitation, 21 out of 27 counties show negative coefficients substantiating the declines in crop yield for excess precipitation. The counties that do not show negative coefficients are Bee, Bosque, Dallam, Deaf Smith, Floyd, and Hansford. Like the MLR without a PCA, most of the counties analyzed mathematically exhibit declining crop yields for excess precipitation showing negative coefficients mostly in maximum 4-day total precipitation and some in growing season total precipitation. Milam is the only county showing a negative coefficient for both the independent variables. Although showing some graphical relationships, Deaf Smith and Floyd are the only counties that did not mathematically exhibit the regression relationship between the independent variables and the dependent variable.

Table 6. Relevance of regression relationships.

County	Significance of Regression without PCA		Significance of Regression with PCA	
	F	Significance F	F	Significance F
Atascosa	3.203	0.057	3.056	0.065
Bailey	3.933	0.032	4.251	0.026
Bee	6.822	0.004	7.333	0.003
Bosque	0.386	0.683	0.386	0.683
Cameron	0.933	0.406	0.933	0.406
Collin	0.841	0.447	0.468	0.631
Cooke	1.595	0.230	1.235	0.308
Coryell	1.617	0.219	1.617	0.219
Dallam	0.295	0.747	0.295	0.747
Deaf Smith	0.669	0.521	0.669	0.521
Fannin	1.244	0.306	1.244	0.306
Floyd	0.226	0.800	0.225	0.799
Gillespie	6.456	0.006	6.465	0.005
Gray	0.739	0.487	0.739	0.487
Hale	0.710	0.508	0.710	0.501
Hansford	0.346	0.710	0.346	0.710
Hunt	0.817	0.453	0.817	0.453
Jackson	1.439	0.255	1.324	0.284
Jones	5.795	0.008	6.100	0.007
Matagorda	1.007	0.379	1.013	0.377
Milam	0.179	0.836	0.182	0.834
Navarro	0.947	0.401	0.914	0.414
Nolan	4.572	0.020	3.109	0.062
Randall	1.473	0.248	1.473	0.248
Wharton	0.097	0.907	0.097	0.907
Wise	0.717	0.497	0.586	0.564
All stations	7.37	0.003	7.272	0.003

Table 7. Results of multiple linear regression analysis (with PCA for removing the relationship between the two independent variables) using annual growing season precipitation, 4-day maximum precipitation, and crop yield.

County	Coefficients for Independent Variables		Intercept	Regression Analysis with PCA
	Variable (X1)	Variable (X2)		Calculated (R^2)
Atascosa	1.156	−2.986	38.517	0.203
Bailey	1.345	−2.153	30.925	0.262
Bee	1.753	5.296	53.441	0.379
Bosque	0.093	0.968	42.508	0.031
Cameron	0.325	−1.869	54.574	0.072
Collin	0.389	−0.232	53.336	0.038
Cooke	0.418	−1.637	53.976	0.093
Coryell	1.331	−3.275	53.757	0.119
Dallam	0.185	1.816	31.922	0.024
Deaf Smith	1.068	2.032	38.246	0.053
Fannin	−0.036	−2.458	54.716	0.094
Floyd	0.381	0.931	39.602	0.018
Gillespie	1.526	−4.232	45.486	0.350
Gray	0.171	−4.769	37.397	0.058
Hale	0.463	−4.351	38.201	0.056
Hansford	−0.320	3.827	41.515	0.028
Hunt	−0.363	−1.926	51.713	0.064
Jackson	−0.150	−2.231	73.163	0.099
Jones	1.718	−1.643	30.740	0.337

Table 7. *Cont.*

County	Coefficients for Independent Variables		Intercept	Regression Analysis with PCA
	Variable (X1)	Variable (X2)		Calculated (R^2)
Matagorda	0.448	−1.116	74.051	0.078
Milam	−0.253	−1.003	60.859	0.015
Navarro	0.167	−4.396	52.597	0.071
Nolan	1.041	−1.337	30.958	0.206
Randall	1.366	−1.252	36.707	0.105
Wharton	0.051	−0.674	76.854	0.008
Wise	0.133	−1.738	41.469	0.047
All stations	1.402	−6.034	47.693	0.377

A comparison of the R^2 values of regression relationships with and without PCA are presented in Table 8 which pointed out that the PCA did not offer a significant improvement in identifying relationships between excess precipitation and rainfed sorghum yield. However, there is some difference in the regression analysis results. In the regression without a PCA, only one county (Milam) did not mathematically show any declining crop yields with excess precipitation. In the regression with PCA, six out of 27 counties analyzed (Bee, Bosque, Dallam, Deaf Smith, Floyd, and Hansford) did not show declining crop yields with excess precipitation. However, the results analyzed in all four different levels point out the existence of crop yield declines with excess precipitation.

Table 8. R^2 with and without PCA.

County	(R^2) without PCA	(R^2) with PCA
Atascosa	0.204	0.203
Bailey	0.239	0.262
Bee	0.363	0.379
Bosque	0.031	0.031
Cameron	0.072	0.072
Collin	0.086	0.038
Cooke	0.151	0.093
Coryell	0.119	0.119
Dallam	0.024	0.024
Deaf Smith	0.053	0.053
Fannin	0.094	0.094
Floyd	0.018	0.018
Gillespie	0.350	0.350
Gray	0.058	0.058
Hale	0.056	0.056
Hansford	0.028	0.028
Hunt	0.064	0.064
Jackson	0.103	0.099
Jones	0.317	0.337
Matagorda	0.075	0.078
Milam	0.014	0.015
Navarro	0.070	0.071
Nolan	0.268	0.206
Randall	0.105	0.105
Wharton	0.008	0.008
Wise	0.054	0.047
All stations	0.371	0.377

3.5. Substantiation of Crop Yield Declines with Excess Precipitation

In the previous section, the existence of crop yield decline with excess precipitation was identified based on separate graphical relationships between crop yield and growing season total precipitation, and crop yield and maximum 4-day total precipitation. The presence of crop yield decline for excess precipitation are substantiated by the graphical plot of crop yield, growing season total precipitation, and maximum 4-day total precipitation together for Hunt county in TX. The thin green rectangle outlined in Figure 10 identifies the hotspots that substantiate our findings described in the previous section(s).

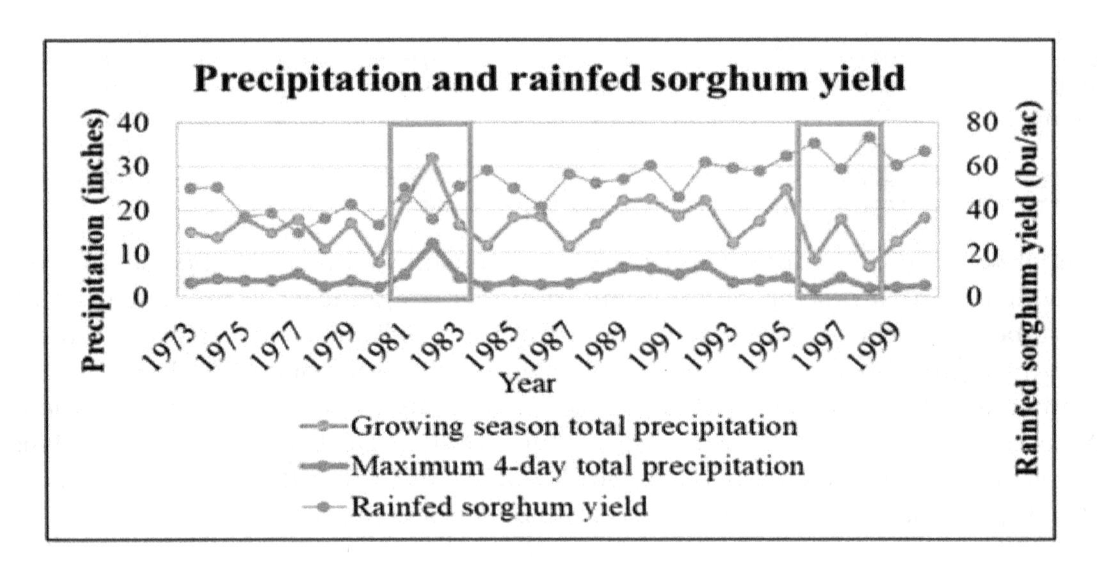

Figure 10. Graph showing rainfed sorghum yield, maximum 4-day total precipitation, and growing season total precipitation for Hunt County.

3.6. Spatial Variation of Declines in Crop Yield for Excess Precipitation

Counties and climate regions in Figures 11 and 12, respectively show the spatial variation of declines in yield of sorghum for excess precipitation. Based on both growing season total precipitation and maximum 4-day total precipitation, the North Central region of Texas appears to be more vulnerable to rainfed sorghum yield declines than other parts of Texas. The other regions showing some crop yield decline for excess precipitation are the High Plains and Southern regions. The large variation of precipitation within the region (Figure 13) and precipitation patterns appear to be the probable reason that can be attributed. However, we need more evidence to substantiate this finding.

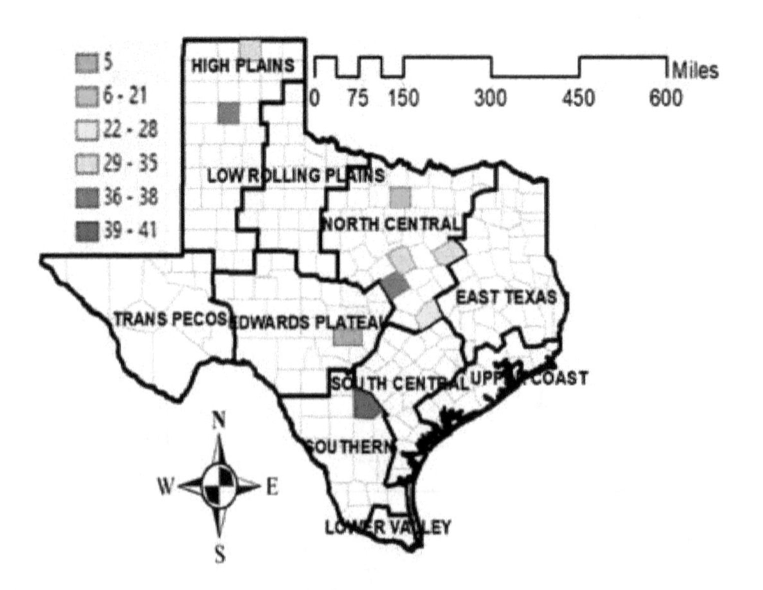

Figure 11. Percent reduction in rainfed sorghum yield between the year with excess precipitation and average crop yield from 1973 to 2000 (based on growing season total precipitation).

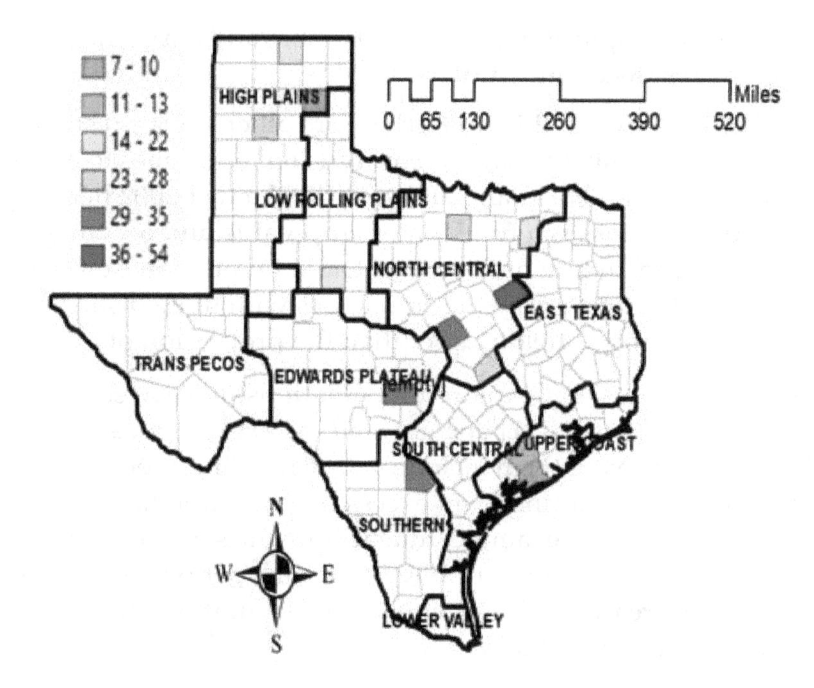

Figure 12. Percent reduction in rainfed sorghum yield between the year with excess precipitation and average crop yield from 1973 to 2000 (based on maximum 4-day total precipitation).

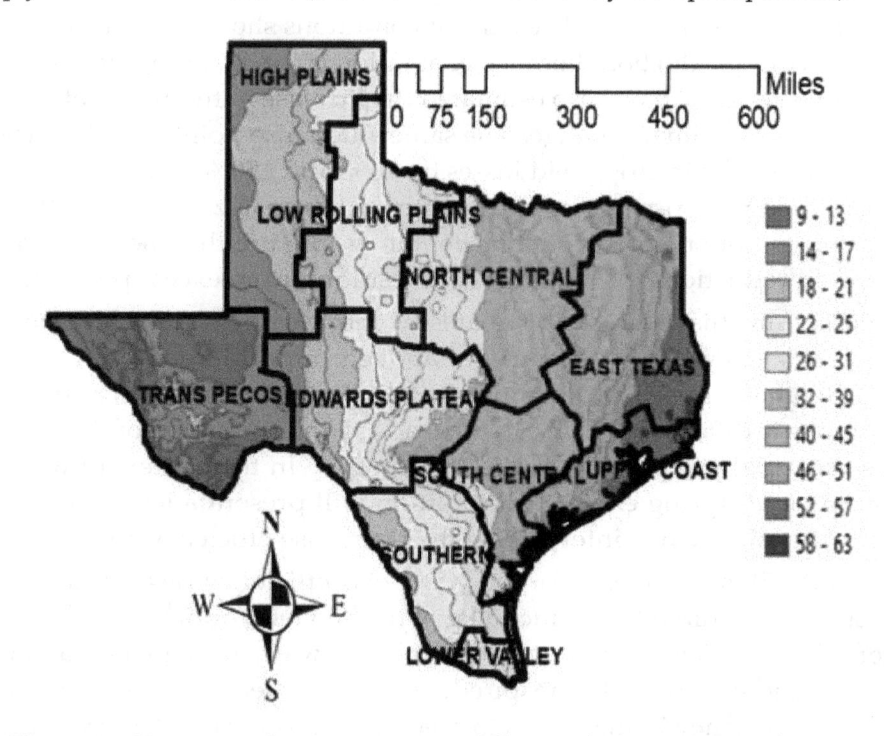

Figure 13. Variation of precipitation in different climate regions of Texas [19].

4. Discussion

For estimating crop yield losses, our study considered the quantity of precipitation alone leaving out another important aspect of precipitation, the timing with respect to the sorghum-growing season. In addition to excess precipitation, there are other contributing factors to yield losses such as high/low temperature (higher than optimum temperature for crop growth and lower than the crop base temperature), wind speed (high winds can dislodge the crop), humidity levels (excess would cause fungal problems), quality of soil (pH, drainage characteristics, depth), human decisions (e.g., whether or not going for pesticide application, irrigation, etc.), human errors in timing of land management operations (fertilizer or pesticide application, tillage, irrigation, and harvest). Therefore, care should be taken when interpreting the results of our study.

In addition to the approach used in this study, there are other ways of estimating crop yield losses by excess precipitation. The possibility of using remote sensing techniques to estimate crop yield losses by flooding was explored in Tapia-Silva et al. [20] using the August 2002 flooding event in Germany. In their approach, the flood crop loss is a function of crop value and a damage factor. The damage factor is a function of type of crop, timing of flood event, and inundation duration. When compared to field observations, they were able to estimate the crop losses with limited success. Their analysis dealt with flood inundation area of cropped fields rather than the proportion of yield loss.

There were a few other studies that explored the relationship between excess precipitation and crop yield reductions. Rosenzweig et al. [21] documented the extreme weather events that occurred in the US between 1977 and 1998; many of them include severe flooding events that resulted in reductions in crop yield. Increased moisture resulting from excess precipitation helps to spread epidemics and prevalence of leaf fungal pathogens, for example, fungal epidemics in corn, soybean, alfalfa, and wheat reported to have occurred in the US Midwest in 1993. The same period also saw incidences of soybean sudden death and mycotoxin increases [21]. Continuous soil saturation causing crazy top and common smut are also documented in the same study.

Corn yield reductions due to excess soil moisture (resulting from high precipitation) during current conditions and future conditions (under climate change) were estimated by Rosenzweig et al. [9] using CERES-maize model for the US Midwest. The current conditions showed a 3% reduction in corn yield ($600 million for the US corn production) because of aeration stress resulting from excess precipitation in the US Midwest. However, they have also estimated the increase in frequency of excess precipitation events in the future because of climate change. The same study also points out that when compared to the present, 90% more decreases in crop yield losses by 2030 and 150% more yield losses are expected by 2090. Winter wheat yield response to many parameters were analyzed in the Netherlands including excess precipitation. Except for one precipitation event in week 31 of the calendar year, they could not find any noticeable yield reductions for winter wheat resulting from excess precipitation [22].

The topic discussed in this manuscript relates to the idea of water use efficiency and water footprint. Water-use efficiency [23] is the ratio of aboveground biomass production to the water evapotranspired. The biomass is usually determined as dry weight rather than as fresh weight because moisture content of crops is different, which can mislead the interpretation of the water-use efficiency results. The results are usually expressed in kg L^{-1} or t m^{-3}. In the context of water-use efficiency, the reductions in crop yield during excess precipitation will present a less water efficient scenario. Therefore, care should be taken when interpreting the water-use efficiency results.

Water footprint [24,25] is the inverse of the water-use efficiency described above. The typical units are L kg^{-1} (L of water required to produce a kg of useful yield) or m^3 t^{-1} (m^3 of water required to produce a metric ton of useful yield). Green water footprint is water from precipitation that is stored in the root zone of the soil and evaporated, transpired, or incorporated by plants [24]. For rainfed crops, the inverse of water-use efficiency is analogous to green water footprint. The reductions in crop yield during excess precipitation will produce a relatively large green water footprint. Therefore, care should be taken when interpreting the water footprint results for crops that underwent an excess precipitation scenario like what is discussed in our study. The simplest way to avoid misleading water-use efficiency and green water footprint results are to use the average values from multiple crop growing years capturing a range of climatic scenarios.

The results of this study and other similar studies have applications in payment of crop insurance claims, parameterization of computer models (estimating crop yield reductions based on aeration stress), policy level decisions on rainfed crop selection, yield forecasting, estimating threats to food production, and water footprint analysis.

5. Conclusions

We collected historical crop yield data for Texas by county for grain sorghum from 1973 to 2000 and the corresponding daily precipitation data from weather stations within the counties. After estimating the crop growing season for sorghum in different parts of Texas, we estimated the growing season total precipitation and maximum 4-day total precipitation for each county growing rainfed grain sorghum. Using the two parameters mentioned above as independent variables, and crop yield of sorghum as the dependent variable, we tried to find out relationships between excess precipitation and decreases in crop yields using both graphical and mathematical relationships. We carried out a multiple linear regression (MLR) analysis with and without the use of a principal component analysis (PCA). Based on the results obtained, we can conclude that:

- Excess precipitation during crop growing season can cause yield reduction in rainfed grain sorghum.
- Total precipitation during the growing season and maximum 4-day total precipitation during the growing season are potential indicators of yield reductions in grain sorghum.
- Yield reductions could be in the range of 18% to 38% for rainfed grain sorghum in Texas because of excess precipitation during the growing season.
- When analyzed spatially, the north-central climate region of Texas appears to be more vulnerable to rainfed sorghum yield reductions because of excess precipitation.

Author Contributions: O.P.S. carried out most parts of the study. N.K. conceptualized the overall study, S.C. designed and carried out the principal component analysis; regression analysis was carried out by O.P.S. under the supervision of S.C. and N.K. B.K.P. analyzed the results of the study. Most of the GIS analysis was carried out by C.M. All the authors contributed to the development of this manuscript.

Acknowledgments: The authors acknowledge Tarleton State University for supporting this research.

Appendix A

Table A1. List of counties in Texas that have rainfed sorghum yield data is available.

Station Number	County	Latitude	Longitude
2	Atascosa	28.92	−98.74
4	Bailey	34.21	−102.73
6	Bee	28.45	−97.70
9	Bosque	32.01	−97.61
16	Cameron	25.91	−97.42
20	Collin	33.03	−96.48
23	Cooke	33.48	−97.15
24	Coryell	31.27	−97.88
26	Dallam	36.23	−102.24
28	Deaf Smith	34.93	−102.98
36	Fannin	33.43	−96.33
39	Floyd	33.98	−101.33
41	Gillespie	30.18	−99.15
42	Gray	35.55	−100.97
43	Hale	34.18	−101.7
45	Hansford	36.19	−101.18
46	Hunt	33.36	−96.06
47	Jackson	28.96	−96.68

Table A1. *Cont.*

Station Number	County	Latitude	Longitude
48	Jones	32.94	−99.8
50	Matagorda	28.68	−95.97
52	Milam	30.61	−97.2
53	Navarro	31.96	−96.68
54	Nolan	32.44	−100.52
58	Randall	34.95	−102.1
66	Wharton	29.31	−96.08
67	Wise	33.35	−97.39

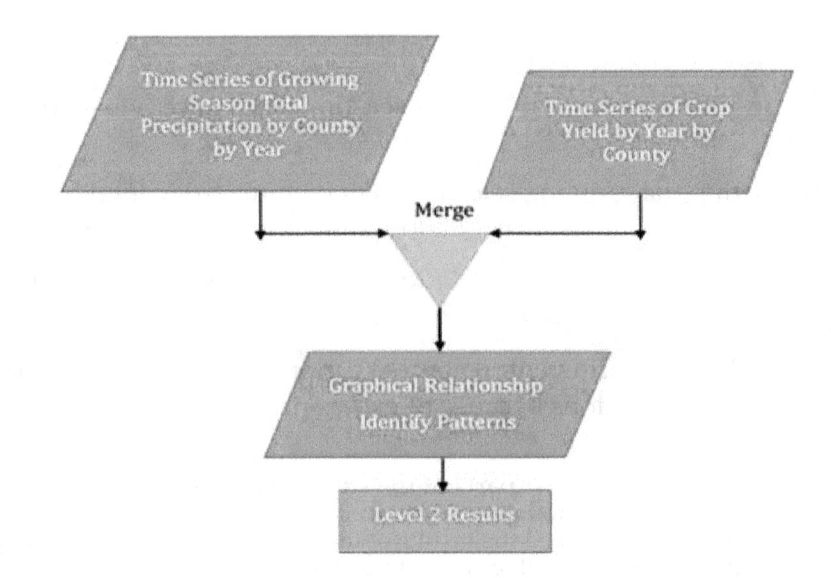

Figure A1. Generation of the graphical relationship between rainfed sorghum yield and growing season total precipitation (level 2 results).

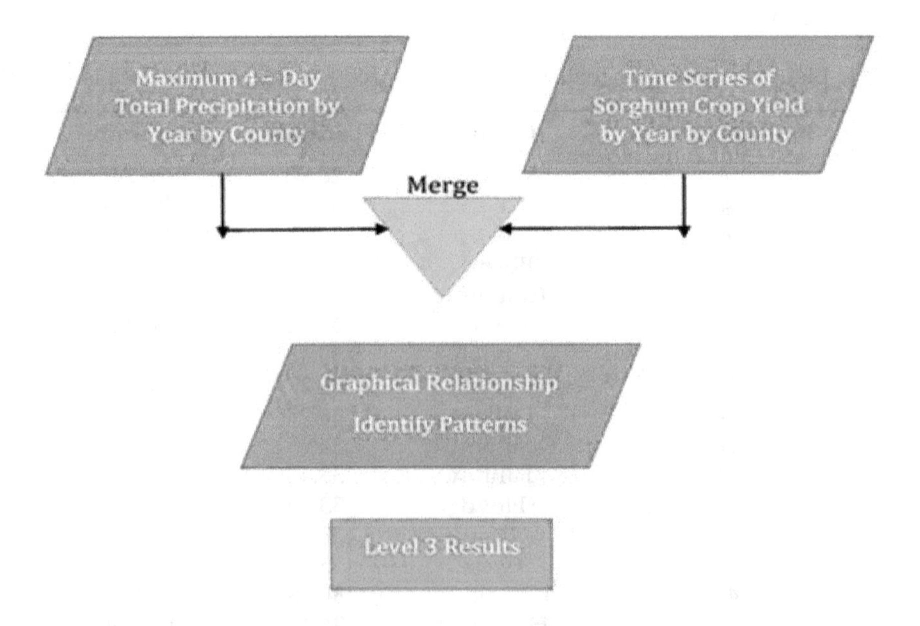

Figure A2. Generation of the graphical relationship between rainfed sorghum yield and maximum 4–day total precipitation (level 3 results).

References

1. Tsuchihashi, N.; Goto, Y. Year-round cultivation of sweet sorghum [*Sorghum bicolor* (L.) Moench] through a combination of seed and ratoon cropping in Indonesian Savanna. *Plant Prod. Sci.* **2008**, *11*, 377–384. [CrossRef]
2. Sorghum Program. Available online: https://www.sorghumcheckoff.com/all-about-sorghum (accessed on 31 July 2019).
3. Eggen, M.; Ozdogan, M.; Zaitchik, B.; Ademe, D.; Foltze, J.; Simane, B. Vulnerability of sorghum production to extreme, sub-seasonal weather under climate change. *Environ. Res. Lett.* **2019**, *14*, 045005. [CrossRef]
4. New, L. *Grain Sorghum Irrigation*; Texas A&M AgriLife Extension: Amarillo, TX, USA, 2004.
5. Arkansas Sorghum Quick Facts. Available online: https://www.uaex.edu/farm-ranch/crops-commercial-horticulture/grain-sorghum/2014-Arkansas-Grain-Sorghum-Quick-Facts.pdf (accessed on 31 July 2019).
6. Promkhambut, A.; Younger, A.; Polthanee, A.; Akkasaeng, C. Morphological and physiological responses of Sorghum (*Sorghum bicolor* L. Moench) to Waterlogging. *Asian J. Plant Sci.* **2010**, *9*, 183–193.
7. Lobell, D.; Burke, M.B.; Tebaldi, C.; Mastrandrea, M.D.; Falcon, W.P.; Naylor, R.L. Prioritizing climate change adaptation needs for food security in 2030. *Science* **2008**, *319*, 607–610. [CrossRef] [PubMed]
8. Grossi, M.C.; Justino, F.; Rodrigues, R.; Andrade, C.L.T. Sensitivity of the sorghum yield to individual changes in climate parameters: Modelling based approach. *Bragantia* **2015**, *74*, 341–349. [CrossRef]
9. Rosenzweig, C.; Tubiello, F.N.; Goldberg, R.; Mills, E.; Bloomfield, J. Increased crop damage in the US from excess precipitation under climate change. *Glob. Environ. Chang.* **2002**, *12*, 197–202. [CrossRef]
10. Balling, R.C., Jr.; Goodrich, G.B. Spatial analysis of variations in precipitation intensity in the USA. *Theor. Appl. Climatol.* **2011**, *104*, 415–421. [CrossRef]
11. Kunkel, K.E.; Karl, T.R.; Brooks, H.; Kossin, J.; Lawrimore, J.H.; Arndt, D.; Bosart, L.; Changnon, D.; Cutter, S.L.; Doesken, N.; et al. Monitoring and understanding trends in extreme storms: State of knowledge. *Bull. Am. Meteorol. Soc.* **2013**, *94*, 499–514. [CrossRef]
12. Texas Natural Resources Information Systems (TNRIS). Available online: https://data.tnris.org/ (accessed on 31 July 2019).
13. Jin, S.; Yang, L.; Danielson, P.; Homer, C.; Fry, J.; Xian, G. A comprehensive change detection method for updating the National Land Cover Database to circa 2011. *Remote Sens. Environ.* **2013**, *132*, 159–175. [CrossRef]
14. United States Department of Agriculture-National Agricultural Statistical Service (USDA-NASS). *Field Crops Usual Planting and Harvesting Dates*; Agricultural Handbook Number 626; USDA-NASS: Washington, DC, USA, 2010.
15. Mishra, A.K.; Singh, V.P. Changes in extreme precipitation in Texas. *J. Geophys. Res. Atmos.* **2010**, *115*. [CrossRef]
16. Nielsen-Gammon, J.W.; Zhang, F.; Odins, A.M.; Myoung, B. Extreme rainfall in Texas: Patterns and predictability. *Phys. Geogr.* **2005**, *26*, 340–364. [CrossRef]
17. Jolliffe, I.T. *Principal Component Analysis*, 2nd ed.; Springer: New York, NJ, USA, 2002.
18. Bluman, A.G. *Elementary Statistics: A Step by Step Approach*, 7th ed.; McGraw Hill Publishers: New York, NJ, USA, 2009.
19. Texas Water Development Board (TWDB). GIS Data. Available online: www.twdb.texas.gov/mapping/gisdata.asp (accessed on 10 June 2019).
20. Tapia-Silva, F.; Itzerott, S.; Foerster, S.; Kuhlmann, B.; Breibich, H. Estimation of flood losses to agricultural crops using remote sensing. *Phys. Chem. Earth* **2011**, *36*, 253–265. [CrossRef]
21. Rosenzweig, C.; Iglesias, A.; Yang, X.B.; Epstein, P.R.; Chivian, E. Climate change and extreme weather events Implications for food production, plant diseases, and pests. *Glob. Chang. Hum. Health* **2001**, *2*, 90–104. [CrossRef]
22. Powell, J.P.; Reinhard, S. Measuring the effects of extreme weather events on yields. *Weather Clim. Extremes* **2016**, *12*, 69–79. [CrossRef]
23. Kirkham, M.B. Water use efficiency. In *Encyclopedia of Soils in the Environment, Reference module in Earth Systems and Environmental Sciences*; Elsevier: Cambridge, MA, USA, 2005; pp. 315–322.
24. Hoekstra, A.Y. *The Water Footprint of Modern Consumer Society*; Routledge: London, UK, 2013.

11

Multi-Crop Production Decisions and Economic Irrigation Water use Efficiency: The Effects of Water Costs, Pressure Irrigation Adoption and Climatic Determinants

Yubing Fan [1,*], Raymond Massey [2] and Seong C. Park [1]

1 Texas A&M AgriLife Research, Vernon, TX 76384, USA; scpark@ag.tamu.edu
2 Division of Applied Social Sciences, University of Missouri, Columbia, MO 65211, USA; masseyr@missouri.edu
* Correspondence: yubing.fan@ag.tamu.edu

Abstract: In an irrigated multi-crop production system, farmers make decisions on the land allocated to each crop, and the subsequent irrigation water application, which determines the crop yield and irrigation water use efficiency. This study analyzes the effects of the multiple factors on farmers' decision making and economic irrigation water use efficiency (*EIWUE*) using a national dataset from the USDA Farm and Ranch Irrigation Survey. To better deal with the farm-level data embedded in each state of the U.S., multilevel models are employed, which permit the incorporation of state-level variables in addition to the farm-level factors. The results show higher costs of surface water are not effective in reducing water use, while groundwater costs show a positive association with water use on both corn and soybean farms. The adoption of pressure irrigation systems reduces the soybean water use and increases the soybean yield. A higher *EIWUE* can be achieved with the adoption of enhanced irrigation systems on both corn and soybean farms. A high temperature promotes more the efficient water use and higher yield, and a high precipitation is associated with lower water application and higher crop yield. Intraclass correlation coefficients (ICC) suggest a moderate variability in water application and *EIWUE* is accounted by the state-level factors with ICC values greater than 0.10.

Keywords: climate variability; water use efficiency; multi-crop production; pressure irrigation systems; water costs; corn; soybeans

---

1. Introduction

In many countries, agricultural production relies heavily on water resources [1]. Most of the cropland is irrigated and some traditionally rain-fed agriculture systems have seen growing irrigation to increase production and mitigate climate risks. Accounting for more than 80–90% of the total water withdrawals, irrigated agriculture needs to contribute an increasing share of food production to meet the growing demands of a rising population [2]. Faced with the dramatic impacts of climate change, many arid and semiarid areas are suffering from severe water shortages, for instance, the Western U.S. [3] and Northwestern China [4]. At the same time, some areas that were not facing water deficiencies are experiencing more, frequent droughts, for instance, the Midwestern U.S. [5,6], thus, increasing the stress on current water resources. In addition, in many areas, the water demand from other sectors is expected to grow faster. Though a large proportion of water demand could be satisfied through new investments in water supply and irrigation systems, and the expansion of water supply could be met with some non-traditional sources, the shrinking water availability increases both economic and environmental costs of developing new water supplies [2,7,8]. Therefore, investments in water systems and developing new water sources to meet growing demands will not be a sufficient solution.

As a more practical path to achieve the sustainability of water resources, water can be saved in current uses through increasing the irrigation water use efficiency (total yield per unit of land divided by irrigation water applied) in agricultural production [9]. The traditional flood (also called furrow or gravity) irrigation systems have been reported to lose 50–70% of the water applied as soil evaporation, seepage, and deep drainage [10,11]. Potential improvements in irrigation water use efficiency can be realized by adopting enhanced pressure irrigation systems.

Most of the studies on irrigation water use efficiency are conducted at the field level based on experiments [12,13]. Two foci of field experiments include the comparison of irrigation water use efficiency at different water application levels and utilizing various irrigation methods, and the interaction and compatibility of improved irrigation systems and other farm-related management practices that are considered the best (e.g., film or straw mulching, irrigation scheduling, and soil testing) [14–17]. Previous studies on irrigation water use efficiency (IWUE) typically used experimental data in one field, collected over multiple years. Because of limited research funding, heterogeneity of experimental fields, and the diversity of cropping systems and farming structures, the available farm-level data are limited. As a result, the evaluation of crop IWUE in multiple fields is very challenging. At the farm level, producers usually plant two or more crops in one growing season. In addition to making adoption decisions regarding different irrigation systems, farmers also need to make decisions on land allocation and irrigation water application for each crop that they choose to plant. These decisions can determine whether the water is used efficiently or not.

The farm-level irrigation and production decisions to improve irrigation efficiency in a multi-crop system are understudied, in particular, across regions with different cropping patterns and climatic conditions [18]. In addition, production decisions in irrigated agriculture may be affected by other factors like water sources, input costs, and the farming area [19]. Analysis of irrigation decisions and crop irrigation water use efficiency, as affected by these and other factors, could help farmers and policymakers adapt to potential climate risks, better manage the irrigation water application, and achieve the sustainable use of limited water resources. Furthermore, given the heterogeneity of farms and states, multi-level models (MLMs) can be readily utilized to deal with the hierarchical nature of the farm-level data and to extract the percentage of variability in each response accounted for by farm- and state-level factors. The multilevel model has been applied in social science research [20,21] and agricultural sciences. To analyze the hierarchically structured data, Neumann et al. [22] adopted the multilevel model to investigate the global irrigation patterns using country-level data, and Giannakis and Bruggeman [23] studied the labor productivity in agricultural system in Europe. However, MLMs have never been used to analyze crop production decisions or farm irrigation efficiency. Given the data structure of the United States Department of Agriculture Farm and Ranch Irrigation Survey (USDA FRIS)—i.e., farms are embedded in states—we explore the applicability of the MLMs to multiple equations relating to production decisions in irrigated multi-crop agriculture.

Therefore, the objective of this study is to better understand the production decisions for irrigated agriculture and economic irrigation water use efficiency of major crops in the U.S., as well as the effects of water costs, the adoption of pressure irrigation systems, and the climatic determinants in a multi-crop production system.

Specifically, this research aims to answer the following fundamental questions:

(1) Are enhanced irrigation systems conserving water and are they more efficient than the traditional systems under diverse farm conditions?
(2) How does climate variability affect production decisions in an irrigated agriculture?
(3) What are the major influential factors and how are the multi-crop production decisions affected by these factors at the farm and state levels?

The layout of the analyses in this paper is presented in Figure 1. Focusing on irrigated farms in a multi-crop production system, four equations on land allocation, water application, crop yield, and economic irrigation water use efficiency are estimated using multilevel models. Intensive and extensive margins of water use to water price and energy costs are calculated. Intraclass correlation coefficients (ICC) as defined later are calculated to find out the proportion of variability in each response is accounted for by each level. Econometric results from the multilevel models are provided to show the effects of exogenous variables on each response variable.

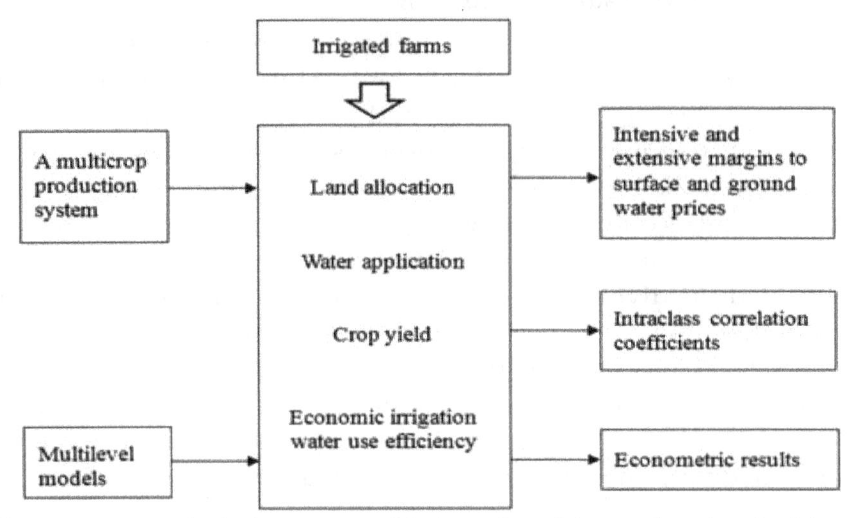

Figure 1. An analytical structure of irrigated multi-crop farming decisions.

2. Literature Review

Crop Water Use Efficiency

In general, water management includes issues relating to five sub-systems existing on most irrigated farms: supply systems, on-farm storage systems, on-farm distribution systems, application systems, and recycling systems [11]. In a report on the Australian cotton industry, Dalton et al. [11] defined water use efficiency at the farm level by focusing on three dimensions: agronomic efficiency, economic efficiency, and volumetric efficiency. The agronomic water use efficiency includes a gross production water use index (yield/total water applied), an irrigation water use index (yield/irrigation water applied), a marginal irrigation water use index (marginal yield due to irrigation/irrigation water applied), and a crop water use index (yield/evapotranspiration). The economic water use efficiency includes a gross production economic water use index (total value/total water applied), an economic irrigation water use index (value/irrigation water applied), a marginal economic irrigation water use index (value due to irrigation/irrigation water applied), and a crop economic water use index (value/evapotranspiration). The volumetric water use efficiency includes the overall project efficiency, conveyance efficiency, distribution efficiency, and field application efficiency, which emphasize irrigation uniformity to avoid over- and under-irrigation issues (reducing the water use efficiency and yield, respectively). Moreover, Pereira [24] discussed various measurements for both distribution uniformity and application efficiency in various irrigation systems.

From a multi-disciplinary perspective, Nair et al. [25] reviewed the efficiency of irrigation water use. Among all the measures of WUE, agronomists defined it as yield per unit area divided by the water used to produce the yield. The yield can be grain yield or the total aboveground biomass depending on the use of the crop produced, and the water can refer to crop evapotranspiration, soil water balance, or precipitation plus irrigation. However, from an economist's perspective, the efficient level of irrigation water occurs "when the marginal revenue (the price of the crop produce in a perfectly competitive market) is equal to the price of water" [25]. The water application level at Stage II in the classical production function was identified as the economically efficient water use amount. Stage II ranges from the point where the marginal physical product (MPP) equals the average physical product (APP), i.e., $w/p = Y/X$ ($MPP = APP$) with w being the water cost, p being the output price, Y being the output quantity, and X being the input quantity, to the yield maximizing point, where $dY/dX = w/p = 0$ (i.e., $MPP = 0$). Other researchers proposed an operating profit water use index to evaluate the water use efficiency, which is defined as $(R - VC - OC)/WU$ with R being the gross return, VC being the variable costs, OC being the overhead costs, and WU being the total amount of water used [26].

Comparing the WUE measures from perspectives of agronomists and economists, a major difference is whether to consider the output price. For example, the economic irrigation water use index (value of crop or grains/irrigation water applied) is the product of the irrigation water use index (yield/irrigation water applied) and the crop price. Because producers are price takers in a competitive market, different farmers growing the same crop will sell it for the same price in the same market. Thus, exogenous variables affecting economic irrigation efficiency and agronomic irrigation efficiency will have the same effects in terms of signs and significance levels, though the magnitude will be different proportionally. To make the analyses easier and follow the mainstream of decision-making on land allocation and water use in order to maximize the expected profit as formulated in the model section below, this study uses the economic measure of irrigation water use efficiency (*EIWUE*) (crop value/irrigation water use), incorporating state-average crop prices in the econometric estimation.

Various approaches have been explored to conserve irrigation water use, such as developing new irrigation techniques [27]; increasing investment in irrigation infrastructure such as canals, wells and drip systems [16]; and designing water conservation policies [28]. Water-conserving irrigation systems have been proposed and applied to various crops in many farming areas around the world. For instance, in eastern Australia [29], arid and semi-arid areas in China [16,30], and southern and southwestern U.S. [15,31,32]. Examples include pressure (or pressurized) irrigation systems (versus gravity irrigation methods), such as linear move, center pivot, sprinkler, and drip irrigation methods. Field experiments with sprinkler and drip irrigation and their comparison with traditional flood or furrow irrigation have been conducted on various crops worldwide [14,33–36]. As a result, a substantial amount of water could be saved using enhanced irrigation systems and crop irrigation water use efficiency can be improved.

3. Hypotheses

In this section, factors affecting farmers' adoption behaviors and irrigation decisions are reviewed, and hypotheses are constructed. Farmers' irrigation decisions are hypothesized to be a function of the expected profit, costs, perceived barriers, information availability, farm and farmer characteristics, and their environmental attitudes and perceptions of climate variability.

Literature reviews on agricultural production and economics show that many changes in socioeconomic, agronomic, technical, and institutional aspects can have considerable positive/negative effects on water use, crop yield, and crop water use efficiencies, and thus diverse effects on the profitability of crop production [37,38]. Farm management practices including controlling the amount and timing of irrigation water, fertilizer/manure use, mulching, and tillage can affect farm returns and profits [39]. Through analyzing various measurements of water use efficiency, Pereira [24] recommended combining improved irrigation methods and scheduling strategies to achieve a higher performance. Pressure irrigation systems are thus expected to decrease water application and increase efficiency.

Based on field-level measurements, Canone et al. [40] assessed the surface irrigation efficiency in Italy. The results from both simulated scenarios and monitored irrigation events highlighted the necessary strategies to improve irrigation efficiencies by reducing the flow rates and increasing the duration of irrigation events. Thus, we hypothesize a higher water availability from various sources and more wells decrease crop water use efficiency.

In addition, the diverse effects of physical factors on farm yield and profits have been reported based on farm-level studies. For instance, with carrot farmer interviews in Pakistan, Ahmad et al. [41] found that farm-level yield and profitability were affected by many factors including expenditures on facility and labor investments regarding the application of fertilizer, irrigation, and weeding. In a similar study, Dahmardeh and Asasi [42] evaluated the effects of the costs of fertilizer, seeds, and water on the profitability of corn farms as well as the effects of income sources. Boyer et al. [19] examined the effects of different energy sources, energy prices, and field sizes on corn production. Thus, the facility expenses and labor payment at the farm level are hypothesized to have positive effects on water application and crop yield, but a mixed effect on water use efficiency.

Farmers face many barriers and challenges when making irrigation and production decisions. Using data on 17 western states from the USDA FRIS, Schaible et al. [43] studied the dynamic adjustment of farmers' irrigation decisions and pointed out some major barriers impacting the adoption of enhanced irrigation technologies. The most important barriers were related to investment cost and financing issues. A greater sharing of costs by government or landlords for installation of advanced irrigation techniques can improve their adoption rates especially for beginning farmers with limited resources and social disadvantages [2]. Moreover, uncertainty about future water availability and farming status could influence farmers' willingness to adopt. Hence, uncertainties regarding potential costs and future benefits will limit the adoption of water conservation practices, and thus discourage farmers to use water more efficiently [44,45].

Information availability and its sources can affect farm irrigation decisions [46]. On the one hand, limited information can be an obstacle to using water efficiently. Rodriguez et al. [47] pointed out that a lack of information on irrigation, crop management, the effectiveness of practices and government programs could be common obstacles for resource-limited farmers when facing the uncertainty of changing to something unknown. On the other hand, effective information can facilitate optimal irrigation decisions by farmers [48]. Frisvold and Deva [49] studied water information used by irrigators and the relationship of information acquisition and irrigation management. Their study indicated that appropriate information use could benefit irrigation management and crop production for farmers with varying acreage. Thus, more information on how to conserve water and use water more efficiently is expected to decrease water use, increase crop yield, and improve irrigation efficiency [37,44].

Regional variables could capture differences in climate, water institutions, and supporting infrastructure [50], as well as farming systems. More generally, which irrigation decisions are appropriate will vary spatially. For example, western states tend to have concentrated irrigation acreage and their irrigation institutions are well established [50]. Eastern and southern states receive moderate amounts of rainfall to support agriculture and do not rely as heavily on irrigation. Thus, we hypothesize that compared with those in the High Plains states, farmers in western states will irrigate more, while farmers in midwestern and southern states will irrigate less.

Furthermore, farmers are also motivated to respond facing varying weather conditions. Climate conditions can influence farm yield and revenue, and irrigation can be considered as a strategy to mitigate the adverse effects and increase profits [51]. Specifically, an awareness of climate change (e.g., drought and heat waves) could motivate farmers to prepare for and take actions to adapt to future risks to production [4,52]. Olen et al. [18] found that farmers were more likely to irrigate crops to mitigate and adapt to various weather and climate impacts including frosts, heats, and droughts. Li et al. [53] reported diverse effects of climate change on corn yield in the United States and China. Therefore, farmers are hypothesized to increase water application rates and decrease irrigation water use efficiency if they perceive or experience less precipitation, higher temperature, or more grain losses due to droughts. This is proxied by changes of weather conditions in 2011, 2012, and 2013.

4. Methods

In this section, we adopt a model of profit maximization [54] and then turn to the maximization of economic irrigation water use efficiency to deal with market failure in water management. In multi-crop irrigated agriculture, producers make decisions on land allocation to each crop, and the amount of water for irrigation [55,56]. Choosing from common crops, a typical producer may plant two or more crops on a farm. Then decisions on land allocation and water supply can be made to maximize the expected total profit [57].

Following a multi-crop production model by Moore et al. [54], the expected profit functions of the multi-crop system and specific crop i can be represented by $\Pi(\boldsymbol{p},\ \boldsymbol{r},\ b,\ N;\boldsymbol{x})$ and $\pi_i(p_i,\ \boldsymbol{r},\ b,\ n_i;\boldsymbol{x})$, respectively. $\boldsymbol{p}$ is a vector of crop prices; p_i is the price of crop i, $i = 1, \ldots, m$; $\boldsymbol{r}$ is a vector of variable input prices excluding water prices; b is the water prices; N is the total farming area as a constraint; n_i is the land allocation for crop i; $\boldsymbol{x}$ represents other exogenous variables including land characteristics, water sources, the adoption of various irrigation systems, and climate perceptions. Each crop-specific profit function π_i is assumed to be convex and homogeneous of degree one in output prices, water price, and other prices of variable inputs, nondecreasing in output price and land allocation, and non-increasing in water prices and other variable input prices.

We extend the model of Moore et al. [54] by adding crop irrigation water use efficiency. A single producer makes production and irrigation decisions to maximize profits. While to achieve sustainability of the water resource, the total profit function of the whole society needs to consider the marginal user cost and higher pumping cost externality of extracting water by every farmer. Thus, in addition to the decision-making on conserving water use and increasing crop yield, the way to achieve higher crop irrigation water use efficiency should be explored. Following the discussion on indicators of water use performance and productivity by Pereira et al. [58], the following definition can be used to calculate the farm-level crop-specific economics irrigation water use efficiency.

$$EIWUE = \frac{Crop\ yield\ \times\ P}{Total\ amount\ of\ irrigation\ water\ applied} \tag{1}$$

where *EIWUE* is the economic irrigation water use efficiency, crop yield is the marketable grain yield, P is the crop price, and irrigation water application is measured based on all irrigation water sources, including well, on- and off-farm surface water. The greater the *EIWUE* value [59], the higher the efficiency due to irrigation water application.

To analyze the effects, *EIWUE* can be a function of the exogenous variables affecting both yield and water application.

$$EIWUE_i = h_i(\boldsymbol{p},\ \boldsymbol{r},\ b,\ N;\boldsymbol{x})\ i = 1, \ldots, m \tag{2}$$

In addition, the farm-level water application can be decomposed to analyze the role of water price on production decisions regarding each crop [54]. The crop-specific water application can be decomposed into an extensive margin of water use (an indirect effect on water use due to land allocation change) and an intensive margin of water use (a direct effect on water use due to water application).

The farm-level total water application (W) equals the sum of water application for each crop grown on the farm with the optimal land allocation [54,60]:

$$W = \sum_{i=1}^{m} w_i(p_i,\ \boldsymbol{r},\ b,\ n_i^*(\boldsymbol{p},\ \boldsymbol{r},\ b,\ N;\boldsymbol{x});\boldsymbol{x})\ i = 1, \ldots, m \tag{3}$$

Taking the derivative of the equation with respect to water price gives

$$\frac{\partial W}{\partial b} = \sum_{i=1}^{m}\left(\frac{\partial w_i}{\partial b} + \frac{\partial w_i}{\partial n_i^*} \times \frac{\partial n_i^*}{\partial b}\right) \tag{4}$$

where $\frac{\partial w_i}{\partial b}$ is the intensive margin, and $\frac{\partial w_i}{\partial n_i^*}\frac{\partial n_i^*}{\partial b}$ is the extensive margin. The total effect can be obtained by summing the effects on all the crops. The intensive margin will decrease in price and $\frac{\partial w_i}{\partial b}$ should have a negative sign for each crop. The sign of the extensive margin depends on $\frac{\partial n_i^*}{\partial b}$. The total farm-level effect on water use should be negative, which indicates a decreasing water application as water price increases. This decomposition of the total marginal effect has been lately employed by Hendricks and Peterson [61], and Pfeiffer and Lin [62].

Multilevel Models

Multilevel models have the advantage of examining individual farms embedded within states and assess the variation at both farm and state levels. The multilevel regression model is commonly viewed as a hierarchical regression model [63]. A multilevel linear modeling technique is utilized to analyze the effects of influential factors on land allocation, water application, crop yield, and *EIWUE*.

For the research questions, we have N individual crop-specific farms ($i = 1, \ldots, N_j$) in J states ($j = 1, \ldots, J$). The X_{ij} represent a set of independent variables at the farm level, and a series of state-level independent variables are represented by Z_j. The model estimation includes two steps. For the first step, a separate regression equation can be specified in each state to predict the effects of independent variables on dependent variables.

$$\mathbf{Y}_{ij} = \beta_{0j} + \boldsymbol{\beta}_{1j}\mathbf{X}_{1ij} \tag{5}$$

For the second step, the intercepts, β_{0j}'s are considered parameters varying across states as a function of a grand mean (γ_{00}) and a random term (u_{0j}). The β_{1j}'s are also assumed to be varying across states and are presented as a function of fixed parameters (γ_{10}) and a random term (u_{1j}).

$$\boldsymbol{\beta}_{0j} = \gamma_{00} + \boldsymbol{\gamma}_{01}\mathbf{Z}_j + u_{0j} \tag{6a}$$

And

$$\boldsymbol{\beta}_{1j} = \boldsymbol{\gamma}_{10} + u_{1j} \tag{6b}$$

Combining Equations (5), (6a) and (6b), we have

$$\mathbf{Y}_{ij} = \gamma_{00} + (\boldsymbol{\gamma}_{10} + u_{1j})\mathbf{X}_{ij} + \boldsymbol{\gamma}_{01}\mathbf{Z}_j + u_{0j} \tag{7}$$

The model is called a random-intercept and random-slope model, as the key features are not only that the intercept parameter in the Level-1 model, β_{0j}, is assumed to vary at Level-2 (state) [64], but that the slope is also random with an error term u_{1j}. The γ_{01} coefficient captures the effects of the state-level variables (Z_j) on the β_{0j}'s, whereas γ_{10} predicts the constant parameter, β_{1j}, (with errors).

To analyze the multi-crop production, four sequential models are estimated for each decision due to their continuous nature, that is, a unconstrained two-level model with random effects for the intercept only and without any predictors (Model 1); random effects for the intercept and fixed effects for level 2 (Model 2); a random intercept as well as a fixed and random level 1 (Model 3); and a random intercept, fixed and random level 1 as well as a fixed level 2 (Model 4) (see Table S1 in the Supplementary Materials for specifications and comparisons of the four models). To determine how much of the variability in the responses is accounted for by factors at the state level, the intraclass correlation coefficient is usually computed from the null model (Model 1) [65] following:

$$ICC = \frac{\tau_{00}}{\tau_{00} + 3.29} \tag{8}$$

where τ_{00} is the covariance parameter estimate for the intercept, and 3.29 is the estimated level-1 error variance [66].

The data were analyzed using the SAS package in the USDA data lab in St. Louis, Missouri, with official permission.

5. Data and Variables

This study uses a national dataset from the 2013 USDA FRIS. Null models for all equations of 17 crops are estimated to calculate the intraclass correlation coefficient. However, only models in the further steps on land allocation [67], water application, crop yield, and *EIWUE* are estimated and presented in this paper focusing on corn and soybeans as they have the most observations but different distribution patterns across the five regions (specified below).

The lower 48 states are grouped into five regions according to the USDA National Agricultural Statistics Services (NASS) [68], including the Western, Plains, Midwestern, Southern, and Atlantic states [69]. The descriptive statistics of the corn and soybean farms [70] at the national level are presented in Table 1. Of the 19,272 irrigated farms, 6030 farms grow corn for grain with an average area of 357 acres, and 3933 farms grow soybeans with an average area of 341 acres [71]. For corn farms, the mean water application is 1.11 acre-feet/acre; the mean yield is 190 bu/acre; and *EIWUE* is 1311 USD/acre-foot on average. For soybean farms, the mean water application, yield, and *EIWUE* are 0.81 acre-foot/acre, 55 bu/acre, and 1221 USD/acre-foot, respectively.

The independent variables are at two levels. At the farm level, the explanatory variables are related to water sources, costs on surface water and energy, expenditures on irrigation equipment, labor payment, farm characteristics including the farming area, number of wells, irrigation systems, barriers for improvements to conserve water, and information sources related to irrigation. Variables related to water sources, federal assistance, barriers, and information sources are dummy variables (Yes = 1, No = 0), and all other independent variables are continuous.

At the state level, in addition to the dummy variables related to the five regions, six explanatory variables on state-wide weather conditions are included using the data from the United States National Oceanic and Atmospheric Administration. The variables are state average precipitation changes in 2011, 2012, and 2013, and the temperature changes in 2011, 2012, and 2013.

Table 1. The summary statistics of crop-specific dependent variables and state-level weather-related independent variables.

Variables	Description (Unit)	N	Mean	Std Dev	CV	Min	Max
Crop-Specific Dependent Variables							
Corn							
Land allocation	Average farming area (acre)	6030	356.84	1426.66	4.00	-	-
Water application	Average water application (acre-foot)	6030	1.11	1.97	1.77	-	-
Crop yield	Average yield of all farms (bu/acre)	6030	190.29	87.19	0.46	-	-
EIWUE	Average economic irrigation water use efficiency ($/acre-foot)	6030	1310.99	3199.15	2.44	-	-
Soybeans							
Land allocation	Average area of all farms (acre)	3933	340.79	1195.10	3.51	-	-
Water application	Average water application of all farms (acre-feet)	3933	0.81	1.13	1.40	-	-
Crop yield	Average yield of all farms (bu/acre)	3933	54.76	27.89	0.51	-	-
EIWUE	Average economic irrigation water use efficiency ($/acre-foot)	3933	1220.55	2352.57	1.93	-	-
State-Wide Average Weather-Related Variables							
PrecipChange2011	Precipitation in 2011—Average precipitation in 1981–2010 (inch)	43	1.51	8.26	5.46	−15.87	17.61
PrecipChange2012	Precipitation in 2012—Average precipitation in 1981–2010 (inch)	43	−3.66	4.74	1.29	−12.21	10.30
PrecipChange2013	Precipitation in 2013—Average precipitation in 1981–2010 (inch)	43	1.74	5.36	3.08	−15.19	14.26
TempChange2011	Temperature in 2011—Average temperature in 1981–2010 (°F)	43	0.54	1.09	2.03	−2.70	2.10
TempChange2012	Temperature in 2012—Average temperature in 1981–2010 (°F)	43	2.47	1.11	0.45	−1.70	4.00
TempChange2013	Temperature in 2013—Average temperature in 1981–2010 (°F)	43	−0.50	0.75	1.48	−2.20	0.90

6. Results

6.1. Descriptive Statistics

The summary statistics of the farm-level independent variables are presented in Table 2. Four water sources are investigated including groundwater only, on- and off-farm surface water [72] only, and two or more water sources (Yes = 1, No = 0). For corn and soybean farms, about 71% and 81% use groundwater only, respectively. Water from on- or off-farm surface sources only account for about 4.5% of soybean farms (about 10.5% of corn farms only use off-farm surface water). About 12% of both farms get water from two or more sources.

Table 2. The summary statistics of farm-level independent variables and region dummies ($: USD).

Variables	Corn (n = 6030)		Soybean (n = 3933)	
	Mean	Std Dev	Mean	Std Dev
Water sources				
Groundwater only (base)	0.713	1.124	0.808	0.926
On-farm surface water only	0.058	0.579	0.045	0.488
Off-farm surface water only	0.105	0.762	0.031	0.406
Two or more water sources	0.124	0.819	0.116	0.752
Costs				
Cost for off-farm surface water ($/acre-foot)	6.891	113.473	4.215	47.154
Energy expenses ($/acre)	47.047	184.994	35.602	62.841
Facility expenses ($/acre)	37.605	367.721	25.131	293.385
Labor payment ($/acre)	5.237	197.95	1.454	25.398
Farm characteristics				
Number of wells used	5.755	23.632	7.365	23.585
Total acre	1879	13497	1665	5238
Percent of owned land	0.497	0.937	0.448	0.852
Pressure irrigation	0.799	0.996	0.708	1.07
Gravity irrigation (base)	0.201	0.996	0.292	1.07
Federal assistance	0.202	0.998	0.219	0.973
Barriers to improvements				
Investigating improvement is not a priority	0.165	0.921	0.14	0.816
Risk of reduced yield or poorer quality crop	0.089	0.708	0.071	0.605
Limitation of physical field or crop conditions	0.11	0.776	0.104	0.718
Not enough to recover implementation costs	0.172	0.937	0.195	0.932
Cannot finance improvements	0.129	0.834	0.114	0.748
Landlords will not share improvement costs	0.119	0.805	0.137	0.808
Uncertainty about future water availability	0.11	0.776	0.08	0.637
Will not be farming long enough	0.075	0.656	0.059	0.554
Will increase management time or cost	0.079	0.671	0.065	0.579
Information sources				
Extension agents	0.33	1.169	0.401	1.153
Private irrigation specialists	0.354	1.188	0.366	1.133
Irrigation equipment dealers	0.31	1.15	0.308	1.086
Local irrigation district employee	0.082	0.683	0.059	0.555
Government specialists	0.153	0.895	0.146	0.831
Media reports	0.118	0.802	0.122	0.769
Neighboring farmers	0.231	1.047	0.231	0.991
E-information services	0.188	0.972	0.191	0.925
Regions				
West	0.139	0.859	0.005	0.171
High Plains (base)	0.554	1.235	0.532	1.174
Midwest	0.16	0.912	0.182	0.908
South	0.113	0.787	0.242	1.008
Atlantic	0.033	0.445	0.038	0.45

All variables have been weighted using weights provided within the FRIS data.

Water costs are measured by the payment for off-farm surface water and energy expenses for pumping groundwater. The average cost for off-farm surface water is 6.89 and 4.22 USD/acre-foot for corn and soybean farms, respectively. The water price measure frees the irrigator from being bind by water institutions [54]. The average energy expenses are 47.05 and 35.60 USD/acre for corn and soybean farms. The energy expenses are a proxy of groundwater price [54]. The average facility expenses and labor payments in 2013 are 37.61 and 5.24 USD/acre for corn, and 25.13 and 1.45 USD/acre for soybeans. The units of costs measure follow the convention by Moore and others [73].

Regarding the farm characteristics, the average number of wells used to irrigate corn and soybeans are 5.76 and 7.37, respectively. The mean areas of the total land are 1879 and 1665 acres/farm for corn and soybeans, and the percentage of owned land is 50% and 45%. For irrigation systems, about 20% of corn farms use gravity systems and 29% of soybean farms use gravity systems, while those using pressure irrigation account for 80% and 71%, respectively. About 20% of the corn farmers received federal assistance to improve irrigation and/or drainage systems, compared to 22% for soybean farmers.

Regarding the barriers to implementing improvements for the reduction of energy costs or water use, nine barriers are investigated in the national survey. The major ones include the following: investigating improvement is not a priority at this time (17% for corn farmers and 14% for soybean farmers), limitation of physical field or crop conditions (11% for corn farmers and 10% for soybean farmers), not enough to recover implementation costs (17% for corn farmers and 20% for soybean farmers), cannot finance improvements (13% for corn farmers and 11% for soybean farmers), and landlords will not share improvement costs (12% for corn farmers and 14% for soybean farmers).

For the eight sources of irrigation information, the top ones are extension agents (33% for corn farmers and 40% for soybean farmers), private irrigation specialists (35% for corn farmers and 37% for soybean farmers), irrigation equipment dealers (31% for both corn and soybean farmers), neighboring farmers (23% for both corn and soybean farmers), e-information services (19% for both), and government specialists (15% for both).

Regarding location, this study includes more irrigated farms in the Plains states, 55% for corn and 53% for soybeans. Farms in the Midwest and South account for 16% and 11% for corn, and 18% and 24% for soybeans, with fewer farms in the Midwest and South.

The state-wide average weather-related variables are presented in Table 1 for the 43 states planting corn. Compared to the 1981–2010 average precipitation, the changes for 2011, 2012, and 2013 are 1.51, −3.66, and 1.74 inches, respectively. Compared with the 1981–2010 average temperature, the changes for 2011, 2012, and 2013 are 0.54, 2.47, and −0.50 °F. While in 2013, the year covered by the survey, it's more favorable for agricultural production as far as the rainfall.

6.2. Decomposition of Farm-Level Water Application

To decompose the effect of water cost on farm-level water application, extensive and intensive margins are provided in Table 3. This paper takes corn and soybeans as examples [74]. The estimated coefficients on crop acreage and water costs in the water application equation suggest a change in water use given a change in land use ($\frac{\partial w_i}{\partial n_i}$), and a marginal change in water use given a change in water cost ($\frac{\partial w_i}{\partial b}$). The estimated coefficients on water cost in the land allocation equation represent a change in land use given a change in water cost ($\frac{\partial n_i}{\partial b}$). The intensive margin can be obtained with $\frac{\partial w_i}{\partial b}$ while adjusting for the estimated probability that the crop is grown. The extensive margin can be calculated using $\frac{\partial w_i}{\partial n_i}\frac{\partial n_i}{\partial b}$. Summing the intensive and extensive margins for each crop gives the total effect of a change in water cost. Further summing the effects on all crops gives the total effect on a typical farm growing both crops.

Margins on both on-surface water costs and energy costs are calculated. Only water from off-farm surface sources is priced and investigated in the survey. Energy expenses on groundwater pumping are considered as the proxy of water price for groundwater. The results show that only $\frac{\partial n_i}{\partial b}$ decreases in energy expenses for soybeans, and other values of $\frac{\partial n_i}{\partial b}$ and $\frac{\partial w_i}{\partial b}$ are positive, which is contradictory to expectations. This indicates more water is used as water prices increases. This is probably true in practice when the adoption of enhanced irrigation systems increase acreage under irrigation and thus increase the amount of irrigation water, as reported in Kansas [75]. There are many debates regarding the empirical changes in water use as a result of changing prices and increasing the adoption of agricultural irrigation technologies [53,61,75]. A numerical illustration can help understand the effects of water prices. A 1 USD increase in groundwater costs (energy expenses) (Δb = $1) would lead to a decrease of 0.109 acre-feet of water application per acre of soybeans, and an increase of 0.0737 acre-feet of water per acre of corn. In a multi-crop system, a typical farm growing both corn and soybeans would decrease water application by −10.87 acre-feet as a result of a $1 increase in energy expenses. These results show water use is highly inelastic in water cost [54]. While this may be different for regions/states with varying availability of water resources, an in-depth analysis of regional or state effect of water costs on water use can be helpful.

Table 3. The crop-specific extensive and intensive margins to surface water cost and energy expenses.

Variables	dw/dn	dn/db	dw/db	Share of Crop-Specific Farms	Extensive Margin	Intensive Margin	Total Effect (Acre-Feet Per Acre)	Total Effect-Farm (Acre-Feet Per Farm)
Surface Water Cost								
Corn	1.0266	0.1766	0.0030	0.3129	0.0567	0.0009	0.0577	20.5769
Soybeans	1.0040	0.0816	0.0006	0.2041	0.0167	0.0001	0.0168	5.7396
Farm total								26.3165
Energy Expenses								
Corn	1.0266	0.2282	0.0012	0.3129	0.0733	0.0004	0.0737	26.2881
Soybeans	1.0040	−0.5334	0.0012	0.2041	−0.1093	0.0002	−0.1090	−37.1606
Farm total								−10.8725

Following the definitions by Moore et al. [54], $\frac{\partial w_i}{\partial n_i}$ is the estimated coefficient of crop acreage in the water application equations, where w_i is the acre-feet of irrigation water on crop i and n_i is the acres of growing crop i. $\frac{\partial n_i}{\partial b}$ is the estimated coefficient of the water price in the land allocation equations, with b being the water price. $\frac{\partial w_i}{\partial b}$ is the estimated coefficients of the water price in the water application equation. The calculation of both intensive and extension margin should be adjusted by the share of the crop planted.

6.3. Intraclass Correlation Coefficients

The first step in conducting a multilevel model is to calculate the ICC which shows how much of the variability in one response variable is accounted for by level 2. The intraclass correlation coefficients for crop-specific multilevel models are presented in Table 4. To better understand these values, for example, the ICC for the water application equation of corn is 0.2102, which suggests about 21% of the variability in water application decisions is accounted for by the factors at the state level, leaving 79% of the variability to be accounted for by the farm-level factors. A moderate variability in water application and *EIWUE* is accounted by the state-level factors, with an ICC value greater than 0.10. However, a higher variability of land allocation and crop yield is accounted for by farm-level factors. In the following sections, results for each estimated equation are presented for corn and soybeans jointly to facilitate the comparison of the effects on the two crops.

Table 4. The intraclass correlation coefficients for null models of each crop-specific multilevel model.

State-Level Variation	Land Allocation	Water Application	Crop Yield	*EIWUE*
Corn	0.0068	0.2102	0.0270	0.1501
Soybeans	0.0291	0.1365	0.0277	0.1763

EIWUE: economics irrigation water use efficiency. The table only presents results from Model 4 (Model 3 + fixed state level) in each equation. More results can be found in Tables S2–S9 of the Supplementary materials.

6.4. Land Allocation

The estimated coefficients from MLMs for land allocation of corn and soybeans are presented in Table 5. The results are shown compared to groundwater use, water uses from on-, off-farm surface only and more sources have a positive effect on land allocation to corn planting. While water from more sources increases the planting of both crops.

Table 5. The results of multilevel models for the land allocation for corn and soybean.

Variables	Corn		Soybeans	
	Estimate	**Std Err**	**Estimate**	**Std Err**
Fixed Effects				
Intercept	−533.020 ***	55.823	12.756	241.4
Water sources				
On-farm surface water only	78.808 ***	23.286	29.948	24.573
Off-farm surface water only	151.370 ***	22.352	49.96	32.727
Two or more water sources	85.086 ***	15.758	71.824 **	20.262
Costs				
Cost for off-farm surface water($/acre-foot)	0.177	0.116	0.082	0.302
Energy expenses ($/acre)	0.228 ***	0.073	−0.533 *	0.253
Facility expenses ($/acre)	0.170 ***	0.036	0.009	0.037
Labor payment ($/acre)	0.034	0.063	0.168	0.428
Farm characteristics				
Number of wells used	38.663 ***	4.081	18.199 ***	2.566
LN(total acre)	81.107 ***	4.816	92.613 ***	10.983
Percent of owned land	−20.374	14.257	−17.387	14.205
Pressure irrigation	21.763	15.337	5.519	22.995
Federal assistance	−26.108	16.153	−25.000 **	12.067
Barriers to improvements				
Investigating improvement is not a priority	−10.970	13.782	23.756 *	13.675
Risk of reduced yield or poorer quality crop	−0.682	19.7	1.521	19.659
Limitation of physical field or crop conditions	−27.543	17.893	−19.917	16.972
Not enough to recover implementation costs	29.853 **	14.16	−2.248	13.184
Cannot finance improvements	−22.241	15.403	−1.889	15.567
Landlords will not share improvement costs	−9.536	16.863	−29.516 **	14.914
Uncertainty about future water availability	−35.165 **	17.261	−28.533	18.987
Will not be farming long enough	3.877	20.056	13.336	20.68
Will increase management time or cost	3.408	20.145	4.782	20.079
Information source				
Extension agents	−29.179 **	11.727	−18.427 *	10.38
Private irrigation specialists	23.782 **	11.218	8.49	10.429
Irrigation equipment dealers	−3.008	11.816	−4.227	11.089
Local irrigation district employee	−2.399	19.916	23.229	21.612
Government specialists	−5.252	15.564	4.839	14.125
Media reports	−0.350	16.675	−0.632	15.178
Neighboring farmers	−27.456 **	12.756	−17.951	11.69
E-information services	12.732	13.76	11.728	12.817
State-level variables				
PrecipChange2011	−2.820	1.936	−8.475	6.35
PrecipChange2012	−3.283	2.405	16.053	9.861
PrecipChange2013	−7.201 **	2.715	30.293 **	11.885
TempChange2011	2.116	16.396	11.812	80.683
TempChange2012	−0.811	12.082	−28.899	73.028
TempChange2013	1.927	18.476	134.110 *	75.466
West	45.578	34.692	−276.610	217.03
Midwest	64.791 *	33.045	−94.133	141.99
South	131.760 ***	41.441	−994.360 ***	163.42
Atlantic	119.730 **	55.951	−343.380	216.11
Error Variance	**Estimate**	**Std Err**	**Estimate**	**Std Err**
Intercept	<0.0001 ***	<0.0001	<0.0001 ***	<0.0001
Residual	858,667 ***	15,746	433,053 ***	9882
Fit Statistics				
N	6030		3933	
−2 Log Likelihood	93,300		58,421	
AIC	93,378		58,513	
AICC	93,379		58,514	
BIC	93,446		58,571	

Significance levels: * 10%; ** 5%; *** 1%.

Surface water price does not affect land allocation, which is consistent with the expectations as the decision on how much land allocated to grow a crop is made mainly depending on the expected crop price and input costs with little consideration of water price, while energy expenses as a proxy of groundwater price increase corn planting and decrease soybean planting. Higher facility expenses increase corn planting as more acres can be irrigated.

Regarding farm characteristics, more wells on a farm increase the planting of both crops. Larger areas of cropland increase the land allocation for both crops. Federal assistance on farm irrigation and drainage management has a negative effect on soybean planting. Unfortunately, land tenure and the adoption of pressure irrigation systems do not have a significant effect on land allocation for both crops.

Regarding barriers to improvements, uncertainties about future water availability have a negative effect on corn planting, and not enough to recover implementation costs has a positive effect. For soybean, landlords not sharing improvements costs has a negative effect on soybean planting, while investigating improvement is not a priority shows a positive effect. While positive effects are unexpected, a comparison of the negative effects on the two crops indicates that corn farmers are more concerned with future uncertainties, and soybean farmers with the share of improvement costs.

Information from extension agents and neighboring farmers decreases the planting of corn and soybean planting is also negatively affected by the information from extension agents, while information from private irrigation specialists increases the planting of corn. These findings indicate the effectiveness of extension programs in promoting the growth of water-conserving crops.

At the state level, the precipitation change in 2013 is negatively associated with corn planting. Both the precipitation change and temperature change are positively associated with soybean acreage. These findings suggest that given climate variability, a lower water available for crop production probably promotes farmers growing more water-conserving crops (in this case, soybeans), and vice versa. Compared with Plains farmers, those in the Midwestern, Southern and Atlantic states are more likely to plant corn, while farmers in the Southern states are less likely to plant soybeans.

6.5. Water Application

The parameter estimates for water application equations of corn and soybeans are presented in Table 6. The results are shown compared to groundwater use only, the water use from two or more sources has a positive effect on water application of corn. High surface water cost, energy expenses, and labor payment are positively associated with water application on corn. The energy expenses are also positively associated with water application on soybeans. The positive effects of water prices and energy expenses are unexpected, but this may indicate the ineffectiveness of a higher water price on water conservation. A positive effect of labor payment may suggest that these factors are complements; more labor use facilitates more irrigation, or producers who need more irrigation to maximize profits use more labor.

Table 6. The results of multilevel models for the mean water application on corn and soybean farms.

Variables	Corn		Soybean	
	Estimate	Std Err	Estimate	Std Err
Fixed Effects				
Intercept	1.041 ***	0.328	1.151 ***	0.227
Water sources				
On-farm surface water only	−0.037	0.069	0.015	0.078
Off-farm surface water only	−0.075	0.083	−0.015	0.055
Two or more water sources	0.106 **	0.044	0.039	0.036
Costs				
Cost for off-farm surface water($/acre-foot)	0.003 **	0.001	0.001	0
Energy expenses ($/acre)	0.001 *	0.001	0.001 **	0.001
Facility expenses ($/acre)	0	0	0	0
Labor payment ($/acre)	0.002 **	0.001	−0.001	0.002

Table 6. *Cont.*

Variables	Corn		Soybean	
	Estimate	Std Err	Estimate	Std Err
Farm characteristics				
Number of wells used	0.002	0.001	0.002 *	0.001
LN(total acre)	0.026 *	0.015	0.004	0.01
Percent of owned land	−0.003	0.049	−0.024	0.035
Pressure irrigation	−0.057	0.107	−0.174 ***	0.044
Federal assistance	0.029	0.033	0.046 *	0.025
Barriers to improvements				
Investigating improvement is not a priority	0.056 ***	0.02	0.006	0.02
Risk of reduced yield or poorer quality crop	0.064 **	0.029	0.079 ***	0.028
Limitation of physical field or crop conditions	−0.094 ***	0.026	0.039	0.025
Not enough to recover implementation costs	0.001	0.021	−0.004	0.019
Cannot finance improvements	0.139 ***	0.023	0.027	0.022
Landlords will not share improvement costs	−0.012	0.025	−0.060 ***	0.022
Uncertainty about future water availability	−0.074 ***	0.026	−0.084 ***	0.028
Will not be farming long enough	0.055 *	0.03	−0.068 **	0.03
Will increase management time or cost	−0.077 ***	0.03	−0.014	0.029
Information source				
Extension agents	−0.062 ***	0.017	−0.041 ***	0.015
Private irrigation specialists	−0.040 **	0.017	−0.042 ***	0.015
Irrigation equipment dealers	0.055 ***	0.018	−0.044 ***	0.016
Local irrigation district employee	0.098 ***	0.03	0.024	0.031
Government specialists	0.038	0.023	0.054 ***	0.02
Media reports	−0.009	0.025	−0.042 *	0.022
Neighboring farmers	−0.069 ***	0.019	−0.037 **	0.017
E-information services	0.060 ***	0.02	0.037 **	0.019
State-level variables				
PrecipChange2011	−0.043 ***	0.015	−0.008	0.006
PrecipChange2012	−0.064 ***	0.019	−0.018 *	0.009
PrecipChange2013	−0.079 ***	0.022	−0.036 ***	0.011
TempChange2011	0.05	0.109	0.111	0.072
TempChange2012	−0.330 ***	0.095	−0.152 **	0.067
TempChange2013	−0.335	0.146	−0.135 *	0.071
West	0.961 **	0.204	0.809 ***	0.192
Midwest	0.180 ***	0.262	−0.160	0.122
South	−0.101	0.276	−0.131	0.149
Atlantic	0.591	0.437	−0.227	0.195
Error Variance	**Estimate**	**Std Err**	**Estimate**	**Std Err**
Intercept	<0.0001 ***	<0.0001	<0.0001 ***	<0.0001
Residual	1.766 ***	0.033	0.886 ***	0.02
Fit Statistics				
N	6030		3933	
−2 Log Likelihood	14,730		6892	
AIC	14,834		6994	
AICC	14,835		6995	
BIC	14,926		7076	

Significance levels: * 10%; ** 5%; *** 1%.

Regarding farm characteristics, the results show that more wells are positively associated with water application on soybean farms, which is consistent with the hypothesis as mentioned above that more wells provide farmers more and easier access to water. A large farming area has a positive association with the average water application on corn farms. The adoption of pressure irrigation systems reduces irrigation water application for soybean farms, which is consistent with the hypothesis that the enhanced pressure irrigation systems reduce water use. Federal assistance increases water use on soybean farms through improved irrigation and drainage.

Barriers showing a negative effect on water application on corn farms include the limitation of physical field or crop conditions, an uncertainty about future water availability, and increase management time or cost. For soybeans, barriers with a negative effect are landlords will not share improvements costs, uncertainty about future water availability, and will not be farming long enough.

These negative effects are in line with the expectations. However, further investigations are needed on variables showing a positive effect.

Information from extension agents, private irrigation specialists, and neighboring farmers have a negative effect on the water use of both corn and soybeans, and irrigation equipment dealers, and media reports also show a negative effect on soybean water use. However, information from E-information services has a positive effect. These findings indicate that certain groups can be more effective in conserving water use.

The state-level variables on climate variability show a very consistent pattern on both corn and soybean water use. Compared to the average precipitation in 1981–2010, more precipitation in 2012 and 2013 leads to less irrigation water application on corn and soybean farms. Compared to the average temperature in 1981–2010, the higher temperature in 2012 and 2013 is negatively associated with the water application of both corn and soybeans in 2013. This indicates that water use is related to both climate variability based on early experience and current water availability. Compared to the farmers in the Plains, those in the West use more water for both crops, which is consistent with the expectations.

6.6. Crop Yield

The MLMs results for crop yield equations of corn and soybeans are presented in Table 7. The results are shown compared to groundwater use only and water from off-farm sources has a positive effect on soybean yield. Unfortunately, none of the cost variables is significantly for both crop yields.

For farm characteristics, more wells used on soybean farms increase the yield. A larger area of farmed land has a positive effect on corn yield, which indicates the economics of scale on corn production. A larger percentage of land owned decreases the yield for both crops. The adoption of pressure irrigation systems shows a positive effect on soybean yield, indicating that soybean yield is increased under enhanced irrigation systems.

Barriers showing a negative effect on yields of both crops include the limitation of physical field or crop conditions, and lack of financing to make improvements. This suggests that crop yield is more related to physical limitation.

Table 7. The results of multilevel models for the mean crop yield of corn and soybean farms.

Variables	Corn		Soybeans	
	Estimate	Std Err	Estimate	Std Err
Fixed Effects				
Intercept	159.780 ***	22.366	51.233 ***	4.905
Water sources				
On-farm surface water only	−6.41	3.823	0.547	1.416
Off-farm surface water only	−2.328	3.64	5.590 ***	1.21
Two or more water sources	1.011	2.681	1.06	0.835
Costs				
Cost for off-farm surface water($/acre-foot)	0.005	0.013	−0.001	0.011
Energy expenses ($/acre)	0	0.025	0	0.014
Facility expenses ($/acre)	0.003	0.007	0	0.001
Labor payment ($/acre)	0.016	0.022	0.035	0.034
Farm characteristics				
Number of wells used	0.02	0.082	0.130 ***	0.032
LN(total acre)	1.980 ***	0.656	−0.262	0.203
Percent of owned land	−5.814 ***	2.129	−2.669 **	1.028
Pressure irrigation	3.956	3.79	2.401 *	1.203
Federal assistance	1.824	1.887	0.81	0.701
Barriers to improvements				
Investigating improvement is not a priority	−1.614	1.151	−0.241	0.484
Risk of reduced yield or poorer quality crop	10.875 ***	1.636	−0.262	0.696
Limitation of physical field or crop conditions	−3.803 ***	1.485	−1.338 **	0.602
Not enough to recover implementation costs	−0.888	1.187	1.401 ***	0.475
Cannot finance improvements	−6.858 ***	1.298	−1.695 ***	0.552
Landlords will not share improvement costs	−0.976	1.39	−0.572	0.53
Uncertainty about future water availability	−2.009	1.441	−1.052	0.677
Will not be farming long enough	0.955	1.682	2.783	0.73
Will increase management time or cost	−0.777	1.679	−0.462 ***	0.711

Table 7. *Cont.*

Variables	Corn		Soybean	
	Estimate	Std Err	Estimate	Std Err
Information sources				
Extension agents	4.107 ***	0.977	1.732 ***	0.368
Private irrigation specialists	3.528 ***	0.94	1.555 ***	0.371
Irrigation equipment dealers	−1.009	0.988	−0.981 **	0.394
Local irrigation district employee	−1.044	1.677	−0.327	0.761
Government specialists	−3.724 ***	1.31	−0.083	0.5
Media reports	2.022	1.381	1.323 **	0.537
Neighboring farmers	−0.522	1.066	0.907 **	0.413
E-information services	2.574 ***	1.147	0.563	0.454
State-level variables				
PrecipChange2011	0.78	0.802	0.045	0.117
PrecipChange2012	−1.145	1.225	0.697 ***	0.171
PrecipChange2013	−1.679	1.234	0.142	0.245
TempChange2011	−11.005	8.57	−3.625 ***	1.388
TempChange2012	3.181	6.313	3.151 **	1.488
TempChange2013	3.242	9.373	5.494 ***	1.439
West	−10.149	18.285	−11.099 ***	4.055
Midwest	9.455	16.155	3.151	2.285
South	25.86	19.538	1.568	2.825
Atlantic	17.813	23.744	−1.05	4.153
Error Variance	**Estimate**	**Std Err**	**Estimate**	**Std Err**
Intercept	326 **	154	<0.0001 ***	<0.0001
Residual	5636 ***	106	534.360 ***	12.309
Fit Statistics				
N	6030		3933	
−2 Log Likelihood	63,275		32,059	
AIC	63,361		32,149	
AICC	63,362		32,150	
BIC	63,440		32,220	

Significance levels: * 10%; ** 5%; *** 1%.

Irrigation information from extension agents and private irrigation specialists show a positive effect on both corn and soybean yield. E-information services only show a positive effect on corn yield, and information from media reports and neighboring farmers have a positive effect on soybean yield. However, information showing a negative effect include government specialists (on corn yield), and irrigation equipment dealers and local irrigation district employees (on soybean yield).

Regarding state-level variables, the precipitation change in 2012 and the temperature changes in 2012 and 2013 show a positive effect on soybean yield. Given the results from the water application regressions, it seems that farmers who have access to more irrigation are able to offset the effects of weather variability. Compared with the Plains States, farms in the West have a lower soybean yield.

6.7. Economic Irrigation Water Use Efficiency

The parameter estimates for *EIWUE* equations of corn and soybeans are presented in Table 8. The results show that irrigation using water from on-farm surfaces only has a positive effect on corn *EIWUE*, compared to groundwater only. Higher water prices decrease *EIWUE* of corn, and higher energy expenses also decrease *EIWUE* of both crops. Combined with the results on water use and yield, these findings suggest that a higher efficiency cannot be achieved through increasing water prices. Higher labor payment also decreases *EIWUE* of corn.

Regarding farm characteristics, the number of wells shows a negative effect on both corn and soybean *EIWUE*. This indicates that fewer wells available on a farm can encourage an efficient use of irrigation water. The adoption of pressure irrigation increases the water use efficiency of both crops, indicating the effectiveness of achieving higher irrigation water use efficiency with the application of enhanced irrigation systems, and this is consistent with the results of water application and crop yield.

Similarly, irrigation efficiency is limited by factors related to the risk of reduced yield or poorer quality crop (on soybeans), limitation of physical field or crop conditions (on soybeans), cannot finance improvements (on corn), and will not be farming long enough (on corn). These findings can be true if water applications are limited by poor water distribution systems and/or farmers are resource-limited.

Effects of information sources are consistent for the two crops. Media reports show a positive effect, and variables showing a negative effect include local irrigation district employees and government specialists.

Regarding the state-level variables on climate variability, for soybean farms, compared with the average precipitation, a higher precipitation in 2011 and 2012 are positively associated with higher irrigation water use efficiency in 2013. The precipitation change in 2013 is positively associated with water use efficiency of both crops. The temperature change in 2011 decreases the *EIWUE* of corn and the temperature changes in 2013 increase *EIWUE* of both crops. These findings suggest that higher temperatures in the growing season lead to farmers using water more efficiently, while perceptions of precipitation are more effective to increase *EIWUE* than perceptions of temperature. Compared to farms in the Plains, both corn and soybean farms in the West have a lower *EIWUE*, while corn farms in the Midwest, South, and Atlantic states have a higher *EIWUE*.

Table 8. The results of multilevel models for the economic irrigation water use efficiency for corn and soybeans.

Variables	Corn		Soybeans	
	Estimate	Std Err	Estimate	Std Err
Fixed Effects				
Intercept	1601.320 ***	341.43	2381.410 ***	692.38
Water sources				
On-farm surface water only	536.290 ***	196.38	22.336	94.799
Off-farm surface water only	561.91	390.29	248.88	289.24
Two or more water sources	−49.571	45.746	−126.34	120.95
Costs				
Cost for off-farm surface water ($/acre-foot)	−11.042 *	6.483	−6.737	4.378
Energy expenses ($/acre)	−3.339 ***	0.986	−3.189 ***	0.859
Facility expenses ($/acre)	−0.461	0.394	0.172	0.55
Labor payment ($/acre)	−0.519 *	0.278	0.147	3.418
Farm characteristics				
Number of wells used	−8.072 ***	2.181	−4.342 *	2.135
LN (total acre)	−17.208	17.449	−18.973	20.326
Percent of owned land	−4.086	62.276	35.36	81.314
Pressure irrigation	141.810 **	64.876	206.590 **	73.067
Federal assistance	14.624	48.547	−12.277	62.92
Barriers to improvements				
Investigating improvement is not a priority	−108.420 ***	38.485	107.210 ***	40.14
Risk of reduced yield or poorer quality crop	−14.715	54.688	−155.590 ***	57.861
Limitation of physical field or crop conditions	−28.084	49.586	−124.660 **	49.947
Not enough to recover implementation costs	−52.497	39.605	−37.805	39.363
Cannot finance improvements	−206.720 ***	43.527	17.588	45.694
Landlords will not share improvement costs	−12.135	46.402	8.707	43.887
Uncertainty about future water availability	111.800 **	48.486	73.29	57.502
Will not be farming long enough	−98.065 *	56.303	64.175	60.414
Will increase management time or cost	188.680 ***	56.305	23.363	59.68
Information sources				
Extension agents	−15.285	32.67	42.179	30.393
Private irrigation specialists	−20.287	31.278	29.727	30.785
Irrigation equipment dealers	−21.818	33.008	6.813	32.874
Local irrigation district employee	−106.130 **	56.215	−119.260 *	64.125
Government specialists	−173.390 ***	43.837	−89.006 **	41.363
Media reports	170.240 ***	46.276	160.840 ***	44.404
Neighboring farmers	54.146	35.616	84.640 **	34.312
E-information services	−35.648	38.361	60.764 *	37.575

Table 8. *Cont.*

Variables	Corn		Soybean	
	Estimate	Std Err	Estimate	Std Err
State-level variables				
PrecipChange2011	7.837	12.116	57.016 ***	14.839
PrecipChange2012	−2.929	17.994	66.779 **	25.763
PrecipChange2013	64.873 ***	18.85	121.720 ***	30.98
TempChange2011	−330.240 ***	117.88	−41.801	195.77
TempChange2012	45.172	96.719	−91.935	215.37
TempChange2013	348.990 **	131.81	375.790 *	198.71
West	−765.200 ***	259.83	−1404.190 **	558.8
Midwest	893.260 ***	224.72	484.59	313.72
South	717.640 **	283.45	−611.14	389.73
Atlantic	1734.600 ***	361.36	−413.26	471.07
Error Variance	**Estimate**	**Std Err**	**Estimate**	**Std Err**
Intercept	<0.0001 ***	<0.0001	98,568 *	68,609
Residual	6,299,184 ***	118,144	3,607,043 ***	84,572
Fit Statistics				
N	6030		3933	
−2 Log Likelihood	105,657		66,842	
AIC	105,759		66,950	
AICC	105,760		66,952	
BIC	105,849		67,037	

Significance levels: * 10%; ** 5%; *** 1%.

7. Discussion

7.1. Balancing Land Allocation

As an important production input, land oftentimes overshadows irrigation water, and the farmers' decision on land input may determine water use and other inputs [76]. A profit maximizing producer might want to optimally allocate land to planting one or more crops while considering the constraints as well as other real and perceived factors. In an irrigated multi-crop production system, water availability is a serious consideration in the production function of farmers. Consistent with Moore and Dinar [73], our results show a better water availability, with more sources and wells for groundwater extraction, as an input for crop production increases land allocation in a multi-crop system. Compared to dryland production, adequate irrigation has been confirmed to increase output and farm income [39]. In particular, for water-consuming crops, farmers' production decisions prefers more water availability and less variation of water supply in growing seasons. Otherwise the yield can be hurt and producers lose incentives to continue farming.

Not only should water sources be considered in agricultural production, but the prices of inputs are important in driving or limiting factors, including the water price and energy price [62]. Water price is integrated with water availability in irrigated farming decisions [54]. While water prices are typically much lower than their real value. Given the inelastic water demand in farm irrigation, a small increase within the low price range may not be effective to conserve water and subsequently may not show a clear influencing pattern on land allocation [54].

In addition to the costs, other farm and farmer characteristics, nonmonetary motivations, or lack thereof, and information availability are equally important in farmers' decision-making [77]. A large farmland may exhibit economies of scale in crop production, thus, increasing the land allocation to more profitable crops. As the cost of per unit input decreases, the cost advantage can be remarkable and larger production returns can be expected. Additionally, farmers' production incentives can be other nonmonetary motivations. In particular, technical assistance and informational support can facilitate scientific farming decisions [78]. Irrigated production incorporates the adoption of agricultural innovations and the best management practices, which are largely adviser-driven and going beyond the farmers' experience. Therefore, the land use decisions are not free from information access and resources for overcoming the obstacles.

Land use decisions can be affected by both weather conditions in the past and farmers' climate perceptions in the coming growing seasons. Climate variability and risks can be a major threat to the farm output if appropriate coping strategies are not in place [18]. Though insurance helps reduce the potential damage, balancing land allocation according to the experienced and perceived weather variability can optimize input combinations and stabilize the expected farm income [4]. In addition, land use decisions are differing with varying geographical conditions at the state and regional levels. Coupled with climatic conditions, farmers growing the same crops may allocate different portions of land to each crop because they are facing different soil types and land slopes, among others [51]. The state and regional boundaries may also represent the implicit effects of water institutions, which influence land allocation decisions through affecting access to different water sources and the priority of water rights.

7.2. Conserving Water Resources

Along with land allocation, water applied for farm irrigation can be managed at the farm level. More sources for water supply may provide better access to water for irrigation purposes and producers may have more flexibility in irrigating crops while considering the real-time soil and crop conditions [78]. According to the conventional production economic theory, a higher input price decreases the amount of input use. In this case, a higher water price should reduce water use [61,62]. The opposite findings from our analysis definitely need close scrutiny, while they might be plausible as an overall effect given the low values of water prices. There are some facts behind these findings: (1) most of the farmers are groundwater users rather than surface water users; (2) some surface water users may have a fee-based surface water delivery system and do not pay a marginal cost for additional water; and (3) surface water use may be highly dependent on the producers' surface water rights, regardless of whether they pay a fee or an additional cost for additional units of surface water [79]. In addition, an increase in water application on per acre basis may be possible if farmers adjust their mix of crops toward more water-consuming crops or varieties, or because yield or revenue can be increased [75].

Meanwhile, the total price effect of surface and groundwater on water use in the multi-crop system is negative and consistent with the previous literature [80]. Pfeiffer and Lin [62] found an increase in the energy price of $1 would decrease groundwater extraction by 5.89 acre-feet per year for an individual farmer in Kansas. Our overall marginal effect is almost doubled, while the area of a typical farm planting multiple crops in their study is less than half of the total area of planting both corn and soybeans in our study. In another study, Hendricks and Peterson [61] found the total elasticity of water demand was −0.10 based on Kansas farm irrigation. As a comparison, Pfeiffer and Lin [62] found an elasticity of −0.26. Therefore, our findings show a modest overall effect of water price on water conservation since we just included two crops, with one being water-intensive and the other being less water-intensive, in our multi-crop production analysis.

Advanced irrigation systems have been promoted in the past decades as a way to conserve irrigation water, while recent studies have reported mixed effects [75,81]. Jevons' Paradox or the rebound effect [82] of an efficient irrigation technology adoption points to an increased water use as a result of crop choices toward more water-intensive crops and an expansion of irrigated acreage [83]. Balanced by both extensive and intensive margins, the rebound effect can be small, moderate, or even larger than 100%. As a typical issue in an irrigated production system with multiple crops, the rebound effect is a serious consideration and it might counteract the water reduction effect of adopting water conservation technologies.

Producers' experienced and perceived climate variability may have a salient influence on water use [5,18]. Similar to the effects of climate risks, a higher variability in rainfall and temperature may ultimately change the real water demand of different crops [51]. To achieve a certain yield goal, farmers routinely want to satisfy the water demand as an attempt to reduce production risks if possible during dry growing seasons, and this is even seen in arid areas [84]. Though the impacts of climate variability on different crops can be different, the effects of the farmers' perceptions may not be proportional to the water demand of different crops [6]. As a result, the effects of climate variability may be mixed and combining those of rainfall, temperature, and others. Additionally, the threshold of climate variability may be of great significance and the effects can depend on the crop-specific and baseline climate conditions [18].

7.3. Improving Crop Yield

The average grain yield on an acre basis has been approved to be higher with adequate irrigation compared with dryland production or with inadequate irrigation [14,42,51]. In a similar vein, a higher water availability by means of either more water sources or more wells facilitates producers to irrigate at a right time, with an appropriate amount of water and at scientific intervals. In the meantime, a large farm may have a higher production efficiency because of the economics of scale and a better ability to mobilize physical and technical resources [19]. Especially, if a large farm owner has the water rights, he can either use as much water as he wants or have a higher priority of withdrawing water for irrigation purposes, even if he grows water-intensive crops. Furthermore, different from the insignificant effect of owned land by Olen et al. [18], our findings show that a larger proportion of owned land decreases average grain yield. This can be true as empirical studies have found farmland rental enhances land productivity [85] and encourages farmers to be more productive and maximize the output within a limited contract period. Leased land may better motivate farmers to utilize machinery and reduce production costs [23], and generally, farmers who rent more land for growing crops specialize in agricultural production [85].

Regarding the barriers to improvements and information sources, their effects can be better understood while jointly looking at the water application and crop yield estimation results. On the one hand, since the barriers are more related to energy reduction or water conservation, their effects are mixed and more indirect. Financial limitation, physical conditions, a short farming horizon, and uncertainty in the future water supply are among the major barriers that push farmers to rely on outdated, conventional irrigation facilities and techniques, which weakens farmers' enthusiasm on water conservation and undermines their ability and effort to maximize crop yield [38,47]. On the other hand, the patterns of information effects are clearer and more direct. Water use can be reduced by extension agents, private specialists, media reports, and neighboring farmers, and these efforts are relatively consistent in enhancing grain yield. As irrigated agriculture becomes increasingly information-dependent, a wide range of scientific and technical information is required for effective decision-making [86]. The information seeking and acquisition behaviors may be influenced by sociodemographic factors as well as the preference of the farmers towards different information sources [87]. Additionally, the information efforts may help overcome the barriers in realizing a lower water consumption and/or higher farm productivity [2,88].

7.4. Enhancing Water Use Efficiency

Farm- and crop-level water use efficiency has been generously reported for different crops under different tillage systems, irrigation levels, and various farm management practices [30,33,39]. By definition, the efficiency is positively correlated with grain yield and negatively correlated with irrigation water application [14]. The effects of the factors on the efficiency can be better understood by comparing their effects on both water applications and crop yield. Water abundance reflected by groundwater use and more wells provide producers easy access and may motivate them to increase the water amount per acre, thus decreasing water use efficiency. In addition, water use efficiency may not have a linear relationship with water application [16,34]. Compared with dryland production or crop growing under drought stresses, a little more water may significantly increase yield, and as a result, the increase in water efficiency can be remarkable [33,34,39], while a higher than usual water use is unlikely to further increase grain yield, and the efficiency change can be reversed [14]. Especially for water-intensive crops like corn, more water usually means no significant yield increase and a declining water efficiency [31].

The costs of variable inputs were previously found to increase water use while having no effect on the grain yield. As a result, water use efficiency decreased. This is largely because more energy and labor were used to provide more water for the water-consuming crop [51], while the adoption of pressure irrigation systems shows a positive effect. This may be because water is saved for water-intensive corn while the yield is not hurt, and both water conservation and grain yield are maintained for less water-intensive soybeans [6].

The weather-related variables show composite effects on water use efficiency by impacting water use and farm yield [6]. On the one hand, more rainfall reduces the supplemental irrigation amount, which results in a higher irrigation efficiency. In particular, the past experiences of ample precipitation may discourage farmers to use more water for a certain yield level [4,51]. On the other hand, a high temperature may have two types of impacts: (1) a hot growing season in previous years, like the drought events in 2011 and 2012, may promote producers to irrigate more to mitigate potential dry conditions in the current year; (2) in a normal year, like 2013, a slightly higher temperature may not lead to notably more irrigation, while the grain yield may be increased as a result of improved photosynthesis [5,51]. Though more fine-scale explorations are necessary to clarify the effects of climate variability, including the direct effects of rainfall and temperature, the evidence here provides insights on the effects of both experienced and perceived weather changes.

8. Conclusions

Using the 2013 USDA FRIS data, this paper analyzes farmers' production decisions relating to irrigated agriculture in a multi-crop production system. To study the role of water costs, the farm-level water application is decomposed into crop-specific application. For each crop, the total effect can be obtained by summing the intensive and extensive margins of water use. With the aggregate effect at the farm level, we can quantify the effect of a one-unit increase in water price. Furthermore, the effects of exogenous variables are analyzed using a multilevel model approach. Four equations regarding land allocation, water application, crop yield, and economic irrigation water use efficiency are formulated using two-level models.

A fundamental finding from the decomposition of farm-level water application illustrates that the higher costs of surface water are not effective to reduce water use for both corn and soybeans through both intensive and extensive margins, while a proxy of groundwater price has a negative effect on soybean water use. This finding is a surprise, but empirically supported by some evidence. Similar to the mixed effects of water price found by Moore et al. [54], water cost is ineffective in conserving water use once producers have made decisions on crop production. Pfeiffer and Lin [75] found farm-level policies to conserve water use may not be effective. In this case, the surface water price is very low and it may not be effective because the water use is inelastic [80,89]. Comparatively, a much higher groundwater price is effective to conserve water use.

In addition, the results from MLMs allow us to make certain of the relative importance of farm- and state-level factors, and the estimation outcomes present the effects of those exogenous variables at both levels. The adoption of pressure irrigation systems reduces the soybean water use and increases the soybean yield. A higher *EIWUE* due to enhanced irrigation methods can also be achieved on both corn and soybean farms.

The findings from MLMs show that the state-level variables on climate variability have consistent effects. A high temperature promotes more efficient water use and higher yield. A high precipitation is correlated with low water application and higher crop yield. Droughts due to less rainfall or high temperature and their perceptions increase farmers' awareness of potential production risks not only during droughts, but in subsequent years [90]. As a result, farmers can be motivated to change land allocation for different crops and irrigate more to mitigate the adverse effects of climate variability. Contrary to Olen et al. [18], we find the irrigation water use is more responsive to precipitation than to temperature. Given the nonlinear impacts of climatic factors, farmers' responses in adapting to climate risks depend on cropping patterns.

This study also leaves some opportunities for future research. The aggregate effect is estimated for a typical farm growing corn and soybeans taking roughly half of the average farming area. Equations on more crops can be estimated to provide a more complete estimate of the water price effect [80], and regional equations can be estimated to account for structural differences across regions. Ideally, the elasticity with respect to water price can be estimated to quantify the price effect from a different and equally important perspective [60]. Though MLMs are supposed to deal with multiple estimation problems, more empirical and methodological investigations are needed, especially on potential endogeneity problems.

Author Contributions: Y.F. designed the research, analyzed the data and wrote the paper. Y.F., R.M., and S.C.P. reviewed and commented on the manuscript.

Acknowledgments: We thank Brad Parks for his support when the first author analyzed data at the USDA NASS data lab in St. Louis, Missouri. We also appreciate helpful comments by Laura McCann, Hua Qin, and Corinne Valdivia on an earlier version of the paper. The authors are grateful to participants at the 2017 Agricultural & Applied Economics Association Annual Meetings. This research was conducted while Yubing Fan was a doctoral candidate at the University of Missouri-Columbia.

References and Notes

1. Fan, Y.; Massey, R.; Park, S.C. Multicrop production decisions and economic irrigation water use efficiency: Effects of water costs, pressure irrigation adoption and climatic determinants. In Proceedings of the Annual Meeting 2017, Chicago, IL, USA, 30 July–1 August 2017; Agricultural and Applied Economics Association: Milwaukee, WI, USA, 2017; pp. 1–60.
2. Schaible, G.; Aillery, M. *Water Conservation in Irrigated Agriculture: Trends and Challenges in the Face of Emerging Demands*; USDA-ERS Economic Information Bulletin No. 99; USDA-ERS: Washington, DC, USA, 2012.
3. EPA (United States Environmental Protection Agency). Climate Change Indicators in the United States—Weather and Climate, 2014. Available online: https://www3.epa.gov/climatechange/pdfs/climateindicators-full-2014.pdf (accessed on 15 July 2016).
4. Jin, J.; Gao, Y.; Wang, X.; Nam, P.K. Farmers' risk preferences and their climate change adaptation strategies in the Yongqiao District, China. *Land Use Policy* **2015**, *47*, 365–372.
5. Zhang, T.; Lin, X.; Sassenrath, G.F. Current irrigation practices in the central United States reduce drought and extreme heat impacts for maize and soybean, but not for wheat. *Sci. Total Environ.* **2015**, *508*, 331–342. [CrossRef] [PubMed]
6. Zhang, T.; Lin, X. Assessing future drought impacts on yields based on historical irrigation reaction to drought for four major crops in Kansas. *Sci. Total Environ.* **2016**, *550*, 851–860. [CrossRef] [PubMed]
7. Wanders, N.; Wada, Y. Human and climate impacts on the 21st century hydrological drought. *J. Hydrol.* **2015**, *526*, 208–220. [CrossRef]
8. Murray, S.J.; Foster, P.N.; Prentice, I.C. Future global water resources with respect to climate change and water withdrawals as estimated by a dynamic global vegetation model. *J. Hydrol.* **2012**, *448–449*, 14–29. [CrossRef]
9. George, B.A.; Shende, S.A.; Raghuwanshi, N.S. Development and testing of an irrigation scheduling model. *Agric. Water Manag.* **2000**, *46*, 121–136. [CrossRef]
10. Batchelor, C.; Lovell, C.; Murata, M. Simple microirrigation techniques for improving irrigation efficiency on vegetable gardens. *Agric. Water Manag.* **1996**, *32*, 37–48. [CrossRef]
11. Dalton, P.; Raine, S.; Broadfoot, K. Best Management Practice for Maximising Whole Farm Irrigation Efficiency in the Australian Cotton Industry. Final Report for CRDC Project NEC2C. 2001. Available online: http://www.insidecotton.com/xmlui/handle/1/3535 (accessed on 12 November 2016).

12. Qin, W.; Assinck, F.B.T.; Heinen, M.; Oenema, O. Water and nitrogen use efficiencies in citrus production: A meta-analysis. *Agric. Ecosyst. Environ.* **2016**, *222*, 103–111. [CrossRef]
13. Gheysari, M.; Loescher, H.W.; Sadeghi, S.H.; Mirlatifi, S.M.; Zareian, M.J.; Hoogenboom, G. Chapter Three-Water-yield relations and water use efficiency of maize under nitrogen fertigation for semiarid environments: Experiment and synthesis. *Adv. Agron.* **2015**, *130*, 175–229.
14. Ibragimov, N.; Evett, S.R.; Esanbekov, Y.; Kamilov, B.S.; Mirzaev, L.; Lamers, J.P.A. Water use efficiency of irrigated cotton in Uzbekistan under drip and furrow irrigation. *Agric. Water Manag.* **2007**, *90*, 112–120. [CrossRef]
15. Schneider, A.D.; Howell, T.A. Scheduling deficit wheat irrigation with data from an evapotranspiration network. *Trans. ASAE* **2001**, *44*, 1617–1623. [CrossRef]
16. Kang, Y.; Wang, R.; Wan, S.; Hu, W.; Jiang, S.; Liu, S. Effects of different water levels on cotton growth and water use through drip irrigation in an arid region with saline ground water of Northwest China. *Agric. Water Manag.* **2012**, *109*, 117–126. [CrossRef]
17. Nijbroek, R.; Hoogenboom, G.; Jones, J.W. Optimizing irrigation management for a spatially variable soybean field. *Agric. Syst.* **2003**, *76*, 359–377. [CrossRef]
18. Olen, B.; Wu, J.; Langpap, C. Irrigation decisions for major West Coast crops: Water scarcity and climatic determinants. *Am. J. Agric. Econ.* **2016**, *98*, 254–275. [CrossRef]
19. Boyer, C.N.; Larson, J.A.; Roberts, R.K.; McClure, A.T.; Tyler, D.D. The impact of field size and energy cost on the profitability of supplemental corn irrigation. *Agric. Syst.* **2014**, *127*, 61–69. [CrossRef]
20. Dolisca, F.; McDaniel, J.M.; Shannon, D.A.; Jolly, C.M. A multilevel analysis of the determinants of forest conservation behavior among farmers in Haiti. *Soc. Nat. Resour.* **2009**, *22*, 433–447. [CrossRef]
21. Guerin, D.; Crete, J.; Mercier, J. A multilevel analysis of the determinants of recycling behavior in the European countries. *Soc. Sci. Res.* **2001**, *30*, 195–218. [CrossRef]
22. Neumann, K.; Stehfest, E.; Verburg, P.H.; Siebert, S.; Müller, C.; Veldkamp, T. Exploring global irrigation patterns: A multilevel modelling approach. *Agric. Syst.* **2011**, *104*, 703–713. [CrossRef]
23. Giannakis, E.; Bruggeman, A. Exploring the labour productivity of agricultural systems across European regions: A multilevel approach. *Land Use Policy* **2018**, *77*, 94–106. [CrossRef]
24. Pereira, L.S. Higher performance through combined improvements in irrigation methods and scheduling: A discussion. *Agric. Water Manag.* **1999**, *40*, 153–169. [CrossRef]
25. Nair, S.; Johnson, J.; Wang, C. Efficiency of irrigation water use: A review from the perspectives of multiple disciplines. *Agron. J.* **2013**, *105*, 351–363. [CrossRef]
26. Harris, G. *Water Use Efficiency: What Is It, and How to Measure, Spotlight on Cotton Research & Development*; Cotton Research & Development Corporation: Narrabri, Australia, 2007; p. 8. Available online: http://era.daf.qld.gov.au/id/eprint/2986/1/50904_CottonCRC_Final_Report_Harris.pdf (accessed on 5 June 2015).
27. Tanwar, S.; Rao, S.; Regar, P.; Datt, S.; Jodha, B.; Santra, P.; Kumar, R.; Ram, R. Improving water and land use efficiency of fallow-wheat system in shallow Lithic Calciorthid soils of arid region: Introduction of bed planting and rainy season sorghum–legume intercropping. *Soil Tillage Res.* **2014**, *138*, 44–55. [CrossRef]
28. Bozzola, M.; Swanson, T. Policy implications of climate variability on agriculture: Water management in the Po river basin, Italy. *Environ. Sci. Policy* **2014**, *43*, 26–38. [CrossRef]
29. Sadras, V.; Rodriguez, D. Modelling the nitrogen-driven trade-off between nitrogen utilisation efficiency and water use efficiency of wheat in eastern Australia. *Field Crops Res.* **2010**, *118*, 297–305. [CrossRef]
30. Fan, Y.; Wang, C.; Nan, Z. Comparative evaluation of crop water use efficiency, economic analysis and net household profit simulation in arid Northwest China. *Agric. Water Manag.* **2014**, *146*, 335–345. [CrossRef]
31. Salazar, M.R.; Hook, J.E.; Garcia y Garcia, A.; Paz, J.O.; Chaves, B.; Hoogenboom, G. Estimating irrigation water use for maize in the Southeastern USA: A modeling approach. *Agric. Water Manag.* **2012**, *107*, 104–111. [CrossRef]
32. Vories, E.D.; Tacker, P.L.; Lancaster, S.W.; Glover, R.E. Subsurface drip irrigation of corn in the United States Mid-South. *Agric. Water Manag.* **2009**, *96*, 912–916. [CrossRef]
33. Dağdelen, N.; Başal, H.; Yılmaz, E.; Gürbüz, T.; Akçay, S. Different drip irrigation regimes affect cotton yield, water use efficiency and fiber quality in western Turkey. *Agric. Water Manag.* **2009**, *96*, 111–120. [CrossRef]
34. Liu, Y.; Li, S.; Chen, F.; Yang, S.; Chen, X. Soil water dynamics and water use efficiency in spring maize (*Zea mays* L.) fields subjected to different water management practices on the Loess Plateau, China. *Agric. Water Manag.* **2010**, *97*, 769–775. [CrossRef]

35. Salvador, R.; Latorre, B.; Paniagua, P.; Playán, E. Farmers' scheduling patterns in on-demand pressurized irrigation. *Agric. Water Manag.* **2011**, *102*, 86–96. [CrossRef]
36. Usman, M.; Arshad, M.; Ahmad, A.; Ahmad, N.; Zia-Ul-Haq, M.; Wajid, A.; Khaliq, T.; Nasim, W.; Ali, H.; Ahmad, S. Lower and upper baselines for crop water stress index and yield of *Gossypium hirsutum* L. under variable irrigation regimes in irrigated semiarid environment. *Pak. J. Bot.* **2010**, *42*, 2541–2550.
37. Pannell, D.J.; Marshall, G.R.; Barr, N.; Curtis, A.; Vanclay, F.; Wilkinson, R. Understanding and promoting adoption of conservation practices by rural landholders. *Aust. J. Exp. Agric.* **2006**, *46*, 1407–1424. [CrossRef]
38. Knowler, D.; Bradshaw, B. Farmers' adoption of conservation agriculture: A review and synthesis of recent research. *Food Policy* **2007**, *32*, 25–48. [CrossRef]
39. Abd El-Wahed, M.H.; Ali, E.A. Effect of irrigation systems, amounts of irrigation water and mulching on corn yield, water use efficiency and net profit. *Agric. Water Manag.* **2013**, *120*, 64–71. [CrossRef]
40. Canone, D.; Previati, M.; Bevilacqua, I.; Salvai, L.; Ferraris, S. Field measurements based model for surface irrigation efficiency assessment. *Agric. Water Manag.* **2015**, *156*, 30–42. [CrossRef]
41. Ahmad, B.; Hassan, S.; Bakhsh, K. Factors affecting yield and profitability of carrot in two districts of Punjab. *Int. J. Agric. Biol.* **2005**, *7*, 794–798.
42. Dahmardeh, N.; Asasi, H. Determined factors on water use efficiency and profitability in agricultural sector. *Ind. J. Sci. Res.* **2014**, *4*, 48–53.
43. Schaible, G.D.; Kim, C.S.; Aillery, M.P. Dynamic adjustment of irrigation technology/water management in western US agriculture: Toward a sustainable future. *Can. J. Agric. Econ.* **2010**, *58*, 433–461. [CrossRef]
44. Rogers, E.M. *Diffusion of Innovations*, 5th ed.; Free Press: New York, NJ, USA, 2003.
45. Sunding, D.; Zilberman, D. The agricultural innovation process: Research and technology adoption in a changing agricultural sector. *Handb. Agric. Econ.* **2001**, *1*, 207–261.
46. Prokopy, L.; Floress, K.; Klotthor-Weinkauf, D.; Baumgart-Getz, A. Determinants of agricultural best management practice adoption: Evidence from the literature. *J. Soil Water Conserv.* **2008**, *63*, 300–311. [CrossRef]
47. Rodriguez, J.M.; Molnar, J.J.; Fazio, R.A.; Sydnor, E.; Lowe, M.J. Barriers to adoption of sustainable agriculture practices: Change agent perspectives. *Renew. Agric. Food Syst.* **2009**, *24*, 60–71. [CrossRef]
48. Hunecke, C.; Engler, A.; Jara-Rojas, R.; Poortvliet, P.M. Understanding the role of social capital in adoption decisions: An application to irrigation technology. *Agric. Syst.* **2017**, *153*, 221–231. [CrossRef]
49. Frisvold, G.B.; Deva, S. Farm size, irrigation practices, and conservation program participation in the US Southwest. *Irrig. Drain.* **2012**, *61*, 569–582. [CrossRef]
50. Negri, D.H.; Gollehon, N.R.; Aillery, M.P. The effects of climatic variability on US irrigation adoption. *Clim. Chang.* **2005**, *69*, 299–323. [CrossRef]
51. Kresovic, B.; Matovic, G.; Gregoric, E.; Djuricin, S.; Bodroza, D. Irrigation as a climate change impact mitigation measure: An agronomic and economic assessment of maize production in Serbia. *Agric. Water Manag.* **2014**, *139*, 7–16. [CrossRef]
52. Li, C.; Ting, Z.; Rasaily, R.G. Farmer's adaptation to climate risk in the context of China: A research on Jianghan Plain of Yangtze River Basin. *Agric. Agric. Sci. Proc.* **2010**, *1*, 116–125.
53. Li, X.; Takahashi, T.; Suzuki, N.; Kaiser, H.M. The impact of climate change on maize yields in the United States and China. *Agric. Syst.* **2011**, *104*, 348–353. [CrossRef]
54. Moore, M.R.; Gollehon, N.R.; Carey, M.B. Multicrop production decisions in western irrigated agriculture: The role of water price. *Am. J. Agric. Econ.* **1994**, *76*, 859–874. [CrossRef]
55. Producers also need to choose which type of irrigation system(s) to adopt, and this has been examined by much research, for instance, Olen et al. as cited in this paper.
56. Just, R.E.; Zilberman, D.; Hochman, E. Estimation of multicrop production functions. *Am. J. Agric. Econ.* **1983**, *65*, 770–780. [CrossRef]
57. Just, R.E.; Zilberman, D.; Hochman, E.; Bar-Shira, Z. Input allocation in multicrop systems. *Am. J. Agric. Econ.* **1990**, *72*, 200–209. [CrossRef]
58. Pereira, L.S.; Cordery, I.; Iacovides, I. Improved indicators of water use performance and productivity for sustainable water conservation and saving. *Agric. Water Manag.* **2012**, *108*, 39–51. [CrossRef]
59. The calculation of EIWUE (and IWUE) just considers irrigation water applied, while excluding rainfall amounts. The measure of water efficiency is restricted due to data paucity of climate-related variables

at the farm level. To exam their effects, the state-level variation is controlled in the multilevel models presented below.

60. Moore, M.R.; Gollehon, N.R.; Carey, M.B. Alternative models of input allocation in multicrop systems: Irrigation water in the Central Plains, United States. *Agric. Econ.* **1994**, *11*, 143–158. [CrossRef]
61. Hendricks, N.P.; Peterson, J.M. Fixed effects estimation of the intensive and extensive margins of irrigation water demand. *J. Agric. Res. Econ.* **2012**, *37*, 1–19.
62. Pfeiffer, L.; Lin, C.-Y.C. The effects of energy prices on agricultural groundwater extraction from the High Plains Aquifer. *Am. J. Agric. Econ.* **2014**, *96*, 1349–1362. [CrossRef]
63. Hox, J.J. *Applied Multilevel Analysis*; TT-Publikaties: Amsterdam, The Netherlands, 1995.
64. Raudenbush, S.W.; Bryk, A.S. *Hierarchical Linear Models: Applications and Data Analysis Methods*, 2nd ed.; Sage: Thousand Oaks, CA, USA, 2002; Volume 1.
65. Ene, M.; Leighton, E.A.; Blue, G.L.; Bell, B.A. Multilevel Models for Categorical data Using SAS PROC GLIMMIX: The Basics. SAS Global Forum 2015. Available online: https://pdfs.semanticscholar.org/a216/864a2a2de19eb194c6523fb8566e601ffa32.pdf (accessed on 15 June 2015).
66. Snijders, T.A.; Bosker, R.J. *Multilevel Analysis: An Introduction to Basic and Advanced Multilevel Modeling*; Sage: Thousand Oaks, CA, USA, 1999.
67. In the empirical analysis below, land allocation refers to harvested acres from the FRIS data.
68. A Map Can Be Found on the USDA NASS Website. Available online: https://www.nass.usda.gov/Charts_and_Maps/Farm_Production_Expenditures/reg_map_c.php (accessed on 16 July 2016).
69. Ideally, analyses on all the production decisions (i.e., 4 equations regarding all crops (17 crops) can be conducted at the region level (i.e., 5 regions). Given the huge amount of work and the focus of this paper, such analyses are beyond the scope.
70. The crop-specific analyses just focus on farms that are at least partially irrigated, while excluding non-irrigated farms.
71. The USDA FRIS targeted at the irrigated farms. The corn and soybean farms included in this analysis are at least partially irrigated. This study only analyzes the harvested acres and excludes the acres that were planted while not harvested due to crop failure or other reasons.
72. According to the USDA FRIS, the on-farm surface water includes recycled water of surface or groundwater that was previous used for irrigation, and reclaimed water from on-farm livestock wastewater after being treated. The off-farm surface water is surface water from off-farm sources, municipal water, rural water supply, as well as reclaimed water from off-farm sources such as municipal reclaimed water, industrial, off-farm livestock operations, and other off-farm sources.
73. Moore, M.R.; Dinar, A. Water and land as quantity-rationed inputs in California agriculture: Empirical tests and water policy implications. *Land Econ.* **1995**, *71*, 445–461. [CrossRef]
74. Ideally, equations on water application and land allocation for each crop can be estimated to obtain both extensive and intensive margins, and then the aggregate effect can be calculated for a typical farm growing all crops. Equations on production decisions can also be estimated for each region to calculate the aggregate effect for a typical farm growing all crops in each specific region. Given the focus of this paper, such analyses are beyond the scope.
75. Pfeiffer, L.; Lin, C.-Y.C. Does efficient irrigation technology lead to reduced groundwater extraction? Empirical evidence. *J. Environ. Econ. Manag.* **2014**, *67*, 189–208. [CrossRef]
76. Moore, M.R.; Gollehon, N.R.; Negri, D.H. Alternative forms for production functions of irrigated crops. *J. Agric. Econ. Res.* **1992**, *44*, 16–32.
77. Koontz, T.M. Money talks? But to whom? Financial versus nonmonetary motivations in land use decisions. *Soc. Nat. Resour.* **2001**, *14*, 51–65.
78. Fan, Y.; McCann, L.M. Farmers' Adoption of Pressure Irrigation Systems and Scientific Scheduling Practices: An Application of Multilevel Models. In Proceedings of the Annual Meeting 2017, Chicago, IL, USA, 30 July–1 August 2017; Agricultural and Applied Economics Association: Milwaukee, WI, USA, 2017; pp. 1–38.
79. We acknowledge the thoughtful ideas from one anonymous reviewer. This issue would be better addressed if there are more detailed data on observations using fee-based systems and the role of water rights in different regions of the United States.

80. Wang, T.; Park, S.C.; Jin, H. Will farmers save water? A theoretical analysis of groundwater conservation policies. *Water Resour. Econ.* **2015**, *12*, 27–39. [CrossRef]
81. Ward, F.A.; Pulido-Velazquez, M. Water conservation in irrigation can increase water use. *Proc. Natl. Acad. Sci. USA* **2008**, *105*, 18215–18220. [CrossRef] [PubMed]
82. While this is not investigated in our study, we realize the potential effect of Jevons' Paradox, which might offset the effect of pressure irrigation in reducing water application.
83. Li, H.; Zhao, J. Rebound effects of new irrigation technologies: The role of water rights. *Am. J. Agric. Econ.* **2018**, *100*, 786–808. [CrossRef]
84. Fan, Y.; Park, S.; Nan, Z. Participatory water management and adoption of micro-irrigation systems: Smallholder farmers in arid north-western China. *Int. J. Water Res. Dev.* **2017**, *34*, 434–452. [CrossRef]
85. Liu, Y.; Wang, C.; Tang, Z.; Nan, Z. Farmland Rental and Productivity of Wheat and Maize: An Empirical Study in Gansu, China. *Sustainability* **2017**, *9*, 1678. [CrossRef]
86. Ali, J.; Kumar, S. Information and communication technologies (ICTs) and farmers' decision-making across the agricultural supply chain. *Int. J. Inf. Manag.* **2011**, *31*, 149–159. [CrossRef]
87. Ma, W.; Renwick, A.; Nie, P.; Tang, J.; Cai, R. Off-farm work, smartphone use and household income: Evidence from rural China. *China Econ. Rev.* **2018**. [CrossRef]
88. Solano, C.; Leon, H.; Perez, E.; Herrero, M. The role of personal information sources on the decision-making process of Costa Rican dairy farmers. *Agric. Syst.* **2003**, *76*, 3–18. [CrossRef]
89. Wang, C.; Segarra, E. The economics of commonly owned groundwater when user demand is perfectly inelastic. *J. Agric. Res. Econ.* **2011**, *36*, 95–120.
90. Peck, D.E.; Adams, R.M. Farm-level impacts of prolonged drought: Is a multiyear event more than the sum of its parts? *Aust. J. Agric. Res. Econ.* **2010**, *54*, 43–60. [CrossRef]

12

Year-Round Irrigation Schedule for a Tomato–Maize Rotation System in Reservoir-Based Irrigation Schemes

Ephraim Sekyi-Annan [1,2], Bernhard Tischbein [1], Bernd Diekkrüger [3] and Asia Khamzina [4,*]

[1] Department of Ecology and Natural Resources Management, Center for Development Research, University of Bonn, Genscherallee 3, 53113 Bonn, Germany; sekyiannan@yahoo.com (E.S.-A.); tischbein@uni-bonn.de (B.T.)
[2] CSIR-Soil Research Institute, Academy Post Office, Private Mail Bag, Kwadaso-Kumasi, Ghana; sekyiannan@yahoo.com
[3] Department of Geography, University of Bonn, Meckenheimer Allee 166, 53115 Bonn, Germany; b.diekkrueger@uni-bonn.de
[4] Division of Environmental Science and Ecological Engineering, College of Life Science and Biotechnology, Korea University, 145 Anam-Ro, Seongbuk-Gu, Seoul 02841, Korea
* Correspondence: asia_khamzina@korea.ac.kr

Abstract: Improving irrigation management in semi-arid regions of Sub-Saharan Africa is crucial to respond to increasing variability in rainfall and overcome deficits in current irrigation schemes. In small-scale and medium-scale reservoir-based irrigation schemes in the Upper East region of Ghana, we explored options for improving the traditional, dry season irrigation practices and assessed the potential for supplemental irrigation in the rainy season. The AquaCrop model was used to (i) assess current water management in the typical tomato-maize rotational system; (ii) develop an improved irrigation schedule for dry season cultivation of tomato; and (iii) determine the requirement for supplemental irrigation of maize in the rainy season under different climate scenarios. The improved irrigation schedule for dry season tomato cultivation would result in a water saving of 130–1325 mm compared to traditional irrigation practices, accompanied by approximately a 4–14% increase in tomato yield. The supplemental irrigation of maize would require 107–126 mm of water in periods of low rainfall and frequent dry spells, and 88–105 mm in periods of high rainfall and rare dry spells. Therefore, year-round irrigated crop production may be feasible, using water saved during dry season tomato cultivation for supplemental irrigation of maize in the rainy season.

Keywords: AquaCrop model; capillary rise; climate change; rainfall variability; supplemental irrigation

---

1. Introduction

Insufficient water availability, owing to variability in rainfall patterns and frequent dry spells exacerbated by climate change [1,2], threatens food security and rural livelihoods in Sub-Saharan Africa (SSA) [3]. In SSA, more than 95% of arable land is under rainfed crop production, which contributes 81% to the regional food basket [4,5]. Because of variable rainfall and low-input cultivation [6,7], grain yields are only from 1 to 2 Mg ha^{-1}, whereas attainable yields range between 4 and 5 Mg ha^{-1} in SSA [5,8]. Furthermore, risks of crop failure in SSA have increased due to land degradation and soil nutrient depletion [9,10], signified by negative annual NPK balances with -26 kg ha^{-1} N, -7 kg ha^{-1} P_2O_5, and -23 kg ha^{-1} K_2O, as reported in [11]. On a continental scale, annual NPK losses averaged 54 kg ha^{-1} (and ranged between 9 kg ha^{-1} in Egypt and 88 kg ha^{-1} in Somalia), resulting in land degradation in more than 40% of Africa's total farmland [12,13]. These risks have further reduced the already insufficient financial capacity of farmers to invest in sustainable land management (SLM) strategies [3,5]. However, such strategies are key for optimizing trade-offs between food production and other agro-ecosystem services [12]. In water-scarce environments such as the Upper East region of Ghana (UER), sustainable soil-water management has been identified as the most influential among agricultural management practices, including soil fertility management, selection of crop varieties, and control of pests and diseases [5,14], for enhancing food security as well as improving the smallholders' livelihoods [5,15–17].

The reservoir-based irrigation schemes in SSA, which store water (i.e., mostly surface runoff) in the rainy season, were originally designed to supply water for dry season crop irrigation, the livestock sector, fish farming, and domestic use, excluding supplemental irrigation in the rainy season. However, increasing climate variability calls for exploring the feasibility of supplemental irrigation for crop cultivation in the rainy season [3]. Supplemental irrigation has considerable potential to increase grain yield, particularly if provided during the critical stages of the crop growing cycle (i.e., booting and grain filling) [18].

Because of increasing competition for stored water in the dry season, the extra water demand for supplemental irrigation to bridge dry spells is likely to result in a mismatch between water supply and demand in the reservoir-based irrigation schemes. Thus, the requirement for supplemental irrigation might be satisfied with water saved through increased irrigation efficiency as a result of improving dry season irrigation scheduling [19]. As long as increased irrigation efficiency is accompanied by yield increments, this provides incentives for irrigators to engage in SLM [5,15,18,20]. Consequently, crop–water–soil–atmosphere models will be useful to determine the most appropriate irrigation schedules for the prevalent cropping practices and for assessing possible alternative scenarios [21–23]. Among the common models capable of simulating irrigated crop growth, those requiring large inputs of primary data, for instance APSIM [24] and CropSyst [25], and that are not available for free, such as the irrigation scheduling model ISAREG [26], might not be favorable for applications in SSA. The DSSAT model [27] has been commonly used to assess the impact of agronomic inputs on irrigated crop yield but at present is not suitable to evaluate the effectiveness of irrigation practices. Some other models, such as CROPWAT [28] do not distinguish between evaporation (i.e., non-beneficial water consumption) and crop transpiration, and do not provide an estimation of yield or, as with EPIC [29], apply simplified routines to evaluate the groundwater contribution to crop water use. Due to relatively modest data requirements, consideration of all major agro-hydrological processes, and its free availability, the AquaCrop model developed by the Food and Agriculture Organization of the Unites Nations (FAO) [22] has found many applications worldwide, including in SSA [30–32].

Current irrigation schedules in reservoir-based irrigation schemes in SSA are based on locally established rules governing access to water for irrigation, but with little consideration of crop- and site-specific water demands in terms of quantity and timing, resulting in the over-irrigation of crops [19]. For instance, in reservoir-based irrigation schemes in onion fields in the UER, the ratio of total water supply to gross irrigation demand ranged between 2.4 and 5.7 during dry season crop irrigation [33]. The problem of over-irrigation in reservoir-based irrigation schemes was further confirmed by gross irrigation amounts (GIAs) ranging from 380 to 852 mm for dry season tomato production in the UER [34], and between 274 and 838 mm for tomato cropping under groundwater irrigation in the same region [35]. Simulations have suggested that the net irrigation requirement (NIR) for dry season tomatoes ranges from 359 to 372 mm in the reservoir-based Koga irrigation scheme in Ethiopia [36], emphasizing the need as well as the potential to improve water management through irrigation scheduling to reduce water losses and increase productivity.

To the best of our knowledge, no study has attempted to develop an irrigation schedule for dry season cropping systems in the UER. Moreover, the limited number of studies on supplemental irrigation in SSA have not explored the feasibility of using dry season water savings in reservoir-based irrigation schemes. For example, Sanfo et al. [3] investigated the economic value of supplemental irrigation of grain crops using farm ponds of 300 m^3 capacity in south-western Burkina Faso, and reported that in years of low rainfall, supplemental irrigation could be a cost-effective intervention to reduce risks of crop failure and increase farmers' incomes. Fox and Rockström [37] also assessed the effect of supplemental irrigation, based on 150 m^3 capacity farm ponds, on the grain yield of sorghum in northern Burkina Faso and found that supplemental irrigation alone resulted in an approximately 56% increase in grain yield, making it a useful technology to mitigate dry spells and shorten the yield gap. Similarly, Mustapha [16] studied the water productivity of pearl millet under supplemental irrigation applied at five different crop growth stages in Nigeria and reported that the supplemental irrigation amount of 84 mm applied at booting and grain filling stages could result in a 69% increase in yields.

This study aims to improve the traditional dry season irrigation practices in reservoir-based irrigation schemes in the UER, and to assess the potential for introducing supplemental irrigation in the rainy season as an adaptation to climate change. To this end, we (i) parameterized and validated the AquaCrop model to render applications for irrigated crop production in the EUR of Ghana (ii) assessed the appropriateness of current water management in the typical tomato–maize rotational system; (iii) developed an improved irrigation schedule for dry season cultivation of tomato; and (iv) determined the requirement for supplemental irrigation of maize in the rainy season under different climate scenarios.

2. Materials and Methods

2.1. Study Area

The study was conducted between May 2014 and April 2016 in the medium-scale Vea irrigation scheme (VIS, 136 km^2) and the small-scale Bongo irrigation scheme (BIS, 0.98 km^2) in the UER, located between latitudes 10°30′ N and 11°15′ N and longitudes 0° W and 1°30′ W (Figure 1). The UER belongs to the Guinea–Sudano–Savanna agro-ecological zone characterized by a single rainy season starting in April/May and ending in September/October, followed by a dry season from November until April/May. The mean annual rainfall is 970 mm with high intra- and inter-seasonal variability, and the mean annual temperature is 29 °C (Figure 2). The annual potential evapotranspiration (ET_0) is twice as much as the annual precipitation, but evapotranspiration is exceeded by rainfall in the rainy season [34]. Soil types in the UER include Gleyic Lixisols, Ferric Lixisols, Haplic Lixisols, Lithic Leptosols, and Eutric Fluvisols, with loam and sandy loam as the dominating soil textures.

Two schemes were selected to capture the typical scale of irrigation schemes in the region, as well as the differing institutional settings in operations by the parastatal Irrigation Company of the Upper Region (ICOUR, Navrongo, Ghana) in Vea, and a community-based operation in Bongo (Figure 1). Furthermore, water allocation in the VIS is supply-driven, and thus a technician implements water supply schedules for 4–5 days continuously with 3–4 days interval between schedules. However, in the BIS, where water allocation is demand-driven, water can flow for the whole week (8 h per day on average) except on market days which occur twice a week. Irrigators in the UER tend to water their crops with as much water as is available resulting in over-irrigation, hence there is an urgent need for improved schedules which are crop- and site-specific [19]. On average, the total irrigation events for the dry season production of tomato ranges between 20 and 29 in both VIS and BIS.

The storage capacity, the elevation of the reservoir's spillway, and the irrigable area of the BIS are 0.43 MCM, 231 m and 12 ha, respectively, while the values for the VIS are 17.27 MCM, 189 m and 850 ha, respectively. The irrigable area in both schemes is equipped with lined trapezoidal primary canals which convey water by gravity to the cropping fields. Farm sizes range between 0.01 and 0.10 ha in the dry season, and up to 0.31 ha in the rainy season, in both irrigation schemes.

Rainfed crops include maize (*Zea mays*), pearl millet (*Pennisetum glaucum*), sorghum (*Sorghum bicolor*), and rice (*Oryza sativa*; cultivated also in the dry season under irrigation). Tomato (*Solanum lycopersicum*) and leafy vegetables such as roselle (*Hibiscus sabdariffa*), lettuce (*Latuca sativa*) and cowpea (*Vigna unguiculata;* grown primarily for the leaves) are irrigated in the dry season only. Currently, irrigation is not practiced in the rainy season. There are no soil bunds constructed on the cropping fields, except around rice fields. Furthermore, furrows are not blocked during irrigation, resulting in the surface runoff of irrigation water.

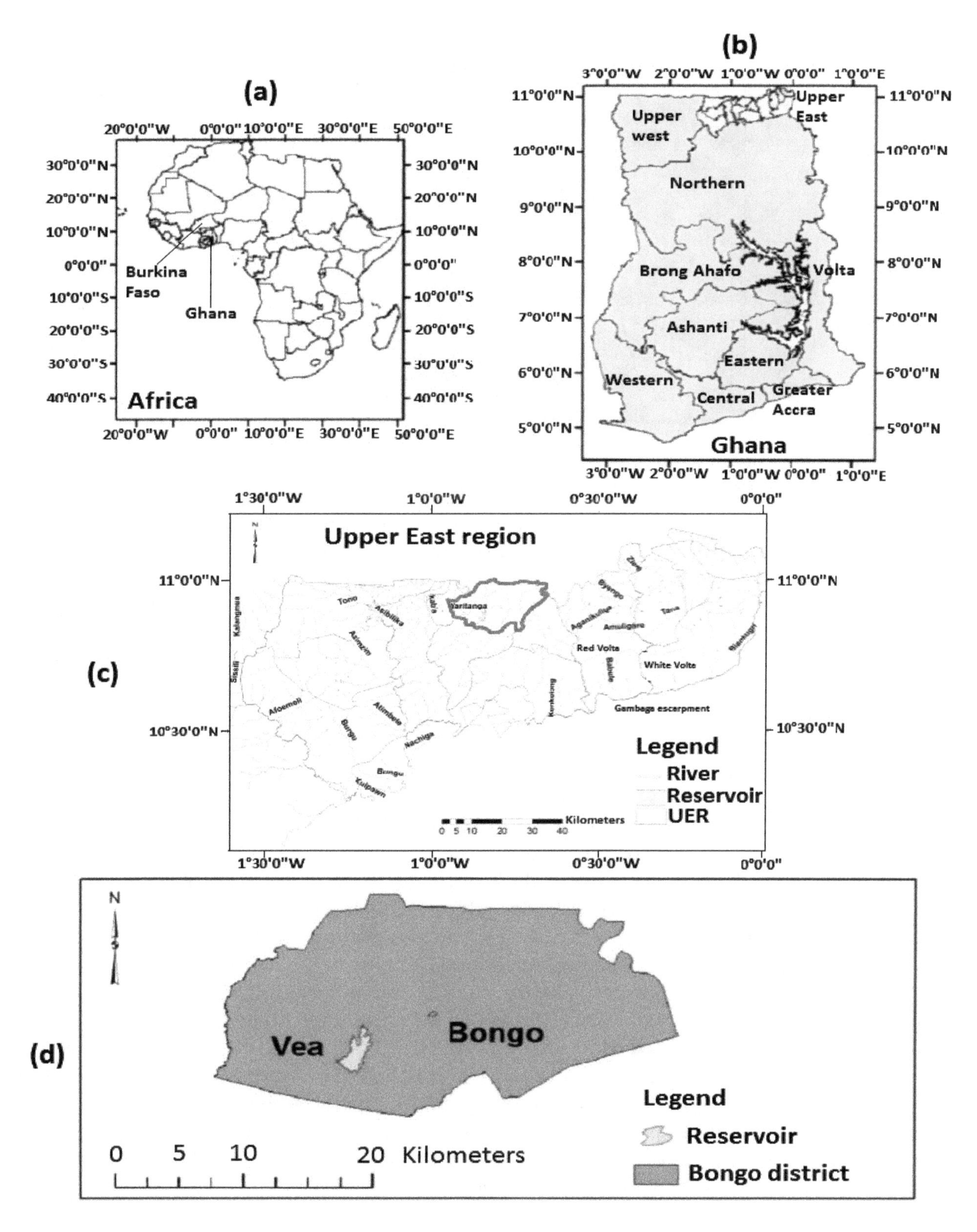

Figure 1. Location of Ghana in West Africa (**a**); the Upper East region (UER) of Ghana (**b**); hydrological network of the UER and location of the study area (**c**); and the study area including the Vea and Bongo reservoirs (**d**).

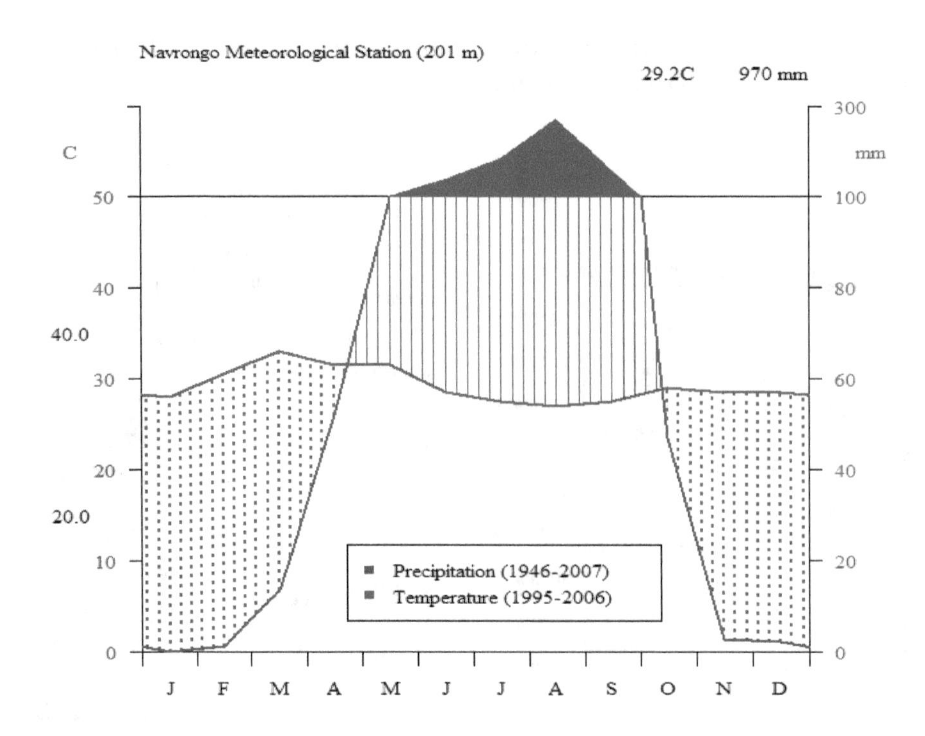

Figure 2. Walter–Lieth climate diagram for the Upper East region of Ghana based on data collected at Navrongo Meteorological Station (latitude 10°54′0″ N and longitude 1°06′0″ W; elevation 201 m above sea level). Precipitation data covered the period 1946–2007, and temperature data were measured between 1995 and 2006. Top of the graph shows the long-term mean annual temperature and rainfall. The value at the top-left of the temperature axis is the mean of the average daily maximum temperature of the hottest month; the value at the bottom of the same axis is the mean of the average daily minimum temperature of the coldest month. Area shaded in blue indicates the moist period and area shaded in red show the arid period. Area filled in blue indicates the period of excess water.

The principal cropping systems include tomato–maize rotation, millet/sorghum–leafy vegetable rotation, and rice mono-cropping under alternate wet–dry irrigation. In the VIS, the shares of the irrigable area in the dry season were 48%, 37%, and 12% for tomato, rice, and leafy vegetables, respectively, and 40%, 5%, and 55% for the same crops in the BIS. In the rainy season, the shares were 50%, 40%, and 10% for millet/sorghum, rice, and maize, respectively, in the BIS, while in the VIS these shares were 34%, 59% and 3%. In this study, the cropping system of dry season irrigated tomato in rotation with maize in the rainy season was selected for detailed analysis due to the socio-economic significance of these crops in the UER and SSA. Tomato was cropped once in the dry season in both the BIS and the VIS. The growing and irrigation period lasted for 113–123 days. In this period, mature tomato fruits were harvested 2–3 times. The duration of tomato seedling development was about 14 days. The growing period of maize sown directly in the field ranged between 84 and 113 days.

The application of cow manure (1 Mg ha^{-1}), NPK (0.21–0.7 Mg ha^{-1}) and ammonium sulfate fertilizer (0.1–0.34 Mg ha^{-1}) for tomato and maize, and Karate (i.e., lambda-cyhalothrin) and DDT insecticides for tomato only was observed in both schemes. On maize fields, manure was applied at ploughing, and mineral fertilizer at a later growth stage. Insufficient application of mineral fertilizer is common due to the high cost involved [9,38]. The fertilizers were applied twice in tomato and maize fields at 2–3 weeks after planting and later at 4–5 weeks after planting.

2.2. Model Description

AquaCrop is a crop water productivity model that simulates the response of crop yield to water supply and is particularly useful where water limits crop production. The model runs in daily time-steps which provides the basis for investigating the appropriateness of irrigation schedules to meet crop-specific demands in practical scheme operation. Consequently, the AquaCrop-based schedules have a high potential to increase crop water productivity [22]. The AquaCrop model can also simulate the effect of climate variability (including variations in temperature, atmospheric carbon dioxide and available water/rainfall) on crop production [22]. Additional useful features of the model are the ability to separate soil evaporation from crop transpiration and to quantify the capillary rise from shallow groundwater.

2.3. Data Collection and Preparation

The input data required for running AquaCrop were collected from two fields under a tomato–maize rotation system in each irrigation scheme (Figure 3).

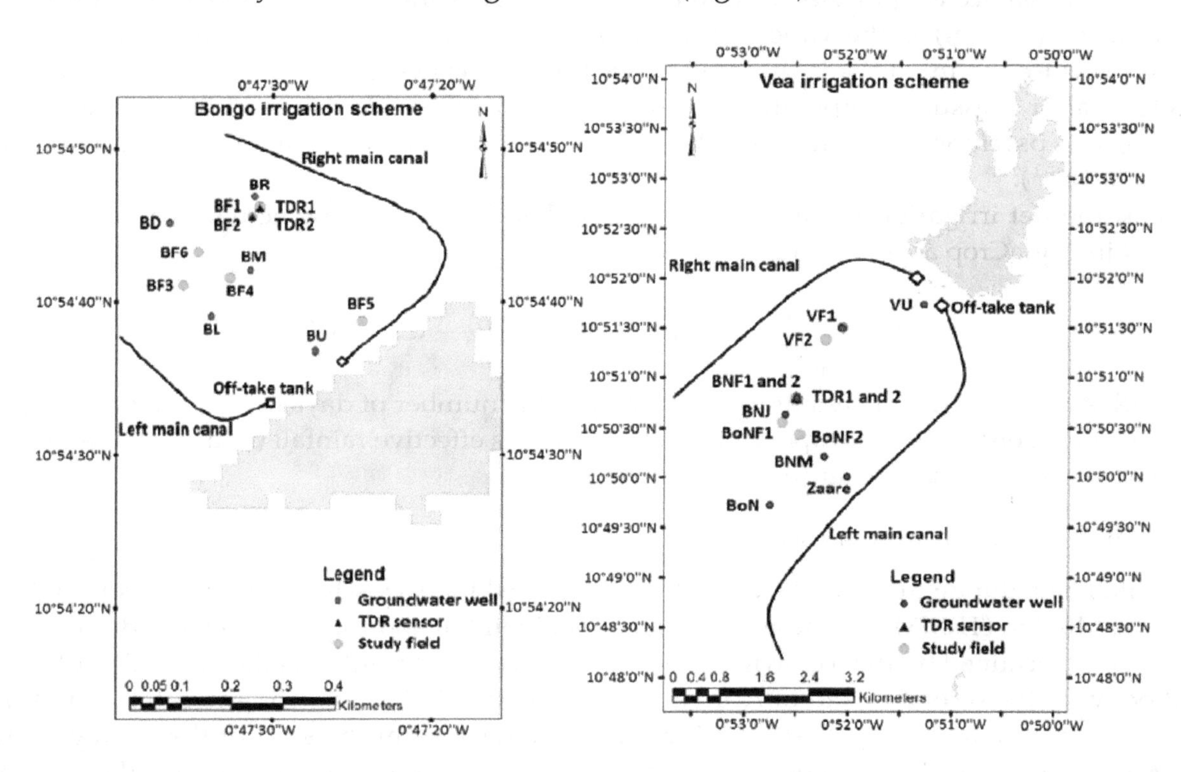

Figure 3. Layout of the irrigation schemes (reproduced from Sekyi-Annan et al. [19]). BF1–6 = Bongo fields, BNF1, 2 = Bongo Nyariga fields, BoNF1, 2 = Bolga Nyariga fields, BR = Bongo right well, BL = Bongo left well, BM = Bongo middle well, BD = Bongo downslope well, BU = Bongo upslope well, VF1, 2 = Vea fields, VU = Vea upslope well, BNM = Bongo Nyariga middle well, BNJ = Bongo Nyariga junction well, TDR1, 2 = Time domain reflectometers.

The model performance, based on the simulation of aboveground dry matter (DM), was assessed with multiple inbuilt statistical indicators including the coefficient of determination (R^2), normalized root mean square error (NRMSE), Nash–Sutcliffe model efficiency coefficient (EF), and Willmott's index of agreement (d). The R^2 indicates the fraction of the variance in observed data explained by the model and ranges from 0 (no agreement) to 1 (perfect agreement) between simulated and observed data. Typically, $R^2 > 0.5$ is acceptable for watershed simulations [39]. The NRMSE signifies the relative difference between the simulated results and the measured data, with NRMSE <10%, 10–20%, 20–30%, and >30% showing excellent, good, fair, and poor model performance, respectively. The EF quantifies the relative magnitude of the residual variance in comparison to the variance of the observed data. EF ranges between 1 and $-\infty$, where 1 signifies a perfect match between predictions and observations, 0 indicates that predictions are as accurate as the observed means, and a negative value indicates poor predictability. The d quantifies the extent to which the measured data are approached by the predictions and ranges from 0 (no agreement) to 1 (perfect agreement).

2.3.1. Estimation of Potential Evapotranspiration and Net Irrigation Requirement

Maximum and minimum air temperatures (T_{max} and T_{min}, °C), average relative humidity (RH, %), wind speed (U, m s^{-1}) and solar radiation (R_s, W m^{-2}) were measured by weather stations located near the study schemes, 10°54′54.1″ N and 0°49′35.3″ W in the BIS, and 10°50′44.6″ N and 0°54′43.9″ W in the VIS. Potential evapotranspiration was calculated based on the Penman–Monteith equation [40]:

$$ET_0 = \frac{1}{\lambda_w}\frac{\Delta(R_n - G) + \rho_a C_p(e_s - e_a)}{\Delta + \gamma_a\left(1 + \frac{r_c}{r_a}\right)} \quad (1)$$

where ET_0 is reference evapotranspiration (mm day^{-1}), R_n is net radiation (W m^{-2}), G is soil heat flux (W m^{-2}), (e_s − e_a) is the vapor pressure deficit of the air (kPa), ρ_a is mean air density at constant pressure (kg m^{-3}), C_p is the specific heat of the air (MJ kg^{-1} $°C^{-1}$), Δ is the slope of the saturation vapor pressure–temperature relationship (kPa $°C^{-1}$), λ_w is latent heat of vaporization (MJ kg^{-1}), γ_a is psychrometric constant (kPa $°C^{-1}$), r_c is crop resistance (s m^{-1}), and r_a is aerodynamic resistance (s m^{-1}).

Next, the net irrigation requirement was calculated based on the actual evapotranspiration simulated in AquaCrop as follows [22,41]:

$$NIR = \sum_{i=1}^{n}[(K_{cb} + K_e)ET_{0_i} - P_{e_i} - CR_i - W_{b_i}] \quad (2)$$

where NIR is the net irrigation requirement (mm), n is the number of days in the crop cycle, K_{cb} is the basal crop coefficient, K_e is the evaporation coefficient, P_e is effective rainfall (mm), CR is capillary rise (mm), and W_b is stored soil water (mm).

2.3.2. Rainfall and Scenario Analyses

Rainfall data during the years 1998–2014 were obtained for each scheme from the Tropical Rainfall Measuring Mission (TRMM) database. The total annual rainfall and total number and duration of dry spells were determined by the following conditions: (i) onset of rainfall is the beginning of a 10 day period between the second dekad of April and the first dekad of May during which the cumulative rainfall is ≥25 mm, and a dry spell ensuing within 30 days from the start of the 10 day period is ≤8 days [1,42]; (ii) cessation of rainfall is the last rainfall event between the third dekad of September and the second dekad of October [42]; (iii) dry spell is two or more consecutive non-rainy days [7], as even a period of two days without rainfall at critical growth stages is detrimental to crop production in savannah environments, particularly during periods of low rainfall; (iv) frequency of dry spells is the number of dry spells during the rainy season in the particular year under focus.

Additionally, the inter- and intra-seasonal variability of rainfall was expressed in the coefficient of variation based on the annual and monthly rainfall data, respectively:

$$CV = \frac{\sigma}{\mu} * 100\ [\%] \quad (3)$$

where CV is the coefficient of variation, σ is the standard deviation, and μ the mean of the rainfall data.

For the estimation of the supplemental irrigation requirement for maize, two climate scenarios (i.e., wet and dry rainfall regimes) were formulated based on rainfall amount and the frequency of dry spells. The first scenario (S1) was a wet year characterized by ≤20% probability of exceedance (i.e., the likelihood of the occurrence of rainfall ≥1057 mm) and by less frequent dry spells [43]. The second scenario (S2) was a dry year characterized by ≥80% probability of rainfall occurrence exceeding 796 mm and by frequent dry spells [43] (Figure 4).

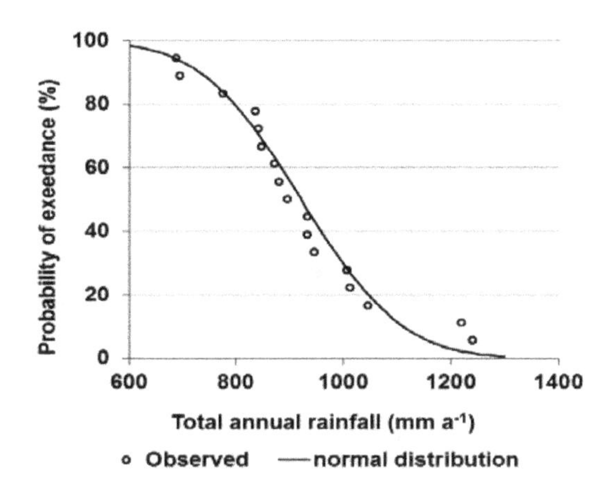

Figure 4. Probability plot of the total annual rainfall for the Vea and Bongo irrigation schemes for 1998–2014.

2.3.3. Soil Characteristics

Three soil profile pits were dug (at BF1 and BF4 in the BIS and at BNF2 in the VIS) to depths of 1.2–1.3 m in the Akrubu and Yaratanga soil series identified in the irrigation schemes. Soil samples were collected from the morphological soil horizons. The soil chemical properties, bulk density, soil moisture at saturation, field capacity, and permanent wilting point were determined in the laboratory [44] (Appendix A). Saturated hydraulic conductivity (K_{sat}) was determined in the laboratory by the falling head method [45] using undisturbed soil cores from the three soil pits. Comparison of the measured K_{sat} values with those determined from the pedo-transfer function based on soil texture and organic matter content [39,46] revealed significantly lower values from the laboratory measurements. This is likely caused by incomplete saturation of the undisturbed soil samples (especially samples with high clay content) before the test, leakage along the metal cylinder during the test, and the impact of soil structure and macropores. Consequently, the K_{sat} values determined from the pedo-transfer functions were used in the analysis.

2.3.4. Crop Growth and Yield Parameters

Sampling areas were demarcated within all selected fields for the collection of total aboveground biomass (AGB) [47]. Three rows were defined for bi-weekly AGB sampling, i.e., four times during the vegetative and reproduction stages. On each sampling day, three samples were collected per field from each defined row by cutting all plants along a 1 m rod. At harvest, two 8 m row sections in each of the selected fields were demarcated for AGB measurement. AGB was sampled and weighed as the yield components, i.e., maize grains and tomato fruits. The samples were weighed, oven-dried at 70–90 °C until a constant weight for at least 72 h, and subsequently, the DM and the yield components were weighed. The planting dates differed from one farmer to another, hence the growth stages of the crops at the time of sampling were not the same. Due to the late start of the field data collection in 2014, the AGB of maize during the vegetative stage was not measured. The harvest index (HI) was estimated as the ratio of the dry yield component to total aboveground DM. The tomato yields in BF1, VF1 and BNF1 could not be assessed during the 2015–2016 season owing to the early onset of the rainy season in 2016, leading to waterlogging and failed tomato yields. Tomato yield measurements were therefore conducted in the neighboring fields characterized by similar soil conditions and farming practices. In the 2014 rainy season, only the BF1 maize field was monitored in the BIS, as the BF6 maize field was not cropped by the farmer. Crop data could not be collected in 2015 in both schemes owing to technical challenges.

Plant density (PD) was determined in all sampling fields. Row spacing was measured as the average distance between two adjacent rows at five random locations in the field [47]. Leaf area index (LAI) was measured bi-weekly using the SunScan probe (SS1-UM-2.0) at five random locations at each field. LAI was converted into canopy cover (CC) using Equation (4), which was developed for maize and soybean but is also applicable to other crops with a similar leaf shape [21,22]:

$$CC = 1.005[1 - \exp(-0.6LAI)]^{1.2} \quad (4)$$

LAI measurements were interrupted in the 2014 rainy season and in the 2015–2016 dry season because of technical challenges. Maximum rooting depth (RD) was measured by manual excavations of at least three plants per crop at harvest time. A summary of all crop growth and yield parameters measured and details of their measurements each season is provided in Appendix B.

2.3.5. Gross Irrigation Amount

At the inlet of each of the selected fields, a Cipoletti weir, or a PVC pipe and a metallic staff gauge (50 cm) with metric graduation, was installed in the canal to measure the water inflow. A discharge equation for flow through the pipes during irrigation events was developed from in-situ measurements through a 'volumetric approach', using a bucket of a known volume (17.5 L) and a stopwatch. The time required to fill the bucket was recorded for seven different water depths read from the staff gauge. Discharges corresponding to the seven measured water depths were computed and, subsequently, discharge (Q, $m^3\ s^{-1}$) was related to water depth (h, m) as follows:

$$Q = 0.073h^{1.334} \quad (5)$$

with $R^2 = 0.972$ and the standard error = 0.001.

The actual water abstraction rate was measured using the volumetric approach in VF1, where pump irrigation was practiced. The discharge was summed up over the irrigation event for the estimation of the gross irrigation amount (GIA) per event.

2.3.6. Groundwater and Capillary Rise

Groundwater was monitored from 1 October, 2014 to 11 May, 2016 to analyze the impact of the groundwater table on water fluxes. Seven georeferenced wells were installed in the irrigable area of the VIS and five in the BIS (Figure 2) at characteristic locations, such as valley bottoms, lateral sites, sites near the dam, and in the middle of the schemes. PVC pipes perforated up to 1 m from the base were used. The depths of the wells ranged from 2.7 to 5.5 m in the VIS and from 2 to 4.9 m in the BIS. An electric contact meter (Seba KLL 077) was used to measure the depth to groundwater table weekly throughout the 2014–2016 observation period. However, measurements could not be carried out between 3 June, 2015 and 15 July, 2015 owing to technical challenges. Because of the late start of the groundwater monitoring in 2014, measurements from the 2015 rainy season were used for the simulation of rainfed maize for 2014.

Capillary rise was estimated in AquaCrop based on soil type and hydraulic characteristics [48] as follows:

$$CR = \exp\left(\frac{\ln(z) - b}{a}\right) \quad (6)$$

where CR is the expected capillary rise in mm day^{-1}, z is the depth to groundwater table in m, and a and b are coefficients specific to the soil type and the hydraulic characteristics.

2.4. Model Parameterization and Validation

The 2014 rainy season dataset from VF1 was used to parameterize the AquaCrop model for maize and the 2014 maize dataset from BF1 was used to validate the model (i.e., inter-farm validation). The 2014 maize crop data from BNF1 were found to be unreliable owing to the effects of waterlogging, and thus were excluded from the analysis. For tomato, the 2014–2015 dry season data from BF1 were used for the parameterization, and inter-farm model validation was performed using 2014–2015 data from BF6. The inter-seasonal validation employed datasets from tomato BF1 collected in 2015–2016. Data from the other tomato fields (VF1 and BNF1) were either unavailable or incomplete owing to technical and environmental (i.e., crop disease attack) challenges in 2014–2015, and the early onset of rainfall destroying the crops in 2016.

The parameters modified in the model were climate, soil characteristics, and agronomic practice (Tables 1 and 2). All the default crop-specific parameters (i.e., yield response factor) for the study crops were used. The climate file in daily time-steps for the period 21 May, 2014 (i.e., beginning of the rainfed farming season in 2014) to 24 May, 2016 (i.e., end of the 2015–2016 dry season farming) was created using the AquaCrop ET_0 file, maximum and minimum temperature file, and a rainfall file.

Table 1. Modified parameters and field data used for the parameterization and validation of the AquaCrop model for maize.

Data Required	Model Parameterization	Inter-Farm Model Validation
Site conditions		
Cropping field	VF1 (2014)	BF1 (2014)
Crop variety	'Obatanpa'	'Obatanpa'
Growing cycle	3 July, 2014–25 August, 2014	24 May, 2014–14 August, 2014
Planting method	Direct sowing	Direct sowing
Soil fertility in relation to biomass	Poor	Poor
Initial canopy cover	High canopy cover	High canopy cover
Maximum canopy cover	Fairly covered	Fairly covered
Maximum rooting depth	0.30 m	0.36 m
Harvest index	0.51	0.53
Crop development	In growing degree days	In growing degree days
Field management		
Soil surface cover	No mulch	No mulch
Soil physical characteristics	Field capacity, wilting point, soil moisture, texture, and thickness of soil layer from soil pit 3 in BNF1	Field capacity, wilting point, soil moisture, texture, and thickness of soil layer from soil pit 1 in BF1
Groundwater level	Weekly depth to groundwater table from VF1 well	Weekly depth to groundwater table from BR well
Simulation period	Calendar of growing cycle	Calendar of growing cycle
Field data file	Aboveground dry matter from VF1	Aboveground dry matter from BF1

Table 2. Modified parameters and field data for the parameterization and validation of the AquaCrop model for tomato.

Data Required	Model Parameterization	Inter-Farm Model Validation	Inter-Seasonal Validation
Site conditions			
Cropping field	BF1 (2014–2015)	BF6 (2014–2015)	BF1 (2015–2016)
Crop variety	'Buffalo'	'Buffalo'	'Buffalo'
Growing cycle	22 October 22, 2014–11 February, 2015	11 November, 2014–6 March, 2015	23 November, 2015–18 March, 2016
Planting method	Transplanting	Transplanting	Transplanting
Soil fertility in relation to biomass	Moderate	Moderate	Moderate
Initial canopy cover	Very small cover	Very small cover	Very small cover
Maximum canopy cover	Fairly covered	Fairly covered	Fairly covered
Maximum rooting depth	0.35 m	0.37 m	0.28 m
Harvest index	0.29	0.29	0.21
Crop development	In growing degree days	In growing degree days	In growing degree days
Field management			
Soil surface cover	No mulch	No mulch	No mulch
Irrigation practice	Irrigation amount per event in mm from BF1	Irrigation amount per event in mm from BF6	Irrigation amount per event in mm from BF6
Soil physical characteristics	Field capacity, wilting point, soil moisture, texture, and thickness of soil layer from soil pit 1 in BF1	Field capacity, wilting point, soil moisture, texture, and thickness of soil layer field from soil pit 2 near BF6	Field capacity, wilting point, soil moisture, texture, and thickness of soil layer from soil pit 1 in BF1
Groundwater level	Weekly depth to groundwater table from BR well	Weekly depth to groundwater table from BD well	Weekly depth to groundwater table from BR well
Simulation period	Calendar of growing cycle	Calendar of growing cycle	Calendar of growing cycle
Field data file	Aboveground dry matter from BF1	Aboveground dry matter from BF6	Aboveground dry matter from BF1

2.5. *Supplemental Irrigation Requirement for Maize*

Irrigation scheduling was simulated for the maize fields (VF1 and BF1) by selecting the 'Net irrigation water requirement' option in AquaCrop, and 50% allowable root zone depletion. The simulation was run to determine the supplemental irrigation requirement under the two aforementioned climate scenarios.

2.6. *Improved Irrigation Scheduling for Tomato*

Datasets from tomato fields (BF1 and BF6) in 2014–2015 were used to optimize the irrigation schedule. Irrigation files for each of the fields were created for the furrow irrigation method. The time criterion selected was 'Allowable depletion of 80% of readily available water' and the irrigation depth criterion used was 'Back to field capacity'. The irrigation water quality was specified as 'excellent' assuming a negligible salinity of irrigation water.

3. Results

3.1. *Rainfall Variability*

Rainfall data revealed a high inter-seasonal variability of rainfall (i.e., 17%) and frequent dry spells lasting for 2–16 days (Figure 5). From 1998 to 2014, the frequency of dry spells in Vea and Bongo ranged between 18 and 28 occurrences. Furthermore, the analysis indicated increasing intra-seasonal rainfall variability in both schemes during the observation period, most likely due to climate change.

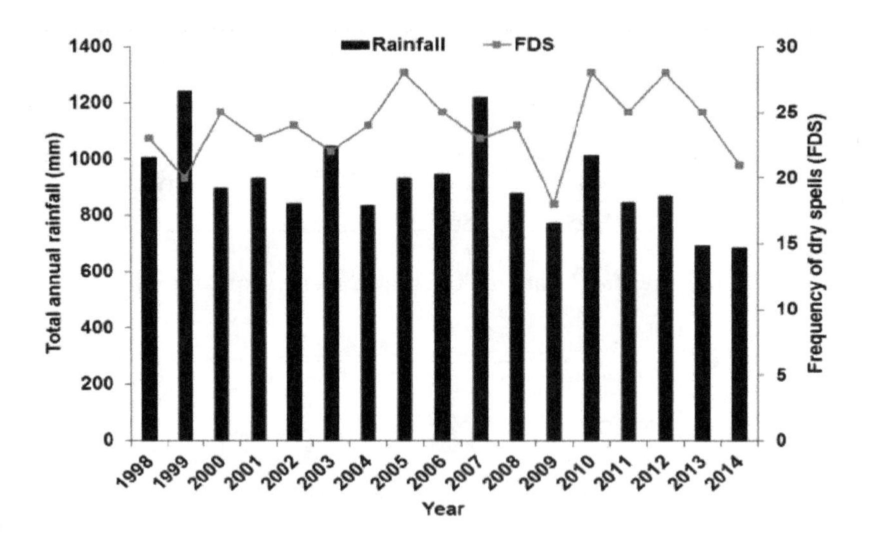

Figure 5. Total annual rainfall and frequency of dry spells (FDS) in the Vea and Bongo irrigation schemes during the years 1998–2014.

3.2. *Crop Growth Parameters*

The PD of maize ranged between 4.1 and 5.5 plants m^{-2} across the schemes (Table 3). The decline in the maize aboveground DM in VF1 and BNF1 in the VIS was attributed to the effects of late planting (i.e., 3 July, 2014) and waterlogging, respectively (Figure 6a). The PD of tomato was generally higher in the BIS (3.3–3.5 plants m^{-2} in 2014–2015, and 3.6–4.2 plants m^{-2} in 2015–2016) than in the VIS (2.6 plants m^{-2} in 2014–2015, and 3.3–3.5 plants m^{-2} in 2015–2016) (Table 3). The difference was partly due to the narrower inter-row spacing observed in the BIS (0.28–0.35 m) compared to that in the VIS (0.25–0.54 m). The remarkably low tomato DM in the Vea BNF1 field in 2014–2015 was due to the impact of plant root disease (Figure 6b).

Table 3. Crop growth and yield components in the Bongo and Vea irrigation schemes during the 2014–2016 observation period.

Crop Type	Farm Label	Plant Density (plants m^{-2})	Maximum Rooting Depth (m)	Fresh Yield (Mg ha^{-1})	Dry Yield (Mg ha^{-1})	Harvest Index
Bongo irrigation scheme				2014 rainy season		
Maize	BF1	4.4	0.36	n.d.	2.9	0.53
				2014–2015 dry season		
Tomato	BF1	3.5	0.35	49.2	2.3	0.29
Tomato	BF6	3.3	0.37	34.3	2.5	n.d.
				2015–2016 dry season		
Tomato	BF1	3.6	0.28	42.8	1.4	0.22
Tomato	BF6	4.2	n.d.	39.6	1.6	0.21
Vea irrigation scheme				2014 rainy season		
Maize	VF1	5.5	0.30	n.d.	2.6	0.51
Maize	BNF1	4.1	0.35	n.d.	1.2	0.41
				2015–2016 dry season		
Tomato	VF1	3.3	0.24	35.3	1.6	0.29
Tomato	BNF1	3.5	0.29	51.3	2.2	0.30

n.d. = not determined/applicable.

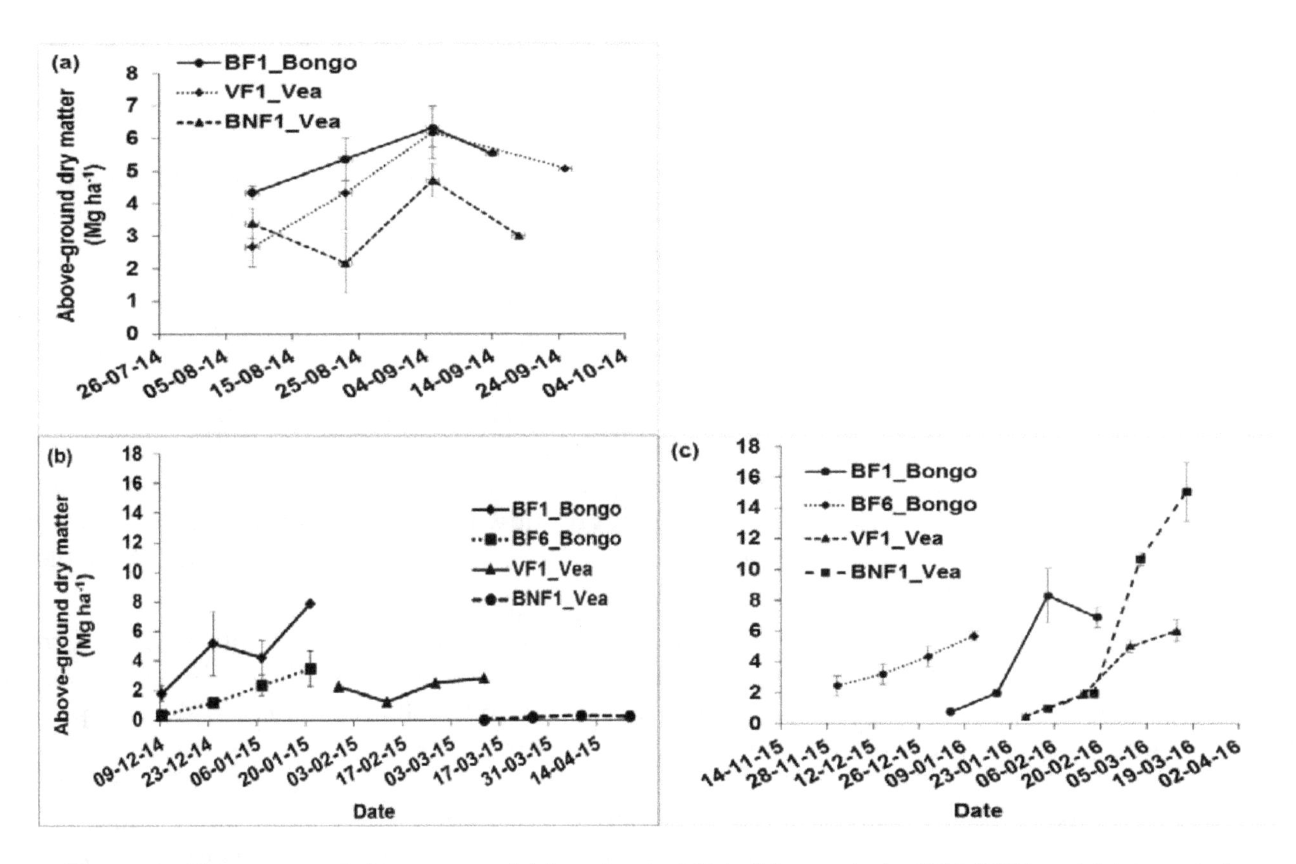

Figure 6. Aboveground dry matter of (**a**) maize in 2014; (**b**) tomato in 2014–2015 and (**c**) tomato in 2015–2016 in the Bongo and Vea irrigation schemes.

The higher tomato DM observed in the 2015–2016 dry season compared to the previous dry season could be due to excessive field-level water application in that season, when an increased water availability was recorded (Figure 6b,c). The downward trend of the LAI of tomato observed in the BIS in 2014–2015 might be due to an insufficient water supply in the later part of the dry season. In contrast, the upward trend of the LAI in the VIS reflects an adequate water supply (Figure 7). The maximum RD of tomato and maize ranged between 0.28 and 0.37 m (Table 1), a result of the shallow soil depth, not exceeding 0.4 m in the UER.

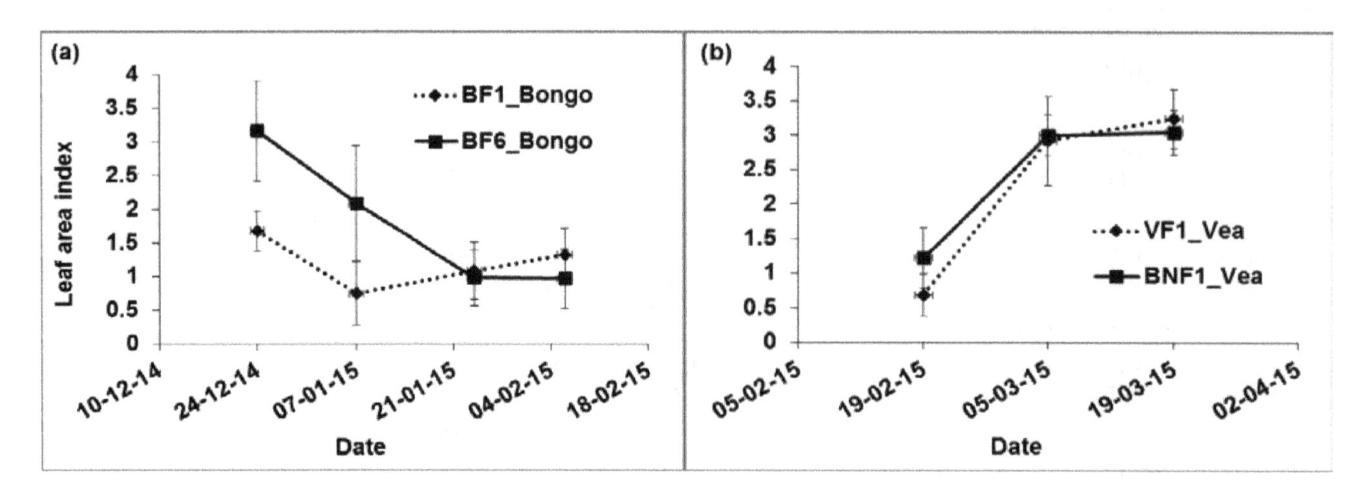

Figure 7. Leaf area index of tomato during the 2014–2015 dry season in the (**a**) Bongo and (**b**) Vea irrigation schemes.

3.3. Crop Yield Components

In 2014, the maize yield ranged between 1.2 and 2.9 Mg ha^{-1}, and the HI ranged from 0.41 to 0.53 across the irrigation schemes (Table 3). A relatively low yield was observed in the BNF1 maize field, in the VIS, possibly due to the combined effect of late planting and waterlogging that occurred in this farm. A relatively high maize yield was recorded in the BIS. The overall range for the annual fresh yields of tomato was 34.3–51.3 Mg ha^{-1} across the irrigation schemes and monitoring periods, and the tomato HI ranged between 0.21 and 0.3.

3.4. Groundwater Level and Capillary Rise

The average depth to groundwater table varied between 0.7 and 2.8 m in the VIS and between 0.6 and 1.3 m in the BIS during 2014–2016 (Figure 8). In the BIS, the soil waterlogging (detected in the BM well) occurred in August and the deepest level (3 m measured in the BD well) was observed in May. The BNF2 well in the VIS recorded the shallowest groundwater level (0.1 m) in August, while the VF1 well measured the deepest groundwater level (3.4 m) in May. The rise in the groundwater table in August most likely resulted from rainfall recharge.

Furthermore, the groundwater level was influenced by nearby streams, reservoirs, and fish ponds. For example, the BU well in the BIS and the VU, BNM and BoN wells in the VIS exhibited stable and relatively shallow groundwater levels due to their proximity to the Bongo reservoir, Vea fish ponds, and streams, even when deep groundwater levels were recorded at other wells (Figure 8). Irrigation events also impacted on the water table. For instance, the groundwater level in the BF1 well in the tomato field increased steadily from the beginning of the dry season and declined from 4 March, 2015, when 2014–2015 dry season irrigation was over. However, the VF1 and BNF2 wells in the VIS in the tomato and leafy vegetable fields, respectively, exhibited rather variable groundwater levels even during the irrigation period and a downward trend after the end of the irrigation period.

The simulated capillary rise into the root-zone of maize was 43–147 mm in 2014, while in the tomato fields it was 18–157 mm in 2014–2015, and 27–263 mm in the 2015–2016 across the irrigation schemes (Figure 9).

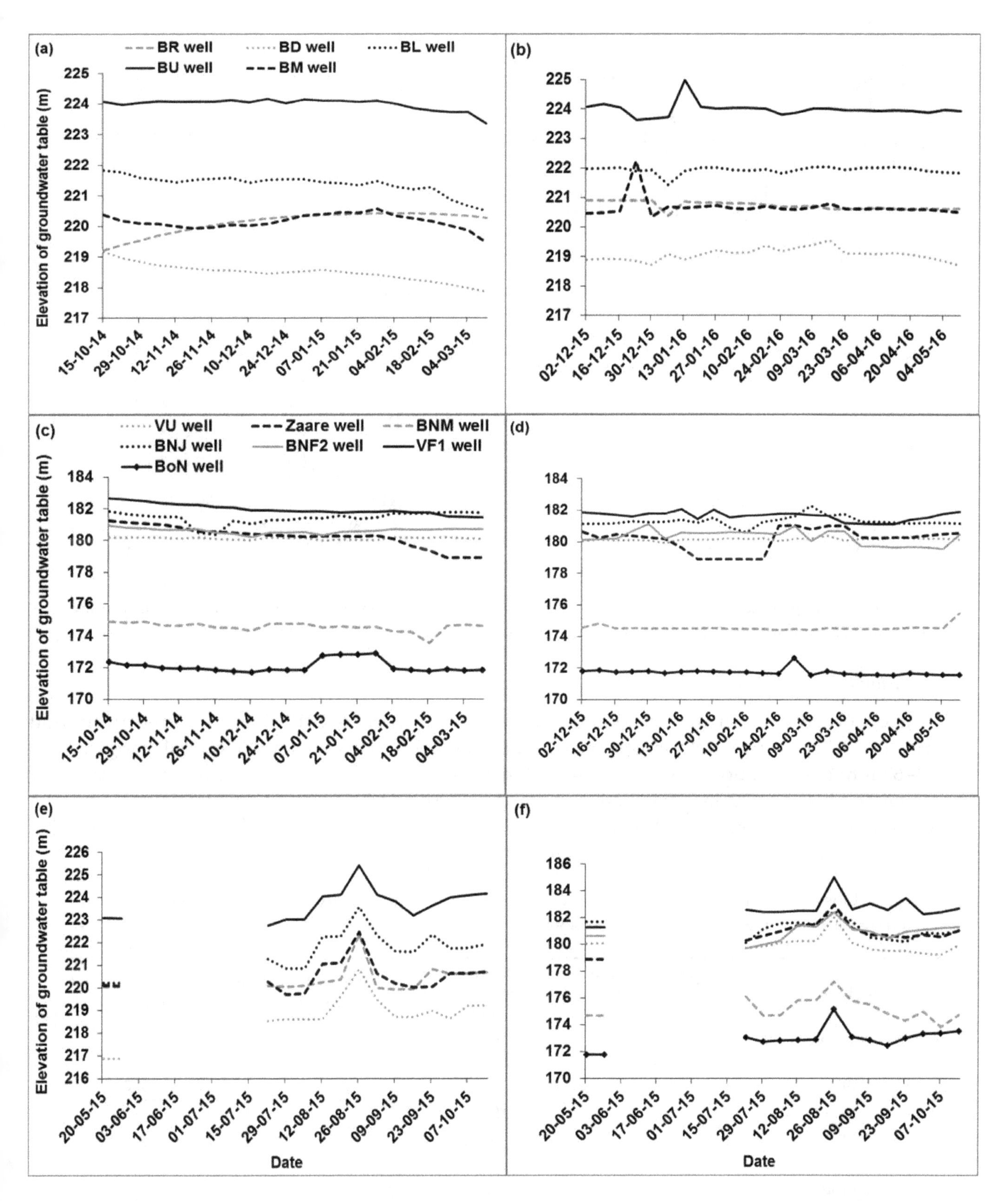

Figure 8. Elevation of the groundwater table during (**a**) 2014–2015 dry season in Bongo (**b**) 2015–2016 dry season in Bongo (**c**) 2014–2015 dry season in Vea (**d**) 2015–2016 dry season in Vea (**e**) 2015 rainy season in Bongo and (**f**) 2015 rainy season in Vea.

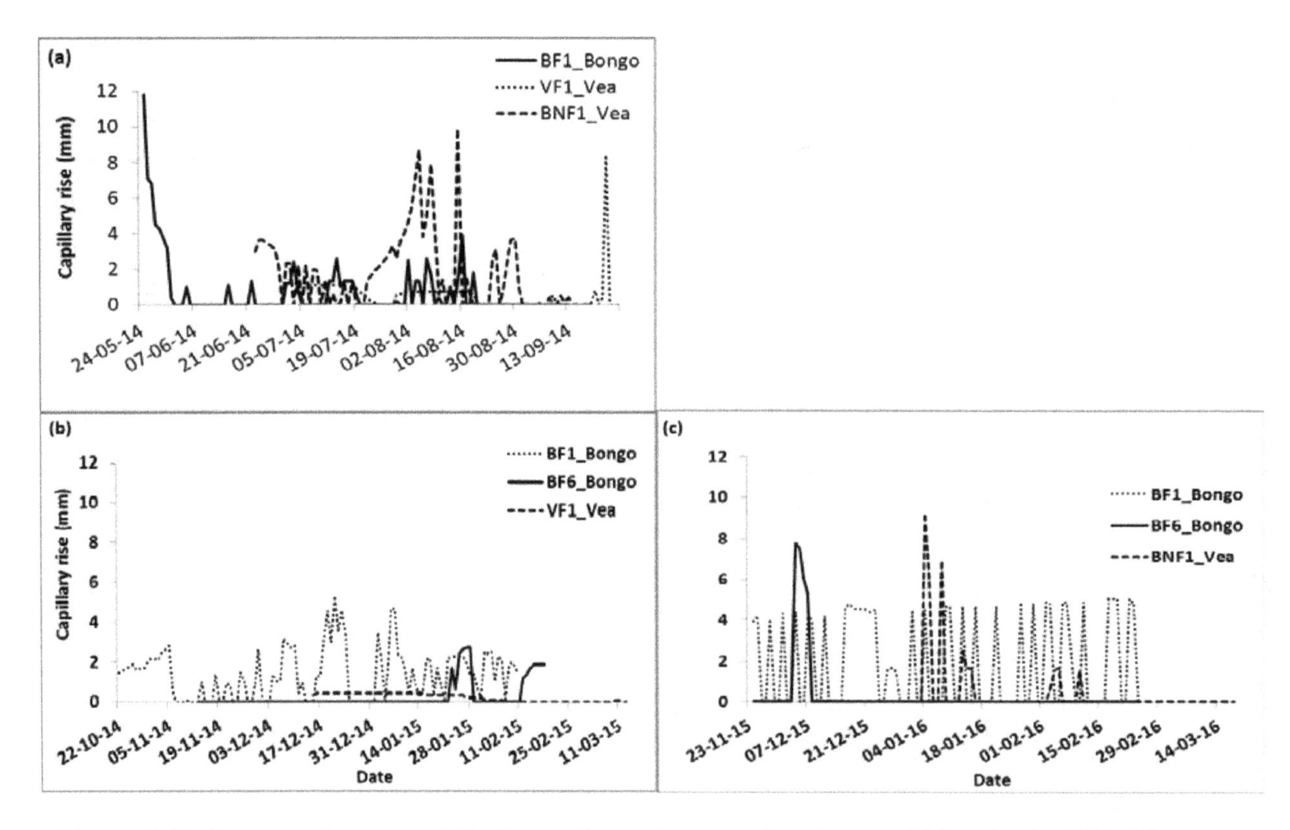

Figure 9. Daily groundwater contribution to the root-zone soil moisture of (**a**) maize in 2014; (**b**) tomato in 2014–2015 and (**c**) 2015–2016 computed by AquaCrop.

3.5. *Traditional Irrigation Scheduling*

The observed GIA for dry season tomato was lower in 2014–2015 than in 2015–2016 in both schemes (Table 4). Across both irrigation schemes and both dry seasons, the overall range of GIA was 21–67 mm per irrigation event and 584–2559 mm per season. The number of irrigation events for tomato ranged between 20 and 29 in both dry seasons. Particularly in the BIS, the irrigation interval in the tomato fields was generally shorter in 2015–2016 than in the previous dry season, owing to the increased availability of water in the Bongo reservoir.

Table 4. Observed field-level irrigation practices and water productivity for tomato in the Bongo and Vea irrigation schemes during the dry seasons.

Field Label	Gross Irrigation Amount Per Season (mm)	Gross Irrigation Amount Per Event (mm)	Average Irrigation Interval (day)	Water Productivity (kg m^{-3})
Bongo irrigation scheme		2014–2015		
BF1	586	19–50	4	8.4
BF6	1247	17–137	5	2.7
		2015–2016		
BF1	1719	20–93	3	2.5
BF6	2559	14–133	2	1.5
Vea irrigation scheme		2014–2015		
VF1	615	13–35	5	n.d.
BNF1	584	21–42	5	n.d.
		2015–2016		
BNF1	1137	33–79	4	4.5

n.d. = not determined.

3.6. Model Performance

The results of model evaluation for tomato DM indicated good agreement (EF = 0.65–0.83, and d = 0.87–0.96) and acceptable error margins (NRMSE = 17.7–42%) (Figure 10c–e).

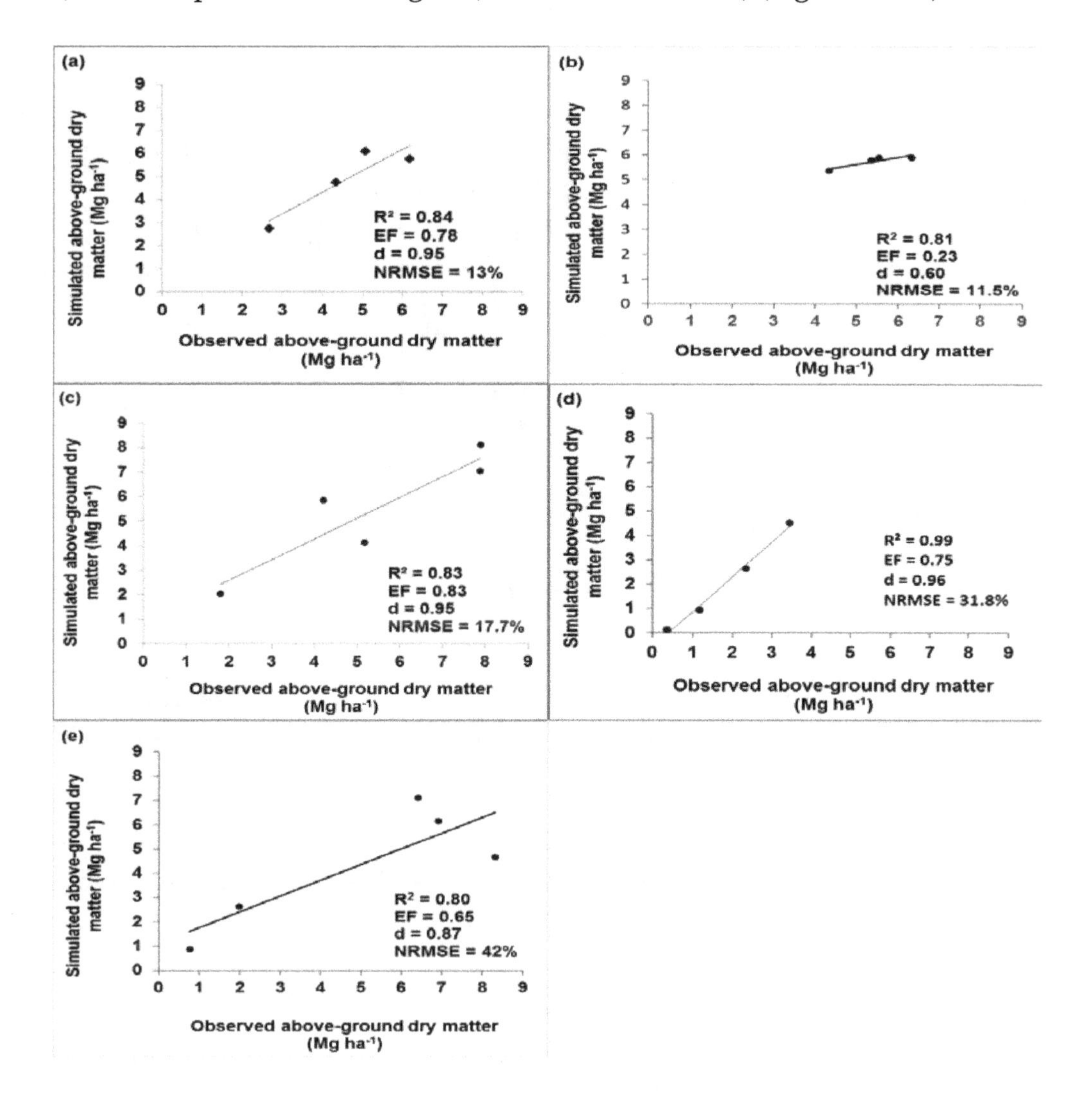

Figure 10. Evaluation of simulated and observed aboveground dry matter of (**a**) maize in VF1 in 2014; (**b**) maize in BF1 in 2014; (**c**) tomato in BF1 in 2014–2015; (**d**) tomato in BF6 in 2014–2015; and (**e**) tomato in BF1 in 2015–2016, in the Vea and Bongo irrigation schemes.

The model evaluation for maize DM also suggested a good agreement (EF = 0.23–0.78, and d = 0.60–0.95) and an acceptable error margin (NRMSE = 11.5–13%). However, the low EF (0.23) for maize in the BF1 field (Figure 10b) could be due to the missing biomass data for the vegetative stage due to the late start of data collection.

3.7. Improved Irrigation Schedule for Tomato

The optimized irrigation schedule for the dry season tomato cropping indicated the need for longer irrigation intervals (6–13 days) in the early crop growth stage and during ripening. In contrast, irrigation intervals should be shorter (2–8 days) in the flowering and yield formation stages (Figure 11). The simulated NIR for tomato ranged from 21 to 29 mm per irrigation event and from 311 to 495 mm per season. The GIA for tomato was estimated as 38–52 mm per irrigation event and 566–900 mm per season, assuming a 55% application efficiency.

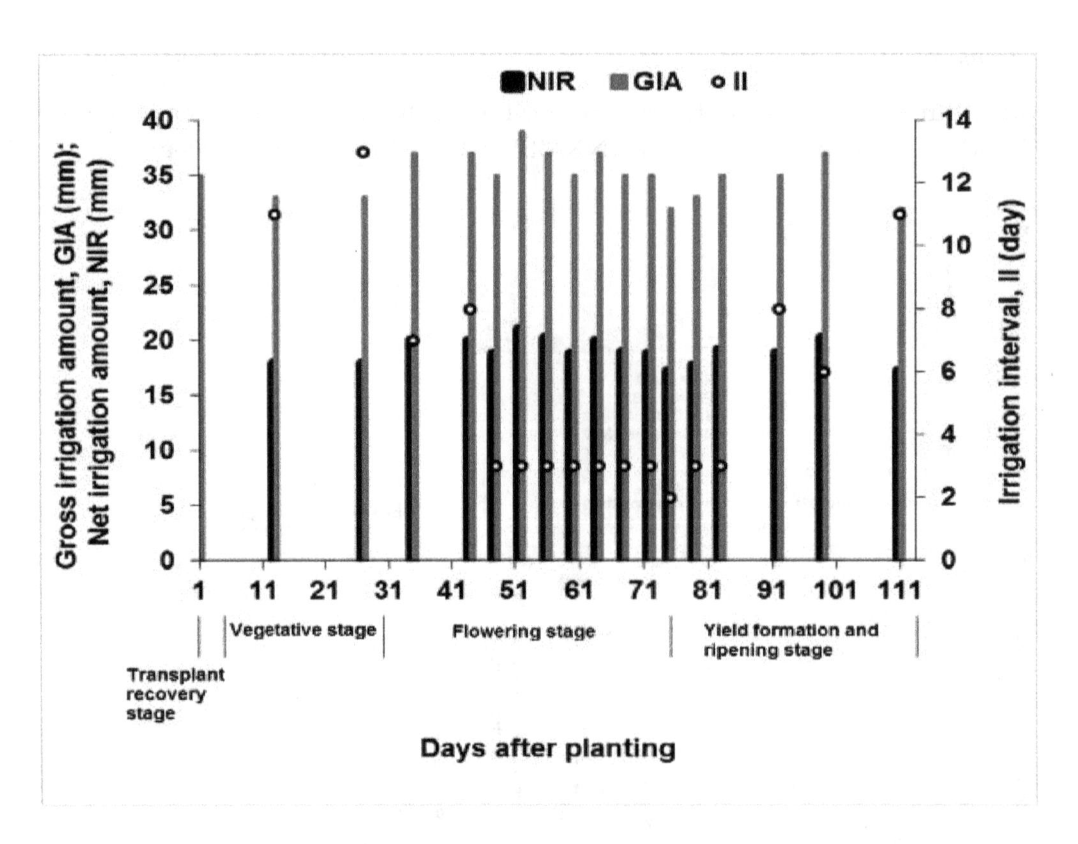

Figure 11. Improved irrigation schedule for tomato cultivation, based on the example of BF1 in the Bongo irrigation scheme. Gross irrigation amount was estimated using a field application efficiency of 55% [19].

The improved irrigation schedule would result in 4–14% yield increment while saving 130–1325 mm (22–52% of GIA) of water, which is otherwise lost through percolation beyond the root zone under the traditional irrigation practice in either scheme (Table 5).

Table 5. Potential water saving and yield increase under the improved irrigation schedule as simulated in AquaCrop.

Field Label	Potential Water Saving (mm)	Tomato Yield under Traditional Irrigation (Mg ha^{-1})	Tomato Yield under Improved Irrigation (Mg ha^{-1})	Potential Yield Increase (%)
BF1 (2014–2015)	130	2.30	2.40	4
BF6 (2014–2015)	775	2.01	2.30	14
BF1 (2015–2016)	1,325	1.58	1.79	14

3.8. Supplemental Irrigation Requirement for Maize

S1 was observed in 1999 when 1240 mm of rainfall and 20 dry spells were recorded in the rainy season, and S2 was observed in 2012 with 871 mm of rainfall and 28 dry spells, the highest frequency of dry spells during the 17 year observation period (Figure 5). Notably, although 2014 recorded the lowest rainfall (687 mm), it was not considered the driest year due to the lower frequency of dry spells (21) compared with 2012. The supplemental irrigation requirement for rainfed maize in the favorable climate scenario S1 was predicted in the range of 88–105 mm (i.e., 25–29% of NIR of maize). The values predicted for S2, the scenario of low rainfall and frequent dry spells, ranged between 107 and 126 mm (i.e., 30–35% of NIR of maize) (Figure 12). The simulated increase in maize yield under supplemental irrigation ranged between 5% and 14%.

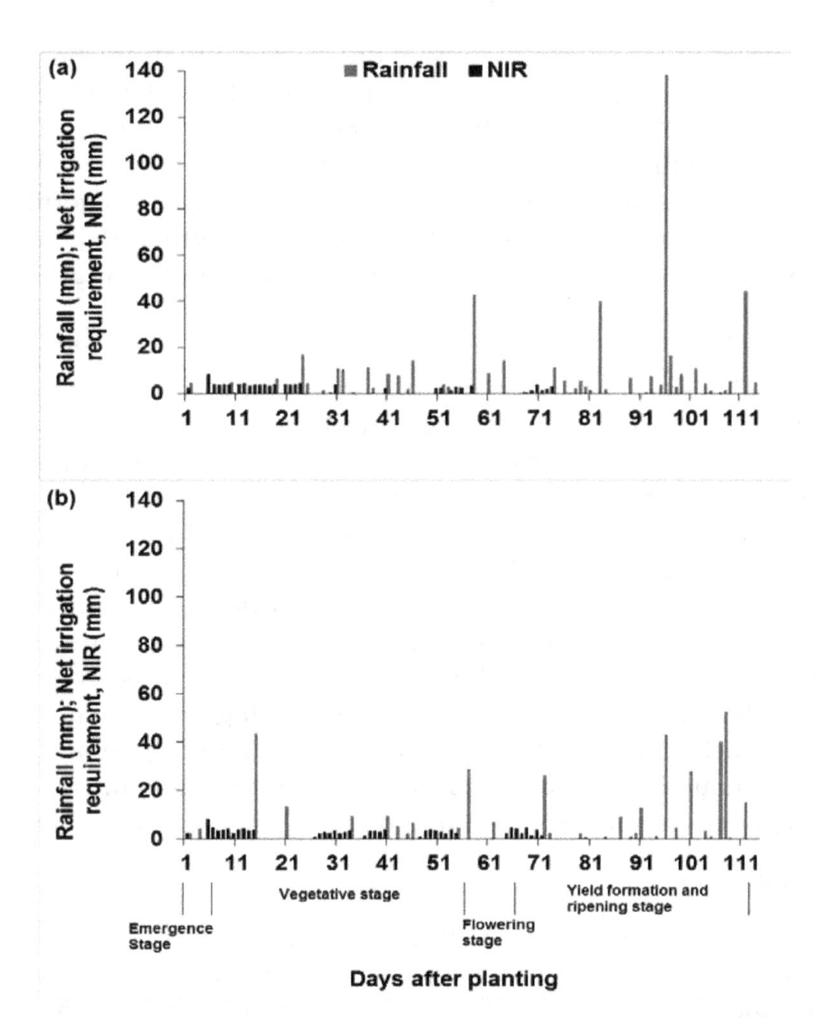

Figure 12. Daily rainfall and net irrigation requirement for supplemental irrigation of maize in the (**a**) 1999 wet year and (**b**) 2012 dry year, as simulated in AquaCrop based on the example of maize BF1 in the Bongo irrigation scheme.

4. Discussion

4.1. Crop Yields

The yields of fresh tomato fruits were similar across schemes but varied remarkably between fields owing to differences in field-level agronomic and irrigation practices and constraints. For instance, the tomato root disease in the VIS field impacted negatively on yield, as the application of insecticides protected only the aboveground biomass. Over-irrigated tomato in 2015–2016 showed high fresh yields and the higher water content of these fruits (i.e., 95% vs. 93% in water-scarce 2014–2015). The values of tomato fresh yield corresponded to the upper ranges measured by [35], reporting 20–36.8 Mg ha^{-1} of fresh tomato yields in the UER, and were greater than the 18 Mg ha^{-1} reported by [49] for rainfed tomato production in the Ashanti and Brong Ahafo regions of Ghana, reflecting the positive impact of irrigation. However, the lower HI of tomato observed in our study (0.21–0.3), compared to values (0.5–0.65) reported by [22] for rainfed tomato in drylands, could be partly due to over-irrigation. The excessive water use was reflected in the low field application efficiencies (30–59%) characteristic of almost all the examined fields [19].

Late planting and waterlogging, due to the lack of drainage facilities, reduced maize grain yield to only 1.2 Mg ha^{-1} in the affected field in the VIS. This observation confirms late planting as one of the causes of sub-optimal yield levels of rainfed maize as rainfall typically declines towards the end of the rainy season. Sallah et al. [50] reported a 30% loss in maize yields due to late planting in northern

Ghana. The observed range of maize grain yields in our study (1.2–2.9 Mg ha^{-1}) was similar to that reported for the fertilized 'Obatanpa' maize variety in Ghana (1.3 to 2.7 Mg ha^{-1}; [38]). The values of [8] for SSA (1.3–1.4 Mg ha^{-1}) were within the lower range of our results. However, Sugri et al. [51] reported the yield potential of the 'Obatanpa' maize variety to be 5.5 Mg ha^{-1} in Ghana. Variations in practices of soil nutrient management and often insufficient applications of fertilizer in the examined fields could also have contributed to the variability in yields. Folberth et al. [9] emphasized that even modest additions of N and P fertilizer might double maize production in most of SSA.

4.2. Irrigation Practice

The examination of field-level irrigation practices during the dry season revealed inappropriate, and in turn, ineffective water application for crop production resulting in over-irrigation in both schemes, mainly due to the lack of consideration of the crop growth stages and water storage characteristics of the soil. Over-irrigation was signified by the high GIA in the water-abundant 2015–2016 season, when farmers in both schemes used more water by shortening irrigation intervals (Table 4), leading to lower water productivity than in the previous, water-scarce season. Because of the lack of appropriate irrigation scheduling and the absence of flow measuring devices in the canals, farmers applied as much water as possible to the tomato crop, and further increased the water application rate with increasing water availability in the reservoir. Faulkner et al. [33] also observed the tendency for excessive water use in response to increasing water availability and attributed this phenomenon to the lack of knowledge of efficient and effective water application at field level. Moreover, the GIAs of tomato in our study were 100–400% larger than the range of values, 274 and 852 mm, previously reported for the UER [34,35], confirming the need for water saving.

4.3. Improved Irrigation Scheduling

The need to adjust irrigation schedules to local hydro-geological conditions is suggested by the modelling analysis, for example, a significant contribution of capillary rise from the groundwater was shown to satisfy the NIR of crops. The groundwater contribution to the NIR of the study crops was highly variable, reflecting the spatial variability in hydro-geological characteristics of the cropping fields. The need to account for this variability complicates the development and application of improved farmer irrigation scheduling in the UER. According to [52], there could be varying contributions of shallow groundwater ($\leq$3 m) to the root-zone soil moisture in fine-textured soils such as those mostly found in the Bongo and Vea irrigation schemes.

The observed increases in tomato yield (i.e., 4–14%) under the improved irrigation schedule most likely resulted from the reduction of the negative effect of over-irrigation on crop yield, as the over-irrigated cropping fields showed the highest potential (14%) to increase yields under the improved irrigation schedule. The simulated magnitude of water saving in the reservoir-based irrigation schemes, which was 22–52% of the GIA under the current irrigation practices, indicates that improving irrigation schedules offers considerable potential for water saving in the dry season in the UER irrigation systems. Overall, however, the improvement of field-level irrigation scheduling alone might not be sufficient for optimizing water productivity and availability in the schemes [16]. To achieve full benefits, equipping irrigation infrastructure with discharge-measuring and dosage structures, and reparation of the decaying water conveyance and distribution sub-systems in the UER would be necessary [18–20]. These interventions to upgrade infrastructure would need to be accompanied by the training of irrigators to handle these facilities, and by further development of water management institutions towards reliably implementing advanced irrigation schedules in order to utilize the full potential of improvements.

4.4. Feasibility of Supplemental Irrigation

The observed temporal variability in rainfall across the irrigation schemes highlights the urgent need for water management strategies to ensure a reduction of the associated risks in rainfed crop production. The intra-seasonal variability of rainfall revealed by the frequency of dry spells was found to be more influential on water demand for crop growth than the total rainfall over the growing season. In the VIS, for example, the supplemental irrigation requirement for maize simulated with the 2014 rainfall was 29 mm, whereas in the wetter year 2012, this value was 107 mm due to the higher frequency of dry spells (28). Similarly, although 1999 was recognized as the wet year for S1 according to the aforementioned criteria, the simulated NIR for supplemental irrigation of maize was 88 mm due to the higher intra-seasonal variability (69%) in that year than in 2014 (64%).

The supplemental irrigation requirement for maize estimated by AquaCrop (29–126 mm) was within the range of values, 20–240 mm, determined by [5] in semi-arid Mwala in Kenya. Furthermore, the temporal rainfall variability was consistent with the findings of [37], who observed high rainfall variability of >25% during the years 1923–1995 in the Sahelian region. Likewise, [7] estimated dry spells lasting for 2–13 days in the Savanna agro-ecological zone of Ghana.

Overall, considering only the crop irrigation sector, the quantity of water saved through improved irrigation scheduling of dry season tomato is largely sufficient to accommodate supplemental irrigation of maize in the rainy season, and thus adapt to rainfall variability and recurrent dry spells. Even for the dry climate scenario of low rainfall coupled with frequent dry spells, about 126 mm of water at field level would be required for the supplemental irrigation of maize during the rainy season. Furthermore, the simulated increase in maize yield upon the introduction of supplemental irrigation offers an incentive for managers of the Bongo and Vea schemes to explore this strategy. Notably, due to the reservoir losses through evaporation and seepage, some of the water saved in the dry season might not be available for supplemental irrigation in the rainy season. Hence, an effective year-round irrigation schedule is required so that supplemental irrigation in the rainy season does not compromise water availability for dry season crop production.

5. Conclusions

High temporal variability in rainfall and frequent dry spells lasting for 2–16 days are common in the UER, requiring adaptive measures to enhance rainfed crop production. The supplemental irrigation requirement for maize under the dry climate scenario of low rainfall and frequent dry spells was estimated between 107 and 126 mm, whereas for periods of high rainfall and rare dry spells, between 88 and 105 mm would be required. These demands can be satisfied via improved irrigation scheduling for dry season tomato that can potentially save 130–1325 mm of water, which would otherwise be lost through percolation and evaporation. Tomato and maize yield increments in the range of 4–14% and 5–14%, respectively, are predicted under the improved irrigation schedule and supplemental irrigation. The AquaCrop model, parameterized using field data collected in the small- and medium-scale reservoir-based irrigation schemes in the Upper East region of Ghana, can be further utilized to improve the irrigation schedule of other cropping systems in the UER. Given the sub-optimal nutrient management practices observed across the study sites, further research should investigate the potential of both soil fertility and water management practices combined for improving crop yields and year-round food security in sub-Saharan Africa.

Author Contributions: E.S.-A., B.T., B.D. and A.K. conceived and designed the field surveys; E.S.-A. performed the surveys; E.S.-A. and B.T. analyzed the data; E.S.-A., B.T., B.D. and A.K. wrote the paper.

Acknowledgments: This study was supported by the German Federal Ministry of Education and Research (BMBF) under the program WASCAL (West African Science Service Center on Climate Change and Adapted Land Use, project No. 00100218). The additional support by the German Academic Exchange Service (DAAD) and Korea University Grant (No. K1608421) is gratefully acknowledged. This paper was presented at the 23rd ICID Congress on Irrigation and Drainage in Mexico City, Mexico.

Appendix A

Table A1. Chemical properties of soils in the Bongo and Vea irrigation schemes.

Morphological Horizon (cm)					Exchangeable Cations (cmol kg^{-1})								Available-Brays	
	pH (1:1 H_2O)	OC (%)	Total N (%)	OM (%)	Ca^{2+}	Mg^{2+}	K^+	Na^+	TEB (cmol kg^{-1})	EA (cmol kg^{-1})	CEC (cmol kg^{-1})	BS (%)	ppmP	ppmK
Pit 1														
0–12	6.5	1.44	0.16	2.48	3.20	0.80	0.08	0.04	4.12	0.15	4.27	96.49	26.95	31.19
12–30	7.0	0.41	0.06	0.71	4.01	1.07	0.04	0.03	5.15	0.15	5.30	97.17	5.50	13.25
30–56	7.2	0.34	0.04	0.59	2.94	1.07	0.03	0.03	4.07	0.05	4.12	98.79	2.55	10.56
56–75	7.1	0.21	0.04	0.36	1.34	1.07	0.02	0.02	2.45	0.50	2.95	83.05	1.99	8.23
75–100	7.9	0.14	0.03	0.24	2.40	1.07	0.04	0.30	3.81	0.03	3.84	99.22	3.11	11.03
100–125	7.9	0.07	0.02	0.12	2.14	1.60	0.03	0.03	3.80	0.03	3.83	99.22	0.88	12.98
Pit 2														
0–20	7.8	1.75	0.16	3.02	15.49	7.34	0.09	0.06	22.98	0.05	23.03	99.78	34.12	35.14
20–44	8.4	0.51	0.07	0.88	9.08	5.07	0.10	0.06	14.31	0.03	14.34	99.79	2.79	37.12
44–80	8.7	0.34	0.04	0.59	7.08	4.01	0.10	0.06	11.25	0.03	11.28	99.73	0.24	38.62
80–120	8.8	0.31	0.04	0.53	8.28	11.21	0.10	0.06	19.65	0.03	19.68	99.85	1.59	34.21
Pit 3														
0–20	7.3	0.72	0.09	1.24	3.74	1.87	0.15	0.08	5.84	0.05	5.89	99.15	13.87	52.48
20–32	7.7	0.48	0.07	0.83	4.81	3.07	0.07	0.04	7.99	0.05	8.04	99.38	28.70	24.31
32–57	8.5	0.45	0.07	0.78	8.01	6.14	0.08	0.04	14.27	0.05	14.32	99.65	0.40	29.67
57–76	8.2	0.41	0.06	0.71	10.68	8.41	0.10	0.06	19.25	0.03	19.28	99.84	0.48	35.29
76–120/128	8.2	0.31	0.05	0.53	18.16	11.35	0.19	0.08	29.78	0.05	29.83	99.83	0.48	70.13

OC = Organic carbon, OM = Organic matter, TEB = Total exchangeable bases, EA = Exchangeable acidity, CEC = Cation exchange capacity, BS = Base saturation.

Table A2. Physical and hydraulic properties of soils in the Bongo and Vea irrigation schemes.

Morphological Horizon (cm)	Soil Texture	Bulk Density (g cm^{-3})	SAT (%)	FC (%)	PWP (%)	TAW (%)	Ksat (mm day^{-1})
Pit 1 (Bongo irrigation scheme)							
0–12	Sandy loam	1.10	49.7	16.9	6.2	10.6	1744
12–30	Loamy sand	1.27	47.5	19.1	4.3	14.9	1816
30–56	Sandy loam	1.26	45.8	19.1	5.8	13.3	1318
56–75	Sandy loam	1.28	46.7	14.1	4.2	9.9	1641
75–100	Sandy loam	1.36	45.2	18.3	6.4	11.9	1109
100–125	Sandy loam	1.44	44.2	17.8	6.3	11.6	885
Pit 2 (Bongo irrigation scheme)							
0–20	Silt loam	1.07	51.5	34.2	11.0	23.2	632
20–44	Loam	1.32	44.6	31.1	12.6	18.5	363
44–80	Loam	1.53	45.1	40.7	18.7	22.0	261
80–120	Loam	1.40	45.7	44.2	14.8	29.4	192
Pit 3 (Vea irrigation scheme)							
0–20	Sandy loam	1.37	47.3	20.3	5.4	14.9	1473
20–32	Sandy loam	1.59	45.3	22.0	8.8	13.2	625
32–57	Loam	1.57	45.1	32.9	14.8	18.1	226
57–76	Loam	1.56	47.0	38.6	14.0	24.6	159
76–128	Clay loam	1.52	49.7	47.3	17.5	29.8	86

SAT = Water content at saturation, FC = Field capacity, PWP = Permanent wilting point, TAW = Total available water, K_{sat} = Saturated hydraulic conductivity determined from pedo-transfer functions.

Appendix B

Table A3. Summary of crop growth and yield parameters and details of their measurements for each study season.

Parameter	Method of Data Collection	Frequency of Data Collection	Cropping Field (Figure 3)
Maize, 2014 rainy season			
Above-ground biomass	Destructive biomass sampling along a 1 m rod on three selected rows Destructive biomass sampling in two 8 m row sections at harvest	Three times during the crop reproduction stage at two weeks interval, and once at harvest time	BF1, VF1, BNF1
Plant density	Counting of total number of plants along the 1 m rod on the three selected rows Estimation of the sampling area	Three times during the crop reproduction stage at two weeks interval	BF1, VF1, BNF1
Row spacing	The average distance between two adjacent rows at five random locations	Once during the reproduction stage	BF1, VF1, BNF1
Maximum rooting depth	Manual excavations of at least three plants per crop	Once at harvest time	BF1, VF1, BNF1
Crop yield	Harvesting and weighing of total maize grain yield from two 8 m row sections	Once at harvest time	BF1, VF1, BNF1
Tomato, 2014–2015 dry season			
Above-ground biomass	Destructive biomass sampling along a 1 m rod on three selected rows Destructive biomass sampling in two 8 m row sections at harvest	Four times during the vegetative and reproduction stages, and once at harvest time	BF1, BF6, VF1, BNF1
Plant density	Counting of total number of plants along the 1 m rod on the three selected rows Estimation of the sampling area	Four times during the vegetative and reproduction stages	BF1, BF6, VF1, BNF1
Leaf area index	Measurements with the SunScan probe (SS1-UM-2.0) at five random locations	Four times during the vegetative and reproduction stages	BF1, BF6, VF1, BNF1
Row spacing	The average distance between two adjacent rows at five random locations	Once at harvest time	BF1, BF6, VF1, BNF1
Maximum rooting depth	Manual excavations of at least three plants per crop	Once at harvest time	BF1, BF6, VF1
Crop yield	Harvesting and weighing of total tomato fruits from two 8 m row sections	Once at harvest time	BF1, BF6

Table A3. *Cont.*

Parameter	Method of Data Collection	Frequency of Data Collection	Cropping Field (Figure 3)
	Tomato, 2015–2016 dry season		
Above-ground biomass	Destructive biomass sampling along a 1 m rod on three selected rows Destructive biomass sampling in two 8 m row sections at harvest	Four times during the vegetative and reproduction stages	BF1, BF6, VF1, BNF1
Plant density	Counting of total number of plants along the 1 m rod on the three selected rows Estimation of the sampling area	Four times during the vegetative and reproduction stages	BF1, BF6, VF1, BNF1
Row spacing	The average distance between two adjacent rows at five random locations	Once during the reproduction stage	BF1, BF6, VF1, BNF1
Maximum rooting depth	Manual excavations of at least three plants per crop	Once at harvest time	BF1, BF6, VF1, BNF1
Crop yield	Harvesting and weighing of total tomato fruits from two 8 m row sections	Once at harvest time	BF6, and fields close to BF1, VF1, BNF1

References

1. Cook, K.H.; Vizy, E.K. Impact of climate change on mid-twenty-first century growing seasons in Africa. *Clim. Dyn.* **2012**, *39*, 2937–2955. [CrossRef]
2. Sylla, M.B.; Nikiema, P.M.; Gibba, P.; Kebe, I.; Klutse, N.A.B. Climate Change over West Africa: Recent Trends and Future Projections. In *Adaptation to Climate Change and Variability in Rural West Africa*; Springer: Cham, Switzerland, 2016; pp. 25–40.
3. Sanfo, S.; Barbier, B.; Dabiré, I.W.P.; Vlek, P.L.G.; Fonta, W.M.; Ibrahim, B.; Barry, B. Rainfall variability adaptation strategies: An ex-ante assessment of supplemental irrigation from farm ponds in southern Burkina Faso. *Agric. Syst.* **2017**, *152*, 80–89. [CrossRef]
4. McCartney, M.; Smakhtin, V. *Water Storage in An Era of Climate Change: Addressing the Challenges of Increasing Rainfall Variability*; International Water Management Institute Blue Paper; IWMI: Colombo, Sri Lanka, 2010; pp. 1–24. Available online: https://www.agriskmanagementforum.org/sites/agriskmanagementforum.org/files/Documents/water%20storage%20in%20era%20of%20climate%20change%20IWMI%20Blue_Paper_2010-final.pdf (accessed on 20 November 2016).
5. Rockström, J.; Barron, J. Water productivity in rainfed systems: Overview of challenges and analysis of opportunities in water scarcity prone savannahs. *Irrig. Sci.* **2007**, *25*, 299–311. [CrossRef]
6. Adwubi, A.; Amegashie, B.K.; Agyare, W.A.; Tamene, L.; Odai, S.N.; Quansah, C.; Vlek, P. Assessing sediment inputs to small reservoirs in Upper East Region, Ghana. *Lakes Reserv. Res. Manag.* **2009**, *14*, 279–287. [CrossRef]
7. Kranjac-Berisavljevic, G.; Abdul-Ghanyu, S.; Gandaa, B.Z.; Abagale, F.K. Dry Spells Occurrence in Tamale, Northern Ghana–Review of Available Information. *J. Disaster Res.* **2014**, *9*, 468–474. [CrossRef]
8. Dzanku, F.M.; Jirström, M.; Marstorp, H. Yield Gap-Based Poverty Gaps in Rural Sub-Saharan Africa. *World Dev.* **2015**, *67*, 336–362. [CrossRef]
9. Folberth, C.; Yang, H.; Gaiser, T.; Abbaspour, K.C.; Schulin, R. Modeling maize yield responses to improvement in nutrient, water and cultivar inputs in sub-Saharan Africa. *Agric. Syst.* **2013**, *119*, 22–34. [CrossRef]
10. Vlek, P.L.G.; Khamzina, A.; Tamene, L. *Land Degradation and the Sustainable Development Goals: Threats and Potential Remedies*; CIAT Publication No. 440; International Center for Tropical Agriculture (CIAT): Nairobi, Kenya, 2017. Available online: http://hdl.handle.net/10568/81313 (accessed on 9 May 2018).
11. Drechsel, P.; Gyiele, L.; Kunze, D.; Cofie, O. Population density, soil nutrient depletion, and economic growth in sub-Saharan Africa. *Ecol. Econ.* **2001**, *38*, 251–258. [CrossRef]
12. Vlek, P.L.G.; Khamzina, A.; Azadi, H.; Bhaduri, A.; Bharati, L.; Braimoh, A.; Martius, C.; Sunderland, T.; Taheri, F. Trade-offs in multi-purpose land use under land degradation. *Sustainability* **2017**, *9*, 2196. [CrossRef]
13. Henao, J.; Baanante, C. *Agricultural Production and Soil Nutrient Mining in Africa: Implications for Resource Conservation and Policy Development*; International Center for Soil Fertility and Agricultural Development: Muscle Shoals, AL, USA, 2006.

14. Droogers, P.; Aerts, J. Adaptation strategies to climate change and climate variability: A comparative study between seven contrasting river basins. *Phys. Chem. Earth Parts ABC* **2005**, *30*, 339–346. [CrossRef]
15. Molden, D.; Oweis, T.; Steduto, P.; Bindraban, P.; Hanjra, M.A.; Kijne, J. Improving agricultural water productivity: Between optimism and caution. *Agric. Water Manag.* **2010**, *97*, 528–535. [CrossRef]
16. Mustapha, A.B. Effect of Dryspell Mitigation with Supplemental Irrigation on Yield and Water Use Efficiency of Pearl Millet in Dry Sub-Humid Agroecological Condition of Maiduguri. In *Proceedings of the 2nd International Conference on Environment Science and Biotechnology;* IACSIT Press: Singapore, 2012; pp. 46–49.
17. Zwart, S.J.; Bastiaanssen, W.G.M. Review of measured crop water productivity values for irrigated wheat, rice, cotton and maize. *Agric. Water Manag.* **2004**, *69*, 115–133. [CrossRef]
18. Ali, M.H.; Talukder, M.S.U. Increasing water productivity in crop production—A synthesis. *Agric. Water Manag.* **2008**, *95*, 1201–1213. [CrossRef]
19. Sekyi-Annan, E.; Tischbein, B.; Diekkrüger, B.; Khamzina, A. Performance evaluation of reservoir-based irrigation schemes in the Upper East region of Ghana. *Agric. Water Manag.* **2018**, *202*, 134–145. [CrossRef]
20. Pereira, L.S. Relating water productivity and crop evapotranspiration. *Options Méditerr. Ser. B* **2007**, *57*, 31–49.
21. Greaves, G.E.; Wang, Y.-M. Assessment of FAO AquaCrop Model for Simulating Maize Growth and Productivity under Deficit Irrigation in a Tropical Environment. *Water* **2016**, *8*, 557. [CrossRef]
22. Steduto, P.; Hsiao, T.C.; Fereres, E.; Reas, D. *Crop Yield Response to Water*; Food and Agriculture Organization of the United Nations: Rome, Italy, 2012. Available online: http://www.fao.org/docrep/016/i2800e/i2800e.pdf (accessed on 13 November 2016).
23. Sekyi-Annan, E.; Acheampong, E.N.; Ozor, N. Modeling the Impact of Climate Variability on Crops in Sub-Saharan Africa. In *Quantification of Climate Variability, Adaptation and Mitigation for Agricultural Sustainability*; Springer International Publishing: Cham, Switzerland, 2017; pp. 39–70.
24. Gaydon, D.S.; Balwinder-Singh; Wang, E.; Poulton, P.L.; Ahmad, B.; Ahmed, F.; Akhter, S.; Ali, I.; Amarasingha, R.; Chaki, A.K.; et al. Evaluation of the APSIM model in cropping systems of Asia. *Field Crops Res.* **2017**, *204*, 52–75. [CrossRef]
25. Sommer, R.; Kienzler, K.; Christopher, C.; Ibragimov, N.; Lamers, J.; Martius, C.; Vlek, P. Evaluation of the CropSyst model for simulating the potential yield of cotton. *Agron. Sustain. Dev.* **2008**, *28*, 345–354. [CrossRef]
26. Fortes, P.S.; Teodoro, P.R.; Campos, A.A.; Mateus, P.M.; Pereira, L.S. Model tools for irrigation scheduling simulation: WINISAREG and GISAREG. In *Irrigation Management for Combating Desertification in the Aral Sea Basin. Assessment and Tools*; Vita Color Publication: Tashkent, Uzbekistan, 2005; pp. 81–96.
27. Jones, J.W.; Hoogenboom, G.; Porter, C.H.; Boote, K.J.; Batchelor, W.D.; Hunt, L.A.; Wilkens, P.W.; Singh, U.; Gijsman, A.J.; Ritchi, J.T. The DSSAT cropping system model. *Eur. J. Agron.* **2003**, *18*, 235–265. [CrossRef]
28. Surendran, U.; Sushanth, C.M.; Mammen, G.; Joseph, E.J. Modelling the Crop Water Requirement Using FAO-CROPWAT and Assessment of Water Resources for Sustainable Water Resource Management: A Case Study in Palakkad District of Humid Tropical Kerala, India. *Aquat. Procedia* **2015**, *4*, 1211–1219. [CrossRef]
29. Wang, X.C.; Li, J. Evaluation of crop yield and soil water estimates using the EPIC model for the Loess Plateau of China. *Math. Comput. Model.* **2010**, *51*, 1390–1397. [CrossRef]
30. Wellens, J.; Raes, D.; Traore, F.; Denis, A.; Djaby, B.; Tychon, B. Performance assessment of the FAO AquaCrop model for irrigated cabbage on farmer plots in a semi-arid environment. *Agric. Water Manag.* **2013**, *127*, 40–47. [CrossRef]
31. Walker, S.; Bello, Z.A.; Mabhaudhi, T.; Modi, A.T.; Beletse, Y.G.; Zuma-Netshiukhwi, G. Calibration of AquaCrop Model to predict water requirements of African vegetables. *Acta Hortic.* **2013**, *1007*, 943–949. [CrossRef]
32. Mabhaudhi, T.; Modi, A.T.; Beletse, Y.G. Parameterisation and evaluation of the FAO-AquaCrop model for a South African taro (*Colocasia esculenta* L. Schott) landrace. *Agric. For. Meteorol.* **2014**, *192–193*, 132–139. [CrossRef]
33. Faulkner, J.W.; Steenhuis, T.; van de Giesen, N.; Andreini, M.; Liebe, J.R. Water use and productivity of two small reservoir irrigation schemes in Ghana's upper east region. *Irrig. Drain.* **2008**, *57*, 151–163. [CrossRef]
34. Mdemu, M.V. Water Productivity in Medium and Small Reservoirs in the Upper East Region (UER) of Ghana. Ph.D. Thesis, University of Bonn, Bonn, Germany, 2008. Available online: http://hss.ulb.uni-bonn.de/2008/1362/1362.pdf (accessed on 21 October 2016).

35. Barry, B.; Forkuor, G. *Contribution of Informal Shallow Groundwater Irrigation to Livelihoods Security and Poverty Reduction in the White Volta Basin (WVB): Current Status and Future Sustainability*; International Water Management Institute: Colombo, Sri Lanka, 2010. Available online: https://cgspace.cgiar.org/bitstream/handle/10568/3940/PN65_IWMI_Project%20Report_May10_final.pdf?sequence=1&isAllowed=y (accessed on 22 February 2017).
36. Asres, S.B. Evaluating and enhancing irrigation water management in the upper Blue Nile basin, Ethiopia: The case of Koga large scale irrigation scheme. *Agric. Water Manag.* **2016**, *170*, 26–35. [CrossRef]
37. Fox, P.; Rockström, J. Supplemental irrigation for dry-spell mitigation of rainfed agriculture in the Sahel. *Agric. Water Manag.* **2003**, *61*, 29–50. [CrossRef]
38. Srivastava, A.K.; Mboh, C.M.; Gaiser, T.; Ewert, F. Impact of climatic variables on the spatial and temporal variability of crop yield and biomass gap in Sub-Saharan Africa—A case study in Central Ghana. *Field Crops Res.* **2017**, *203*, 33–46. [CrossRef]
39. Raes, D.; Steduto, P.; Hsiao, T.C.; Fereres, E. Chapter 2—Users guide. In *Reference Manual: AquaCrop, Version 4.0*; FAO, Land and Water Division: Rome, Italy, 2012; pp. 1–164.
40. Allen, R.G.; Pereira, L.S.; Raes, D.; Smith, M. *Crop Evapotranspiration: Guidelines for Computing Crop Water Requirements*; Food and Agricultural Organization of the United Nations: Rome, Italy, 1998. Available online: https://www.unirc.it/documentazione/materiale_didattico/1462_2016_412_24101.pdf (accessed on 15 November 2016).
41. Doorenbos, J.; Pruitt, W.O.; Aboukhaled, A.; Damagnez, J.; Dastane, N.G.; Van Den Berg, C.; Rijtema, P.E.; Ashford, O.M.; Frère, M. *Crop Water Requirements*; FAO Irrigation and Drainage Paper No. 24; Food and Agriculture Organization of the United Nations: Rome, Italy, 1992; pp. 70–72.
42. Amekudzi, L.; Yamba, E.; Preko, K.; Asare, E.; Aryee, J.; Baidu, M.; Codjoe, S. Variabilities in Rainfall Onset, Cessation and Length of Rainy Season for the Various Agro-Ecological Zones of Ghana. *Climate* **2015**, *3*, 416–434. [CrossRef]
43. Raes, D. *Frequency Analysis of Rainfall Data*; International Centre for Theoretical Physics (ICTP): Leuven, Belgium, 2004. Available online: http://indico.ictp.it/event/a12165/session/21/contribution/16/material/0/0.pdf (accessed on 3 January 2017).
44. Mbah, C.N. Determining the field capacity, wilting point and available water capacity of some Southeast Nigerian soils using soil saturation from capillary rise. *Niger. J. Biotechnol.* **2012**, *24*, 41–47.
45. Pedescoll, A.; Samsó, R.; Romero, E.; Puigagut, J.; García, J. Reliability, repeatability and accuracy of the falling head method for hydraulic conductivity measurements under laboratory conditions. *Ecol. Eng.* **2011**, *37*, 754–757. [CrossRef]
46. Saxton, K.E.; Rawls, W.J. Soil Water Characteristic Estimates by Texture and Organic Matter for Hydrologic Solutions. *Soil Sci. Soc. Am. J.* **2006**, *70*, 1569–1578. [CrossRef]
47. Bell, M.A.; Fischer, R.A. *Guide to Plant and Crops Sampling: Measurements and Observations for Agronomic and Physiological Research in Small Grain Cereals*; CIMMYT: Mexico City, Mexico, 1994. Available online: http://libcatalog.cimmyt.org/download/cim/53067.pdf (accessed on 11 November 2016).
48. Raes, D.; Steduto, P.; Hsiao, T.C.; Fereres, E. Chapter 3: Calculation procedures. In *Reference Manual: AquaCrop, Version 4.0*; FAO, Land and Water Division: Rome, Italy, 2012; pp. 1–130.
49. Adu-Dapaah, H.K.; Oppong-Konadu, E.Y. Tomato production in four major tomato-growing districts in Ghana: Farming practices and production constraints. *Ghana J. Agric. Sci.* **2002**, *35*, 11–22. [CrossRef]
50. Sallah, P.Y.K.; Twumasi-Afriyie, S.; Kasei, C. Optimum planting dates for four maturity groups of maize varieties grown in the Guinea savanna zone. *Ghana J. Agric. Sci.* **1997**, *30*, 63–69. [CrossRef]
51. Sugri, I.; Kanton, R.A.L.; Kusi, F.; Nutsugah, S.K.; Buah, S.S.J.; Zakaria, M. Influence of Current Seed Programme of Ghana on Maize (*Zea mays*) Seed Security. *Res. J. Seed Sci.* **2013**, *6*, 29–39. [CrossRef]
52. Bos, M.G.; Kselik, R.A.; Allen, R.G.; Molden, D. Capillary Rise. In *Water Requirements for Irrigation and the Environment*; Springer Netherlands: Dordrecht, The Netherlands, 2009; pp. 103–118.

PERMISSIONS

All chapters in this book were first published in MDPI; hereby published with permission under the Creative Commons Attribution License or equivalent. Every chapter published in this book has been scrutinized by our experts. Their significance has been extensively debated. The topics covered herein carry significant findings which will fuel the growth of the discipline. They may even be implemented as practical applications or may be referred to as a beginning point for another development.

The contributors of this book come from diverse backgrounds, making this book a truly international effort. This book will bring forth new frontiers with its revolutionizing research information and detailed analysis of the nascent developments around the world.

We would like to thank all the contributing authors for lending their expertise to make the book truly unique. They have played a crucial role in the development of this book. Without their invaluable contributions this book wouldn't have been possible. They have made vital efforts to compile up to date information on the varied aspects of this subject to make this book a valuable addition to the collection of many professionals and students.

This book was conceptualized with the vision of imparting up-to-date information and advanced data in this field. To ensure the same, a matchless editorial board was set up. Every individual on the board went through rigorous rounds of assessment to prove their worth. After which they invested a large part of their time researching and compiling the most relevant data for our readers.

The editorial board has been involved in producing this book since its inception. They have spent rigorous hours researching and exploring the diverse topics which have resulted in the successful publishing of this book. They have passed on their knowledge of decades through this book. To expedite this challenging task, the publisher supported the team at every step. A small team of assistant editors was also appointed to further simplify the editing procedure and attain best results for the readers.

Apart from the editorial board, the designing team has also invested a significant amount of their time in understanding the subject and creating the most relevant covers. They scrutinized every image to scout for the most suitable representation of the subject and create an appropriate cover for the book.

The publishing team has been an ardent support to the editorial, designing and production team. Their endless efforts to recruit the best for this project, has resulted in the accomplishment of this book. They are a veteran in the field of academics and their pool of knowledge is as vast as their experience in printing. Their expertise and guidance has proved useful at every step. Their uncompromising quality standards have made this book an exceptional effort. Their encouragement from time to time has been an inspiration for everyone.

The publisher and the editorial board hope that this book will prove to be a valuable piece of knowledge for researchers, students, practitioners and scholars across the globe.

LIST OF CONTRIBUTORS

Paola Sánchez-Bravo, Luis Noguera-Artiaga, Esther Sendra and Ángel A. Carbonell-Barrachina
Research Group "Food Quality and Safety (CSA)", Department of Agro-Food Technology, Escuela Politécnica Superior de Orihuela (EPSO), Universidad Miguel Hernández de Elche (UMH), Ctra. de Beniel, km 3.2, 03312 Orihuela, Alicante, Spain

Edgar Chambers V and Edgar Chambers IV
Center for Sensory Analysis and Consumer Behavior, Kansas State University, Manhattan, KS 66502, USA

Zied Hammami and Asad S. Qureshi
International Center for Biosaline Agriculture (ICBA), Dubai, UAE

Ali Sahli, Fatma Ezzahra Ben Azaiez, Sawsen Ayadi and Youssef Trifa
Laboratory of Genetics and Cereal Breeding, National Agronomic Institute of Tunisia, Carthage University, 43 Avenue Charles Nicole, 1082 Tunis, Tunisia

Arnaud Gauffreteau
INRA–INA-PG–AgroParisTech, UMR 0211, Avenue Lucien Brétignières, F-78850 Thiverval Grignon, France

Zoubeir Chamekh
Laboratory of Genetics and Cereal Breeding, National Agronomic Institute of Tunisia, Carthage University, 43 Avenue Charles Nicole, 1082 Tunis, Tunisia
Carthage University, National Agronomic Research Institute of Tunisia, LR16INRAT02, Hédi Karray, 1082 Tunis, Tunisia

Stanslaus Terengia Materu, Andrew Tarimo and Siza D. Tumbo
Department of Engineering Sciences and Technology, Sokoine University of Agriculture, Chuo Kikuu, Morogoro, Tanzania

Sanjay Shukla and Rajendra P. Sishodia
Agricultural and Biological Engineering Department, University of Florida, 2685 State Road 29 N, Immokalee, FL 34142, USA

Pavel Trifonov and Gilboa Arye
French Associates Institute for Agriculture and Biotechnology of Drylands, Jacob Blaustein Institutes for Desert Research, Ben Gurion University of the Negev, Sede Boqer Campus, Midreshet Ben-Gurion 84990, Israel

John Rohit Katuri
French Associates Institute for Agriculture and Biotechnology of Drylands, Jacob Blaustein Institutes for Desert Research, Ben Gurion University of the Negev, Sede Boqer Campus, Midreshet Ben-Gurion 84990, Israel
Department of Agronomy, Directorate of Crop Management, Tamil Nadu Agricultural University, Coimbatore, Tamil Nadu 641003, India

Tamara Avellán
United Nations University Institute for Integrated Management of Material Fluxes and of Resources (UNU-FLORES), Ammonstrasse 74, 01067 Dresden, Germany

Niels Schütze
Institute of Hydrology and Meteorology, Technische Universität Dresden, 01069 Dresden, Germany

Agossou Gadédjisso-Tossou
United Nations University Institute for Integrated Management of Material Fluxes and of Resources (UNU-FLORES), Ammonstrasse 74, 01067 Dresden, Germany
Institute of Hydrology and Meteorology, Technische Universität Dresden, 01069 Dresden, Germany

Arindam Malakar
Nebraska Water Center, part of the Robert B. Daugherty Water for Food Global Institute, 109 Water Sciences Laboratory, University of Nebraska, Lincoln, NE 68583-0844, USA

Daniel D. Snow
School of Natural Resources and Nebraska Water Center, part of the Robert B. Daugherty Water for Food Global Institute, 202 Water Sciences Laboratory, University of Nebraska, Lincoln, NE 68583-0844, USA

Chittaranjan Ray
Nebraska Water Center, part of the Robert B. Daugherty Water for Food Global Institute 2021 Transformation Drive, University of Nebraska, Lincoln, NE 68588-6204, USA

Hakan Kadioglu
School of Natural Resource Sciences, North Dakota State University, Fargo, ND 58105, USA

Harlene Hatterman-Valenti and Halis Simsek
Agricultural & Biosystems Engineering, North Dakota State University, Fargo, ND 58105, USA

Xinhua Jia
Plant Science, North Dakota State University, Fargo, ND 58105, USA

Xuefeng Chu
Civil & Environmental Engineering, North Dakota State University, Fargo, ND 58105, USA

Hakan Aslan
Farm Structure and Irrigation, Ondokuz Mayis University, 55270 Samsun, Turkey

Chantha Oeurng
Faculty of Hydrology and Water Resources Engineering, Institute of Technology of Cambodia, Russian Federation Bd, Phnom Penh 12156, Cambodia

Aurore Degré
BIOSE, Gembloux Agro-Bio Tech, Liège University, Passage des Déportés 2, Gembloux 5030, Belgium

Pinnara Ket
Faculty of Hydrology and Water Resources Engineering, Institute of Technology of Cambodia, Russian Federation Bd, Phnom Penh 12156, Cambodia
BIOSE, Gembloux Agro-Bio Tech, Liège University, Passage des Déportés 2, Gembloux 5030, Belgium

Sarah Garré
TERRA, Gembloux Agro-Bio Tech, Liège University, Passage des Déportés 2, Gembloux 5030, Belgium

Lyda Hok
Department of Soil Science, Faculty of Agronomy, Royal University of Agriculture, Phnom Penh 12401, Cambodia

Juan F. Velasco-Muñoz, José A. Aznar-Sánchez and Ana Batlles-delaFuente
Department of Economy and Business, Research Centre CAESCG and CIAIMBITAL, University of Almería, 04120 Almería, Spain

Maria Dolores Fidelibus
Department of Civil, Environmental, Land, Building Engineering and Chemistry, Polytechnic University of Bari, 70126 Bari, Italy

Om Prakash Sharma
Department of Chemistry, Geosciences and Physics, Tarleton State University, Stephenville, TX 76402, USA

Narayanan Kannan and Bijay Kumar Pokhrel
Texas Institute for Applied Environmental Research (TIAER), Tarleton State University, Stephenville, TX 76402, USA

Scott Cook
Department of Mathematics, Tarleton State University, Stephenville, TX 76402, USA

Cameron McKenzie
Department of History, Sociology, Geography and GIS, Tarleton State University, Stephenville, TX 76402, USA

Yubing Fan and Seong C. Park
Texas A&M AgriLife Research, Vernon, TX 76384, USA

Raymond Massey
Division of Applied Social Sciences, University of Missouri, Columbia, MO 65211, USA

Bernhard Tischbein
Department of Ecology and Natural Resources Management, Center for Development Research, University of Bonn, Genscherallee 3, 53113 Bonn, Germany

Ephraim Sekyi-Annan
Department of Ecology and Natural Resources Management, Center for Development Research, University of Bonn, Genscherallee 3, 53113 Bonn, Germany
CSIR-Soil Research Institute, Academy Post Office, Private Mail Bag, Kwadaso-Kumasi, Ghana

Bernd Diekkrüger
Department of Geography, University of Bonn, Meckenheimer Allee 166, 53115 Bonn, Germany

Asia Khamzina
Division of Environmental Science and Ecological Engineering, College of Life Science and Biotechnology, Korea University, 145 Anam-Ro, Seongbuk-Gu, Seoul 02841, Korea

Index

Printed in the USA
CPSIA information can be obtained
at www.ICGtesting.com
JSHW051409091023
49903JS00006B/350